MOLECULAR BIOLOGY

An International Series of Monographs and Textbooks

Editors: BERNARD HORECKER, NATHAN O. KAPLAN, JULIUS MARMUR, AND HAROLD A. SCHERAGA

A complete list of titles in this series appears at the end of this volume.

SPIN LABELING II

Theory and Applications

SPIN LABELING II
THEORY AND APPLICATIONS

Edited by LAWRENCE J. BERLINER

Department of Chemistry
The Ohio State University
Columbus, Ohio

ACADEMIC PRESS   New York San Francisco London **1979**

A Subsidiary of Harcourt Brace Jovanovich, Publishers

ACADEMIC PRESS, INC.
111 Fifth Avenue, New York, New York 10003

United Kingdom Edition published by
ACADEMIC PRESS, INC. (LONDON) LTD.
24/28 Oval Road, London NW1 7DX

Library of Congress Cataloging in Publication Data
Main entry under title:

Spin labeling.

    (Molecular biology, an international series of mono-
graphs and textbooks)
    Includes bibliographies and index.
    1. Spin labels. I. Berliner, Lawrence J.
II. Series.
QH324.9.E36S64   574.1'9285   75-3587
ISBN 0-12-092352-1 (v. 2)

PRINTED IN THE UNITED STATES OF AMERICA

79 80 81 82   9 8 7 6 5 4 3 2 1

To Barbie

Contents

List of Contributors   xi

Preface   xiii

1. Saturation-Transfer Spectroscopy

James S. Hyde and Larry R. Dalton

    I. Introduction   3
   II. Physics of the Very Slow Tumbling Domain   4
  III. Experimental Methods   14
  IV. Theory   21
   V. Computational Efficiency   27
  VI. Applications of Saturation-Transfer Electron Paramagnetic Resonance 38
 VII. Anisotropic or Special Motions   58
      References   66

2. The Spin-Probe–Spin-Label Method

James S. Hyde, Harold M. Swartz, and William E. Antholine

    I. Introduction   71
   II. Theory   74
  III. Survey of Chemical and Magnetic Properties of
     Paramagnetic Metal-Ion Probes   90
  IV. Applications of the Method   101
      References   109

3. New Aspects of Nitroxide Chemistry

John F. W. Keana

    I. Introduction   115
   II. Structural Features That Make a Nitroxide Stable   116
  III. Characterization of New Nitroxide Spin Labels and
     Nitroxide Spectra   126
  IV. Synthesis of New Stable Nitroxide Spin Labels   129

V.   Chemical Reactions Involving the Nitroxide Grouping   147
VI.   Chemical Reactions Not Involving the Nitroxide Grouping   152
VII. Dinitroxide Spin Labels   152
VIII. Experimental Procedures for the Synthesis of Selected Proxyl and
      Imidazolidine-Derived Nitroxide Spin Labels   154
      References   166

4.   Spin-Labeled Synthetic Polymers

Wilmer G. Miller

I.   Introduction   173
II.   Sample Preparation   174
III. Analysis in Polymeric Materials   185
IV.   Motion in Bulk Polymers   195
V.   Effect of Diluents   204
VI.   Conformation of Polymers Adsorbed at Solid–Liquid Interfaces   214
VII. Conclusion   216
      References   217

5.   Spin Labeling in Pharmacology

Colin F. Chignell

I.   Introduction   223
II.   Drug Absorption, Distribution, and Excretion   224
III. Drug Metabolism   227
IV.   Pharmacological Effects of Drugs   231
V.   Future of Spin-Labeling Studies in Pharmacology   243
      References   243

6.   Spin Labeling in Biomedicine

Lawrence H. Piette and J. Carleton Hsia

I.   Introduction   247
II.   Spin Assay   248
III. Spin Immunoassay   254
IV.   Spin-Membrane Immunoassay   259
V.   Spin Labeling in Carcinogenesis   264
VI.   Spin Trapping   279
VII. In Vivo Electron Spin Resonance   284
      References   287

7.   Applications of Spin Labeling to Nucleic Acids

Albert M. Bobst

I.   Introduction   291
II.   Spin Labeling–Spin Probing of Nucleic Acids   293
III. Electron Spin Resonance Spectra of Nucleic Acids Containing
      Spin Labels or Spin Probes   308

IV. Monitoring Conformational Changes in Nucleic Acids by
Electron Spin Resonance 319
V. Molecular Association Studies with Spin-Labeled Nucleic Acids 329
VI. Conclusion 341
References 342

Appendix I 347
Appendix II Updated List of Commercial Sources of Spin Labels and
Nitroxide Precursors 349
Index 351

List of Contributors

Numbers in parentheses indicate the pages on which the authors' contributions begin.

William E. Antholine (71), National Biomedical ESR Center, Department of Radiology, Medical College of Wisconsin, Milwaukee, Wisconsin 53201

Albert M. Bobst (291), Department of Chemistry, University of Cincinnati, Cincinnati, Ohio 65221

Colin F. Chignell (223), Laboratory of Environmental Biophysics, National Institute of Environmental Health Sciences, Research Triangle Park, North Carolina 27709

Larry R. Dalton (1), Department of Chemistry, State University of New York, Stony Brook, New York 11790

J. Carleton Hsia (247), Department of Pharmacology, Faculty of Medicine, University of Toronto, Toronto, Ontario, Canada

James S. Hyde (1, 71), National Biomedical ESR Center, Department of Radiology, Medical College of Wisconsin, Milwaukee, Wisconsin 53201

John F. W. Keana (115), Department of Chemistry, University of Oregon, Eugene, Oregon 97403

Wilmer G. Miller (173), Department of Chemistry, University of Minnesota, Minneapolis, Minnesota 55455

Lawrence H. Piette (247), Cancer Center of Hawaii, University of Hawaii, Honolulu, Hawaii 96844

Harold M. Swartz (71), National Biomedical ESR Center, Department of Radiology, Medical College of Wisconsin, Milwaukee, Wisconsin 53201

Preface

This volume, which is not intended as part of a continuing series, serves to complement *Spin Labeling: Theory and Applications* published in 1976. As the editor, my personal interest in compiling this second volume stemmed from my awareness that pertinent material was omitted from the first volume due to limitations on page length. In addition, the encouragement to compose *Spin Labeling II* was fostered by my many friends and colleagues in the field of biological magnetic resonance as well as the original contributors and the publisher.

Again, we have been fortunate in obtaining the efforts of expert contributors in the field. While only three of the chapters address biomedical or molecular biological approaches, the general subject of spin-labeled biopolymers is addressed in its importance to both macromolecular modeling as well as some industrially based problems. Two still emerging approaches, saturation transfer spectroscopy, and spin-probe–spin-label interactions, are covered thoroughly in both theory and applications. Since publication of the first volume, many radically new synthetic approaches to nitroxide chemistry have been discovered. These are presented here in the quite well-received format of Gaffney in the original volume; it includes detailed synthetic methods as well as the rationale behind these various reaction schemes. The order of chapters is two theoretical chapters, then chemistry and polymer applications, and finally pharmacology, biomedical aspects, and nucleic acids.

I wish to express my personal gratitude to the people who helped make this and the first volume possible.

Lawrence J. Berliner

1

Saturation-Transfer Spectroscopy

JAMES S. HYDE

NATIONAL BIOMEDICAL ESR CENTER
DEPARTMENT OF RADIOLOGY
MEDICAL COLLEGE OF WISCONSIN
MILWAUKEE, WISCONSIN

and

LARRY R. DALTON

DEPARTMENT OF CHEMISTRY
STATE UNIVERSITY OF NEW YORK
STONY BROOK, NEW YORK

and

DEPARTMENT OF BIOCHEMISTRY
VANDERBILT UNIVERSITY MEDICAL SCHOOL
NASHVILLE, TENNESSEE

I.	Introduction	3
II.	Physics of the Very Slow Tumbling Domain	4
	A. Spectral Diffusion of Saturation	4
	B. The Average Spectral Diffusion Velocity and the Adiabatic Condition	6
	C. Spin–Lattice Relaxation	8
	D. Transverse Relaxation	10
	E. Binding Tightness	12
III.	Experimental Methods	14
	A. Progressive Saturation	14
	B. First-Harmonic Dispersion Out of Phase	15
	C. Second-Harmonic Absorption Out of Phase	17
	D. Electron–Electron Double Resonance	18
IV.	Theory	21
V.	Computational Efficiency	27
	A. Approximation of Inhomogeneous Broadening by Gaussian Convolution	27
	B. Gaussian Preconvolution Followed by Runge–Kutta Solution of the Master Supermatrix Equation	27
	C. A Perturbation Scheme in ω_m and the Differential Approximation	30

1

VI. Applications of Saturation-Transfer Electron Paramagnetic Resonance   38
    A. Actin–Myosin Interactions   39
    B. Hemoglobin S Self-Association   45
    C. Protein–Lipid Interactions   47
    D. Phospholipid Motion in Gel-Phase Lipid Bilayers   54
    E. Molecular Motion on Surfaces   56
VII. Anisotropic or Special Motions   58
    A. Rotational Diffusion of a Spin-Labeled Ellipsoid of Revolution in an
       Isotropic Medium   58
    B. Experimental Methods   60
    C. Theoretical Comments   63
    References   66

Definitions of Symbols

U_m — Dispersion signal component in phase with respect to the applied Zeeman modulation detected at the mth harmonic of the modulation (see Fig. 8)

U_m' — Dispersion signal component out of phase (in phase quadrature) with respect to the applied modulation

V_m — Absorption signal component detected in phase with respect to the applied modulation

V_m' — Absorption signal component detected in phase quadrature to the applied modulation

ω_m — Frequency of the applied modulation in radians per second

m — Modulation harmonic

d_m — $(\frac{1}{2})\gamma_e H_m$, modulation amplitude in radial frequency units

Δ — $\omega_0 - \omega_\lambda$, where ω_0 is the applied microwave frequency, and ω_λ is the transition frequency of the λth EPR hyperfine transition

d_0 — $(\frac{1}{2})\gamma_e H_1$, microwave amplitude in radial frequency units

q — $\hbar/NkT$, where N is the number of spin states

T_1 — Electron spin–lattice relaxation time

T_2 — Electron spin–spin relaxation time

Y_m^+ — $2(Z_{\alpha\alpha}'^{\lambda(m)} - Z_{\beta\beta}'^{\lambda(m)})$, where α and β are electron spin states

iY_m^- — $2(Z_{\alpha\alpha}''^{\lambda(m)} - Z_{\beta\beta}''^{\lambda(m)})$, where α and β are electron spin states

By use of the above notation, the four signal components existing at the mth harmonic of the modulation frequency can be defined as

$$U_m = Z_{\beta\alpha}'^{(m)} + Z_{\beta\alpha}'^{(-m)}$$

$$U_m' = Z_{\beta\alpha}''^{(m)} - Z_{\beta\alpha}''^{(-m)}$$

$$V_m = Z_{\beta\alpha}''^{(m)} + Z_{\beta\alpha}''^{(-m)}$$

$$V_m' = Z_{\beta\alpha}'^{(m)} - Z_{\beta\alpha}'^{(-m)}$$

where Z has been decomposed into real and imaginary parts as $Z = Z' + iZ''$.

I. INTRODUCTION

The major thrust of the spin-label method has been to study motion. The nitroxide radical moiety appears to be essentially unique—no other stable free radical has been synthesized that can compete with it as a general-purpose probe of biological systems. The range of motions that can be studied with the nitroxide radical using the approaches described in the book "Spin Labeling: Theory and Applications" (Berliner, 1976) is an accident of the anisotropy of the magnetic interactions of the label. The motional range is 10^{-7}–10^{-11} sec, and scientists, when using nitroxide labels, were forced to ask in the design of their experiments questions concerning motions that fell in this range. Although there has been little motivation for studying motions faster than 10^{-11} sec, many very interesting experiments could have been designed if only it were possible to study motions slower than 10^{-7} sec. For example, these slow ranges correspond to motions expected for proteins embedded in membranes and to motions of macromolecular assemblies of proteins.

The methods of saturation-transfer spectroscopy (sometimes abbreviated ST-EPR) discussed in this chapter extend the range of motions that can be studied using nitroxide radical spin labels by four additional orders of magnitude to times as long as 10^{-3} sec. There are few, if any, physicochemical methods that compete effectively with the method in this range of 10^{-3}–10^{-7} sec, nor are there other techniques that can probe such a wide range of times (10^{-3}–10^{-11} sec) as is possible using conventional and saturation-transfer techniques in concert.

The theories of magnetic resonance spectroscopy in the solid phase and in the liquid phase are very well developed, and the amount of experimental literature is large indeed. There exists, however, an intermediate range of viscosities between solid and liquid where the theories fail and about which very little has been written. We define this range as the *very slow tumbling domain*. If we introduce three critical frequencies, $\Delta\omega$, τ_R^{-1}, and T_1^{-1}, where $\Delta\omega$ is the order of magnitude of the anisotropy of the magnetic interaction of a spin label, τ_R is the rotational correlation time of the label, and T_1 is the spin–lattice relaxation time of the label, then this intermediate region is characterized by the relationships

$$\Delta\omega\tau_R \gg 1 \qquad 100T_1 > \tau_R > 0.01T_1$$

The first inequality guarantees that the ordinary EPR spectrum is the same as that from a rigid powder. The second inequality leads to spectral diffusion of saturation if an intense microwave field is incident on the sample, since rotational diffusion is comparable to spin–lattice relaxation. Nitroxide radical spin labels fulfill this inequality in the range of correlation times of 10^{-3} to 10^{-7} sec, and it is in this range of conditions that saturation-transfer spectroscopy functions. There are no textbook discussions of motional modes, relaxation mechanisms, and magnetic resonance theory in this domain. Whatever exists has essentially been done, almost in passing, during the development of the methodology of saturation-transfer spectroscopy.

The first section in this chapter, then, is a review of the magnetic resonance physics of the very slow tumbling domain. Readers should recognize that there is very little literature and that there are many opportunities for experimental and theoretical work.

II. PHYSICS OF THE VERY SLOW TUMBLING DOMAIN

A. Spectral Diffusion of Saturation

Saturation-transfer spectroscopy depends on the spectral diffusion of saturation that occurs in a time T_1 because of modulation of anisotropic interactions by rotational diffusion. Saturation-transfer spectra are sensitive in shape to rotational diffusion because rotational diffusion gives rise to spectral diffusion that varies across the spectrum according to the anisotropy of the magnetic interactions. On these two sentences hang all that follows. They should be studied carefully.

These two concepts—namely, that rotational diffusion causes spectral diffusion of saturation and that the effects are different in different parts of the spectrum—were introduced by Hyde et al. (1970) in a study of the saturation characteristics of flavin radicals in flavoproteins.

As an idealized example consider the powder pattern associated with the high-field (-1) nuclear spin configuration. We approximate it by an axial magnetic interaction. Then the resonance condition is

$$H_R = (H_\perp^2 \sin^2 \theta + H_\parallel^2 \cos^2 \theta)^{1/2} + H_0 \qquad (1)$$

The powder pattern, assuming infinitely narrow linewidths, is

$$I(H) \propto \frac{\sin \theta}{\partial H_R / \partial \theta}$$

$$\propto \frac{(H_\perp^2 \sin^2 \theta + H_\parallel^2 \cos^2 \theta)^{1/2}}{\cos \theta (H_\perp^2 - H_\parallel^2)} \qquad (2)$$

This function is plotted in Fig. 1. Each point on the abscissa corresponds to a unique value of θ.

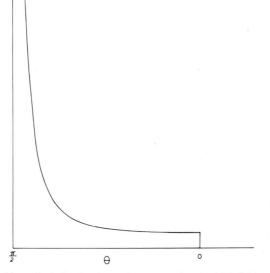

Fig. 1. Nitroxide radical simulated powder pattern for the high-field spectral fragment associated with the (-1) nuclear quantum number assuming infinitely narrow lines and an axial hyperfine tensor with $A_\parallel = 32$ G, $A_\perp = 7$ G.

Consider an ensemble of spins contributing to the intensity of the spectrum at H' corresponding to an orientation θ'. Imagine an intense microwave field which is turned on at time $t = 0$ at a value H' of the resonant condition. The situation is illustrated schematically in Fig. 2. Spins in a band

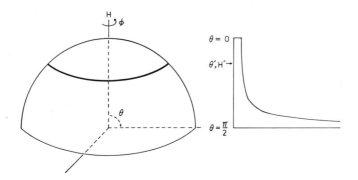

Fig. 2. If we let H_0 coincide with a diameter of a sphere, then each orientation of the nitroxide molecular frame with respect to H_0 corresponds to a point on the surface of the sphere, and rotational diffusion can be modeled by random walk on that surface. For an axial Hamiltonian and (-1) nuclear quantum number, the angle between H_0 and the z axis of the nitroxide (for example, θ') results in magnetic resonance at a unique field (for example, H') and corresponds to a band or ring of orientations on the surface of the sphere.

of orientations (the solid line) corresponding to a unique value of θ' are excited. Let us assume that the microwave field is left on for some time and that the spins simultaneously undergo spin–lattice relaxation and isotropic rotational Brownian diffusion.

Let us approximate this diffusion by assuming that each spin undergoes steps of equal angular size $\Delta\Omega$ on the surface of the sphere and that the steps occur at a constant frequency f. The mean square angular displacement in a time t is

$$\overline{(\theta - \theta')^2} = ft\left(\frac{2}{\pi}\frac{\Delta\Omega}{}\right)^2 \equiv \frac{2}{3}\frac{t}{\tau_R}\left(\frac{2}{\pi}\right)^2$$

or

$$t = \frac{3\pi^2}{8}\tau_R\overline{(\theta - \theta')^2} \tag{3}$$

[We note that if the average step is $\Delta\Omega$, the average step in θ is $(2/\pi)\,\Delta\Omega$.] The equilibrium distribution of saturated spins if $\tau_R/T_1 > 1$ is

$$S(\theta - \theta') = (3\pi^2\tau_R/16T_1)^{1/2}\exp-[(\theta - \theta')^2(3\pi^2/4)(\tau_R/T_1)]^{1/2} \tag{4}$$

This is an analytical expression for the average extent of angular diffusion of saturation that occurs when an ensemble of spins is relaxing with a time constant T_1.

The extent of spectral diffusion varies across the spectrum as given by the expression

$$\sin\theta\,\frac{dH_R(\theta)}{d\theta} = \frac{\sin^2\theta\,\cos\theta(H_\perp{}^2 - H_\parallel{}^2)}{(H_\perp{}^2\sin^2\theta + H_\parallel{}^2\cos^2\theta)^{1/2}} \tag{5}$$

Thus, at the extrema corresponding to $\theta = 0$ or $\theta = \pi/2$, there is *no* spectral diffusion, even though angular diffusion has occurred over an infinitesimal angular distance $\Delta[(\theta - \theta_0)^2]$. The maximal spectral diffusion of saturation occurs at an intermediate angle that can be determined readily from Eq. (5).

In this section, then, we have illustrated the two key concepts: (1) rotational diffusion gives rise to spectral diffusion, and (2) the effect varies spectrally. Equation (5) is, of course, a differential form. If $[(\theta - \theta')^2]^{1/2}$ is not small, then an integral over the diffusing and relaxing ensemble of spins must be written in order to calculate the spectral distribution of saturation.

B. The Average Spectral Diffusion Velocity and the Adiabatic Condition

As we have seen, rotational diffusion causes spins to pass through the resonance condition because of the anisotropy of the magnetic interactions. We can do a simple calculation to determine the average spectral diffusion velocity. Let $[\overline{(H - H')^2}]^{1/2}$ be the average spectral diffusion distance that is associated with an average angular diffusion distance $[\overline{(\theta - \theta')^2}]^{1/2}$. Then,

$$\frac{d[\overline{(H - H')^2}]^{1/2}}{dt} = \frac{dH}{d\theta} \frac{d[\overline{(\theta - \theta')^2}]^{1/2}}{dt} \tag{6}$$

and from Eqs. (3) and (5),

$$\frac{d[\overline{(H - H')^2}]^{1/2}}{dt} = \frac{\sin\theta \cos\theta (H_\perp{}^2 - H_\parallel{}^2)}{(H_\perp{}^2 \sin^2\theta + H_\parallel{}^2 \cos^2\theta)^{1/2}} \left(\frac{2}{3\pi^2}\right)^{1/2} \frac{1}{(t\tau_R)^{1/2}} \tag{7}$$

In the absence of any spin–lattice relaxation, this is the average spectral diffusion velocity as a function of time after an initial coding of the spins as, for example, by the application of microwave power in the circumstance illustrated by Fig. 2. The average spectral diffusion velocity exhibits a $t^{-1/2}$ dependence as shown by Eq. (7). Since the spins are relaxing with a relaxation probability $T_1^{-1} \exp(-t/T_1)$, the average spectral diffusion velocity is given by

$$\left\langle \frac{d[\overline{(H - H')^2}]^{1/2}}{dt} \right\rangle = \frac{\sin\theta \cos\theta (H_\perp{}^2 - H_\parallel{}^2)}{(H_\perp{}^2 \sin^2\theta + H_\parallel{}^2 \cos^2\theta)^{1/2}}$$

$$\times \left(\frac{2}{3\pi^2}\right)^{1/2} \frac{1}{(T_1\tau_R)^{1/2}} \int_0^\infty \frac{e^{-t/T_1}}{t^{1/2}} dt$$

$$= \frac{\sin\theta \cos\theta (H_\perp{}^2 - H_\parallel{}^2)}{(H_\perp{}^2 \sin^2\theta + H_\parallel{}^2 \cos^2\theta)^{1/2}} \left(\frac{2}{3\pi^2}\right)^{1/2} \frac{1}{(T_1\tau_R)^{1/2}} \tag{8}$$

The well-known adiabatic condition

$$dH/dt \ll \gamma H_1{}^2 \tag{9}$$

enters frequently into discussions of rapid-passage phenomena in magnetic resonance. As will be discussed in considerable detail, nearly all saturation-transfer experiments have employed rapid-passage methods. If the sweep of the dc field through resonance is sufficiently slow that the adiabatic condition is fulfilled but is still fast compared to T_1, then in the rotating frame the resultant magnetization vector lies along the effective field throughout the sweep. Rotational diffusion on the average will result in substantial breakdown of the adiabatic condition when, combining Eqs. (8) and (9),

$$\tau_R = \frac{2}{3T_1} \left[\frac{\sin^2\theta \cos^2\theta (H_\perp{}^2 - H_\parallel{}^2)}{H_\perp{}^2 \sin^2\theta + H_\parallel{}^2 \cos^2\theta}\right] \frac{1}{\gamma^2 H_1{}^4} \tag{10}$$

Thomas et al. (1976a) estimated that $T_1 = 6.6 \times 10^{-6}$ sec in the very slow tumbling domain (vide infra) and performed all saturation-transfer experiments with a microwave field of $H_1 = 0.25$ G. Using $H_{\parallel} = 35$ G, $H_{\perp} = 7$ G, and $\theta = 45°$,

$$\tau_R = 4 \times 10^{-5} \text{ sec} \tag{11}$$

In the absence of rotational diffusion, the resultant magnetization traces a complicated path under *adiabatic* rapid-passage conditions, and the projections on the x-y plane in the rotating frame are the experimental observables. If rotational diffusion is now "turned on" and Eq. (11) is approximately satisfied, passage is no longer adiabatic, the magnetization vectors are no longer aligned along the effective fields, and the experimental observable is altered.

We believe that the concept of "average spectral diffusion velocity" is useful in understanding saturation-transfer spectroscopy. The concept makes plausible the experimental result that EPR displays obtained under adiabatic rapid-passage conditions are sensitive in shape to rotational diffusion. The equations developed here certainly do not constitute a theory that predicts the form of the spectral observable. Indeed, despite numerous efforts we have never been able to extract from the complete theory fully satisfactory conceptual models of the phenomena occurring under rapid-passage saturation-transfer conditions. Apparently, spectral diffusion effects are intermixed into the general equations in a most complex manner.

C. Spin–Lattice Relaxation

Two model systems have been extensively used to investigate the physics of very slowly tumbling spin labels: (1) low molecular weight nitroxides in supercooled *sec*-butylbenzene and (2) maleimide-labeled hemoglobin in glycerol–water or sucrose solutions. The early progressive-saturation experiments described by Hyde and Dalton (1972) showed that the spin–lattice relaxation time of the nitroxide radical moiety in the very slow tumbling domain is nearly independent of the motion. It was this discovery that made saturation-transfer spectroscopy a practical technique. If T_1 were to show a strong dependence on rotational diffusion, the entire saturation-transfer approach could well fail.

M. Huisjen and J. S. Hyde (unpublished) measured the spin–lattice relaxation time of the alcohol nitroxide 2,2,6,6-tetramethyl-4-piperidinol-1-oxyl (TANOL) (I) in *sec*-butylbenzene using the saturation-recovery method. For rotational correlation times between 10^{-5} and 10^{-7} sec, T_1 was unchanged and about equal to 1.5×10^{-5} sec (see Fig. 3). Huisjen and Hyde (1974) reported a value of 6.6×10^{-6} sec for maleimide-spin-labeled hemoglobin in water at room temperature. This value is recommended for use in

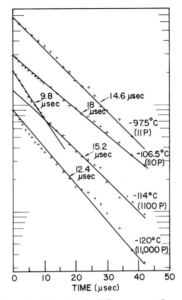

Fig. 3. Spin–lattice relaxation times derived from saturation-recovery signal amplitudes for 3×10^{-3} *M* TANOL (**I**) in supercooled *sec*-butylbenzene. The viscosities were determined from data given by Barlow *et al.*, (1966). Pump power, 100 mW; pump duration, 5 μsec; observing power, 1 mW, observing aperture, 1 μsec. See Huisjen and Hyde (1974).

analysis of saturation-transfer experiments on biological systems in the absence of any information to the contrary.

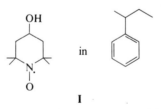

I

Theoretical models for spin–lattice relaxation in the very slow tumbling domain are qualitative, at best. The contribution of the commonly encountered liquid-state spin–lattice relaxation mechanisms (spin–rotational interaction, anisotropic electron–Zeeman interaction, anisotropic hyperfine interaction) can be calculated, as was done by Percival and Hyde (1976), with the result that the contribution from these mechanisms is found to be negligible compared to experimentally observed rates for the very slow motion region. Moreover, the dependence of measured relaxation rates on the ratio of viscosity to temperature is weaker than predicted by the aforementioned liquid-state mechanisms. It is thus reasonable to look elsewhere for significant contributions to electron spin–lattice relaxation.

Measurements (Muromtsev et al., 1975; L. R. Dalton, A. L. Kwiram, and J. A. Cowen, unpublished results) of model system nitroxides (TANOL) at low temperatures (77°–4°K) demonstrate that a two-phonon Raman process is the dominant contribution to spin–lattice relaxation. Simple extrapolation of these results to higher temperatures suggests that the contribution from this mechanism will be between 10^{-4} and 10^{-5} sec in the very slow motion region, which is not insignificant. However, since the predicted $T_1 \propto T^2$ dependence of this mechanism deviates from that observed by Huisjen and Hyde (Fig. 3), one must consider other relaxation models.

The phonon spectra of low-symmetry materials, such as the nitroxide solute and the solvent matrix, are observed to be structured with peaks (known as Einstein spikes), which can be associated with characteristic local modes. Spin modulation by local modes is known to be an effective spin–lattice relaxation mechanism in certain inorganic materials, and the temperature dependence of electron spin-lattice relaxation,

$$T_1^{-1} = 7.65 \times 10^{-6}\tau_R^{-1} + 1.14 \times 10^8 \exp(-1207.75/T) + 0.167T^2 \quad (12)$$

observed in the study of the nitroxide radical model system of TANOL in sec-butylbenzene (Percival and Hyde, 1976) is consistent with such a contribution. However, final characterization of the contributions to electron spin–lattice relaxation must await further experimentation. The study of relaxation as a function of external pressure in order to separate intra- and intermolecular contributions and the investigation of the dependence of relaxation on the detailed nature of radical and solvent structure would seem appropriate.

D. Transverse Relaxation

In the theoretical description of saturation-transfer spectroscopy, which is reviewed in this chapter, there is a parameter T_2 into which has been lumped all mechanisms of transverse relaxation except the loss of phase coherence that occurs because of the diffusion of saturation. Despite the fact that T_2 is an extremely important parameter, very little is known about it.

The electron–nuclear dipolar (END) interaction provides, we suspect, an important mechanism in the very slow tumbling domain for inducing ^{14}N nuclear relaxation. This pseudosecular interaction couples different portions of the powder EPR spectrum that correspond to the same orientations with respect to the magnetic field but different nuclear quantum numbers and is a transverse relaxation mechanism for the electrons. The nuclear resonance frequency of ^{14}N is of the order of 10^7 Hz. A rotational correlation time of 10^{-7} sec is the short time limit for saturation-transfer spectroscopy, and at this limit this END mechanism should be quite effective.

At the other limit of the method, 10^{-3} sec, one would expect that ^{14}N relaxation induced by the spin–lattice relaxation of the unpaired electron (the solid-state effect) would become important. Little is known about the relative importance of these two pseudosecular mechanisms for the ^{14}N nucleus at intermediate correlation times ($\tau_R \sim 10^{-5}$ sec).

The same interactions can be discussed for the protons. In fact, the most crucial and perplexing aspect of phase relaxation (and the analysis of resonance linewidths) is the existence of unresolved or partially resolved proton hyperfine interactions. The effects of proton hyperfine interactions in turn depend on proton relaxation. If proton relaxation is fast, the proton hyperfine lines will be effectively coupled and each of the nitrogen hyperfine lines will behave as a homogeneously broadened line. For such a condition, proton relaxation can be viewed as a phase relaxation mechanism, and the direct relationship between resonance linewidth and phase relaxation rate obtains. If, on the other hand, proton relaxation is slow, the nitrogen hyperfine lines will behave as inhomogeneously broadened lines and no direct relationship between resonance linewidth and phase relaxation rate will exist. This condition is known to obtain for the fast- and intermediate-motion regions for dilute solutions of nitroxides. However, since the magnitude of the proton–nuclear resonance frequency is in the range 10–20 MHz, the greatest contribution from a proton END mechanism would be expected for correlation times near the short side of the saturation-transfer range— namely, 10^{-6}–10^{-7} sec. The detailed computer analysis of saturation-transfer spectra supports the hypothesis of fast proton relaxation in the 10^{-6}–10^{-7} sec region, with characterization of the situation at longer correlation times being more ambiguous. There are, of course, other mechanisms, such as methyl group tunneling–rotation, that could contribute to proton relaxation. Unfortunately, there is currently no information available that would permit estimation of the magnitude of such contributions.

Another possible contribution to phase relaxation is modulation of the electron–electron dipolar interaction—the cross-relaxation mechanism. This is a long-range interaction and can be important even for the dilute radical solutions commonly encountered in spin-label studies.

Two mechanisms can in general be ruled out as significant contributors to relaxation in the very slow motion region: (1) Heisenberg spin exchange and (2) electron spin diffusion. These are unimportant due to the low collisional frequencies existent in the very slow tumbling domain coupled with the fact that the systems are magnetically dilute.

E. Binding Tightness

To the mechanisms responsible for loss of phase coherence and therefore contributing to T_2, as discussed in the preceding section, must be added the effects arising from librational motions of the label. If the nitroxide were more rigidly attached, resulting in a longer T_2, spectral diffusion over a smaller spectral distance could be measured and a longer rotational correlation time could be inferred. Alternatively, analysis of saturation-transfer spectra in order to obtain T_2 could yield useful information on motional modes of a protein to which the label is attached. The subject of binding tightness is of crucial importance.

This can be seen mathematically. One can estimate that

$$\overline{(\theta - \theta')^2_{\text{min}}} \sim 1/(\Delta\omega \, \gamma T_2)^2 \tag{13}$$

where $(\gamma T_2)^{-1}$ is the width arising from the librational motions and $\Delta\omega$ is the anisotropy of the magnetic interactions. Writing, as before, from Eq. (3)

$$\tau_R = \frac{8}{3\pi^2} \frac{T_1}{(\theta - \theta')^2} \tag{14}$$

we have

$$(\tau_R)_{\text{max}} \sim T_1 (\Delta\omega \, \gamma T_2)^2 \tag{15}$$

and as T_2 becomes shorter so does $(\tau_R)_{\text{max}}$ (the longest time that can in principle be measured).

The mathematical treatment of librational motion depends on the type of motion involved. If the motional process can be described by a double well potential, the effect of motion can be computed by superimposing the effects of a two-site jump model on the effects of rotational diffusion. For the computation of a single spectrum, the result of a two-site jump superimposed on isotropic rotational Brownian diffusion is often indistinguishable from a large-angle jump diffusion model. However, by analysis of the spectra as a function of temperature and viscosity the effects of the diffusional and librational mechanisms can be distinguished. For more complex librational models (e.g., such as found in DNA) more complex mathematical models are required for the simulation of spectra.

The literature concerning the tightness of binding of nitroxide labels to proteins is sparse. The only study devoted exclusively to that subject is by Johnson (1978). The present authors have occasionally commented on the subject (see, for example, the review article by Hyde, 1978.) We discuss in the next paragraphs several aspects of the general binding-tightness problem with the hope that the discussion will serve as a guide for future work.

1. Mediation of Free Molecule Intramolecular Modes by Protein Conformational Fluctuations. The nitroxide radical moiety in the piperidine ring system undergoes chair–chair interconversions at a rate that appears to be in the kilohertz range (Brière *et al.*, 1967). It seems likely that this intramolecular motion is too slow to be important in ST-EPR. The methyl groups undergo rotation in the 10^9-Hz range. It remains a possibility that the presence of the protein can alter methyl group rotation in such a way as to affect the ST-EPR spectra. The N—O bond in single crystals is at a 16° angle (Berliner, 1970). Little is known about the rate of inversion of the N—O bond with respect to the C—N—C plane. A particularly probable internal motion would be a torsional oscillation about the C—N central bond of the maleimide spin label. In any event, it seems likely that the normal modes of the label will be quenched to various degrees when the label is embedded in a protein and that fluctuations in the protein can mediate these internal motions.

2. Mediation of Label–Protein Binding by Protein Conformational Fluctuations. Even if the nitroxide moiety were rigid, fluctuations of the label with respect to the protein coordinate system could be expected in a manner that depended on the bonding at the point of attachment, van der Waals forces, and steric effects. Protein conformational changes or fluctuations in atomic coordinates of the protein could be expected to affect these nitroxide fluctuations.

3. Slow-to-Fast Motion about One Axis and Very Slow Motion about the Other Axis. Another example of motion of a spin-label moiety relative to the biostructure of interest is provided by intercalation-labeled DNA. Robinson *et al.* (1978b) showed that the spin-labeled acridines and ethidiums intercalated into DNA are free to rotate in the plane of the base pairs, i.e., are free to rotate about the axis normal to the plane of the spin-labeled dye. Such motion is not, however, observed for labels such as propidium and psoralen derivatives.

These three aspects illustrate the need for caution. Relatively little attention has been paid thus far to design and use of labels that would give improved saturation-transfer results. Such labels might have altered intramolecular motional modes, bifunctional binding, and improved space filling.

We attempt in this chapter to give realistic limits of the saturation-transfer method and to discuss objectively opportunities for future research. The subject of binding tightness is to the authors particularly challenging. However, in a great many practical situations, broadening from these motions is at most 2 or 3 G, and the saturation-transfer method can be used without undue complications from them.

III. EXPERIMENTAL METHODS

In the preceding sections we described spin dynamics and relaxation processes in the very slow tumbling domain. In this section we discuss the various experimental methods by means of which one determines the rotational correlation time. We restrict ourselves at this point in the chapter to consideration of experimental methods that have actually been used; at a later point, ideas for other experimental approaches are suggested.

Four methods have been described in the literature: (1) progressive (or CW) saturation, in which the ordinary EPR absorption is observed as a function of microwave field intensity; (2) observation of the EPR dispersion signal using high-frequency field modulation, detecting the first-harmonic response with the reference phase of the phase-sensitive detector set 90° out of phase with respect to the field modulation; (3) observation of the EPR absorption signal using high-frequency field modulation, detecting the *second*-harmonic response with the reference phase set 90° out of phase; and (4) electron–electron double resonance (ELDOR), in which saturation is induced at one point in the spectrum by a first microwave source and its presence at a second point in the spectrum is detected using a second microwave source.

These four methods are discussed in this section. A summary evaluation of the relative merits of the methods from a practical experimental point of view is as follows. Progressive saturation is seldom used, primarily because of difficulties in interpretation and relatively low sensitivity to motion. First-harmonic dispersion out-of-phase detection has been used by a number of investigators and is the best rapid-passage method from the point of view of theoretical analysis. It suffers from a poor signal-to-noise ratio when employed in the usual reflection-type microwave bridge because of the extreme sensitivity of the bridge in this arrangement to FM noise from the microwave oscillator. Second-harmonic absorption out of phase is by far the most practical and widely used technique. ELDOR is in many ways the most satisfactory method except that a special microwave bridge and cavity are required, and they are not widely available.

A. Progressive Saturation

The observation that EPR spectra of flavin radicals are sensitive to motion when recorded at high microwave power (Hyde *et al.*, 1970) was the historical event that led directly to the development of saturation-transfer spectroscopy. Hyde and Dalton (1972) immediately carried out experimental progressive-saturation studies on nitroxides, and Goldman *et al.* (1973) wrote a theory of saturation in the very slow tumbling domain. Since these early papers, however, nothing more has appeared, and the future of the method is uncertain.

The experimental and theoretical studies (see particularly Fig. 13 of Goldman *et al.*, 1973) show that the greatest sensitivity of the progressive-saturation method to motion is in the range of correlation times of 10^{-6} to 10^{-7} sec. The microwave powers at which EPR signals show maximal intensity vary by only about a factor of 2 over the entire range of rotational correlation times (Fig. 2 of Hyde and Dalton, 1972). The present authors believe that the method could be a reasonable alternative experimental approach in the 10^{-6}–10^{-7} sec range but that it would not be satisfactory for measurement of slower motions.

It should be recognized that no one has done careful experimental studies on model systems undergoing isotropic rotational diffusion, nor have theoretical simulations been done in detail. Thus, to a degree this method suffers by comparison with the others discussed here from a simple lack of development.

The experimentalist should on every sample to be investigated with saturation-transfer spectroscopy perform progressive-saturation experiments as a routine procedure. The experiments should be carried out using a sufficiently low frequency of field modulation (1 kHz, for example) that the field modulation does not affect the saturation results. This is a normal and appropriate procedure that serves as one of the controls on the experiment.

B. First-Harmonic Dispersion Out of Phase

In EPR it is experimentally difficult to sweep the polarizing magnetic field sufficiently rapidly to observe adiabatic rapid-passage effects, but it is quite common to observe effects when high-frequency field modulation is used. Portis (1955) first described these effects. They were studied experimentally by Hyde (1960) and analyzed in considerable detail by Weger (1960). Feher (1959) also gave a helpful discussion.

One can easily understand the 90° out-of-phase adiabatic rapid-passage experiment using a rotating-frame argument. Referring to the left half of Fig. 4, assume that $\omega_m T_1 < 1$, where ω_m is the field modulation frequency, so that the spin system has come to equilibrium when the sinusoidal modulation field is at one end of its excursion with a sweep rate going to zero. The vectors **M** and **H**$_{\text{eff}}$ are parallel and remain so during a sweep, assuming the adiabatic condition [Eq. (9)] holds. The dispersion signal is detected as the projection of **M** on the $+x$ axis. If the system can now come to equilib-

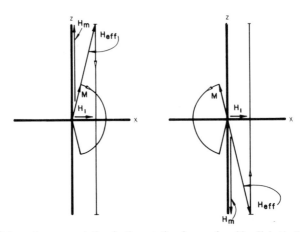

Fig. 4. Schematic representation in the rotating frame of rapid, adiabatic, "not fast" passage for a single spin packet. During passage H_{eff} and M remain parallel, with the ends of these vectors tracing paths indicated by the lines labeled with open arrows.

rium again at the other end of the sweep, then M will initially be along the $+z$ axis, during passage M and H_{eff} will be antiparallel, and the signal will be the projection of M on the $-x$ axis. Thus, the signal from a single spin packet will be $90°$ out of phase with respect to the field modulation. Weger (1960) integrates this result from a single spin packet over the envelope $h(\omega - \omega_0)$ that is the distribution of spins. The result is a signal that has the shape $h(\omega - \omega_0)$ and is $90°$ out of phase.

The qualitative idea in the Hyde and Dalton paper was simply that rotational diffusion in the very slow tumbling domain ought in some way to modify this result. Their experimental spectra are shown in Fig. 5. The spectra were obtained under various combinations of rotational correlation time and field modulation frequency. With the slowest motion and highest field modulation frequency, the spectrum in the lower left corner is obtained. It has the shape expected for a pure absorption. The spectrum in the upper right corresponding to lowest field modulation frequency and fastest rotational diffusion shows absorption only at the turning points, where we have already seen that the average spectral diffusion velocity goes to zero [for example, Eq. (8)].

Comparing spectra on diagonals we see that to a good approximation a particular $\omega_m \tau_R$ product results in a unique lineshape. Thomas et al. (1976a) analyzed this experiment in considerable detail. The various experimental parameters in the theory are H_m, ω_m, H_1, T_2, T_1, and τ_R. It is a remarkable result that the shapes of these spectra are found both experimentally and theoretically to be essentially independent of H_m, H_1, and T_1. They show some dependence on T_2 but are largely determined by the product of $\omega_m \tau_R$.

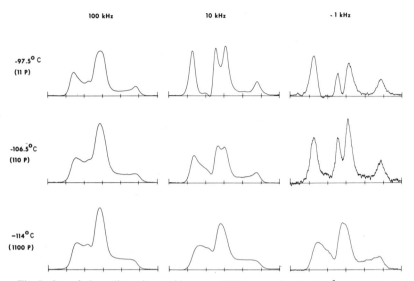

Fig. 5. Out-of-phase dispersion rapid-passage EPR spectra for 3×10^{-3} M TANOL (I) in supercooled *sec*-butylbenzene. The abscissa ticks are at 20-G intervals.

Thus, the experimentalist can probe the motion by varying the field modulation frequency.

Maximal sensitivity of this display to motion lies in the range 10^{-5}–10^{-6} sec. At times as short as 10^{-7} sec, the spectra show sensitivity to motion but the signal intensity is low. At times near 10^{-4} sec, the agreement between simulated and experimental spectra is not quite as good, suggesting that librational motions may be affecting the results in ways not fully incorporated into the theory.

As mentioned previously, dispersion spectra obtained with the usual reflection-type microwave bridge are extremely noisy. One can overcome this noise using an induction-type microwave cavity and bridge (see, for example, Fig. 3 of Huisjen and Hyde, 1974). There has been difficulty finding reliable cavity designs, but it now seems likely, on the basis of projects underway in our own laboratories and those of others, that this problem will be satisfactorily solved. It is predicted that the method of this section will be widely used in the future because of (1) the high signal intensity, (2) optimal sensitivity in the 10^{-5}–10^{-6} sec range, and (3) ease of interpretation.

C. Second-Harmonic Absorption Out of Phase

This unlikely display was announced by Hyde and Thomas (1973) and has become by far the most commonly employed. It resulted from a systematic experimental investigation of eight displays: in-phase and out-of-phase, first- and second-harmonic absorption and dispersion. Weger's (1960) theoretical analysis also contributed to the selection of this display. It has the immediate advantage of being insensitive to FM noise, which causes so much difficulty when one is observing dispersion.

Criteria in the experimental search were, as before in the Hyde and Dalton paper, sensitivity of the shape to the product of $\omega_m \tau_R$ and insensitivity to the other experimental parameters. The success with respect to these criteria was not as great as in the first-harmonic dispersion out-of-phase display. The detailed analysis of Thomas *et al.* (1976a) shows good insensitivity to H_1, H_m, and T_1, as before, but shapes were not uniquely determined by the product of $\omega_m \tau_R$.

There are two types of rapid-passage phenomena that contribute to the observed spectra. At slowest motion and highest modulation frequency, the shape is the same as for the first-harmonic dispersion out of phase—namely, the pure absorption. The signal is all on one side of the baseline, meaning that the signal arises from the middle of the field modulation interval. At faster motions and lower modulation frequencies the signal can be on both sides of the baseline. This change in phase of the detected signal with respect to the field modulation can arise only if there are contributions to the detected signal from the ends or extrema of the field modulation cycle.

It is observed experimentally and confirmed theoretically that second-harmonic absorption out-of-phase line shapes are sensitive to motion over a wider range of rotational correlation times than are first-harmonic dispersion out-of-phase displays. Best sensitivity is in the 10^{-4}–10^{-5} sec range. Thus, the two displays are complementary, and it is becoming increasingly apparent that both should be used whenever possible. Experimental and theoretical comparisons of the two displays are shown in Fig. 6.

The paper by Thomas *et al.* (1976a) is intended as a detailed guide to the experimentalist, and the reader who wishes to perform saturation-transfer experiments is urged to study it carefully. Although this material is not repeated here, one comment seems appropriate. The authors feel that many persons have failed to appreciate the care and attention that must be paid, as discussed at length in this paper, to setting of the reference phase of the phase-sensitive detector.

D. Electron–Electron Double Resonance

In a stationary ELDOR experiment, one portion of a spectrum is saturated by the application of an intense microwave field, and the effects of this saturation at other points in the spectrum are investigated using a second microwave bridge. The magnitude of the transfer of saturation is monitored as a function of both the pumping resonant condition and the

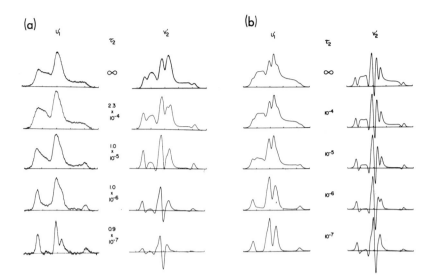

Fig. 6. (a) Representative saturation-transfer spectra obtained from maleimide-spin-labeled (**II**) hemoglobin. From the top, the samples are ammonium sulfate-precipitated Hb; Hb in 90% glycerol at $-12°C$; Hb in 80% glycerol at $5°C$; Hb in 60% glycerol at $5°C$; and Hb in 40% glycerol at $5°C$. The correlation times τ_R (τ_2 in the figure) (in seconds) were calculated from the Debye equation for the Brownian diffusion, $\tau_R = 4\pi\eta r^3/3kT$, employing measured solvent viscosities η and an effective radius of 29 Å for the hydrated protein. The signals displayed are the first-harmonic dispersion out of phase (U_1') and the second-harmonic absorption out of phase (V_2'). The applied modulation frequencies ($\omega_m/2\pi$) are 100 kHz for U_1' and 50 kHz for V_2', with detection of both at 100 kHz. A microwave field intensity (H_1) of 0.25 G and a modulation amplitude (H_m) of 5 G (peak to peak) were employed in recording spectra. Abscissa ticks are at intervals of 20 G. [Reproduced from Thomas *et al.*, (1976a).] (b) Simulated saturation-transfer spectral displays of first-harmonic dispersion out-of-phase (U_1', left) and second-harmonic absorption out-of-phase (V_2', right) signals at selected rotational correlation times τ_R. The magnetic field H_0 (horizontal axis) ranges from 3350 to 3450 G with tick marks at 20-G intervals. Modulation frequencies ($\omega_m/2\pi$) of 100 kHz for U_1' and 50 kHz for V_2' were employed. Input parameters were chosen to model the maleimide-spin-labeled Hb system in Fig. 6a: microwave field intensity $H_1 = 0.25$ G; modulation amplitude $H_m = 5$ G (peak to peak); $T_1 = 6.6 \times 10^{-6}$ sec; $T_2 = 2.4 \times 10^{-8}$ sec; $A_{\parallel} = 35$ G, $A_{\perp} = 7$ G, $g_{\parallel} = 2.00241$, and $g_{\perp} = 2.00741$. [Reproduced from Thomas, *et al.* 1976a).]

observing resonant condition. Since two microwave sources are involved, a bimodal cavity capable of resonating at the two different frequencies is employed. ELDOR is discussed at length in the book "Electron Spin Double Resonance Spectroscopy" by Kevan and Kispert (1976). Very slow rotational diffusion giving rise to transfer of saturation from one part of the nitroxide spectrum to another was investigated as an ELDOR mechanism by Smigel *et al.* (1974) and Hyde *et al.* (1975). These companion papers provide theory and experiments that undergird many of the physical con-

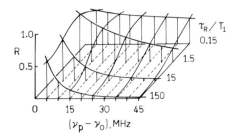

Fig. 7. Three-dimensional theoretical plot of ELDOR reduction of the high-field (-1) turning point of a very slowly tumbling nitroxide radical as a function of the rotational correlation time and the spectral distance separating pumping and observing resonant conditions. See Smigel *et al.* (1974) and Hyde *et al.* (1975). The pumping microwave field was 0.3 G, and T_1 was 6.6×10^{-6} sec.

cepts of saturation-transfer spectroscopy. A particularly important result was that by directly observing the saturation transfer it was possible to establish the validity of the two model systems, maleimide-labeled hemoglobin in glycerol–water and TANOL in supercooled *sec*-butylbenzene, that had been used in the development of the rapid-passage methods.

Figure 7 graphically illustrates spectral diffusion of saturation. The ELDOR effect is expressed in terms of a fractional reduction R of the observed signal intensity due to the presence of the pumping microwave field.

$$R = \frac{\text{(signal without pumping power)} - \text{(signal with pumping power)}}{\text{signal without pumping power}} \qquad (16)$$

This figure is a three-dimensional theoretical plot of the reduction of the high-field turning point of a nitroxide spectrum as a function of the spectral position of the pumping microwave source and the ratio τ_R/T_1. When $\tau_R/T_1 \gg 1$ and the difference frequency between pump and observing sources is large (i.e., the lower right-hand corner of the figure), no effect is observed. When $\tau_R/T_1 \ll 1$, corresponding to the rear spectral diffusion profile in the figure, the entire spectral fragment associated with the -1 nuclear quantum number (see Fig. 1) is saturated.

However, despite the conceptual attractiveness of the ELDOR method, we are concerned here with developing an understanding of the relative merits of the various techniques in analyzing motion in actual biological systems. No such applications of ELDOR have appeared, and our remarks are qualitative and speculative.

The magnitude of the ELDOR effects from the rotational diffusion mechanism is large—of the order of 10–50% change in the EPR signal intensity. In principle, it should be possible to perform ELDOR experiments in the very slow tumbling domain with excellent signal-to-noise ratios. In fact, it generally has not been possible to achieve this predicted result. The primary reason, in the authors' opinion, lies in the fact that the geometry of bimodal cavities used for ELDOR has not been optimized for aqueous samples.

The potential power of the method is very great: spectral diffusion of saturation is observed directly and interpretation is more or less immediate and obvious. A spectral diffusion profile can be measured for every point in the spectrum. Combinations of ELDOR and rapid-passage methods are possible (Percival *et al.*, 1975). Perhaps a proper point of view is that ELDOR is an important potential tool for studying complicated motions and that it is likely to be used more frequently in the future. However, methodological development will be necessary.

IV. THEORY

Either the well-known Bloch equation (Bloch, 1946) or density matrix equation (Karplus and Schwinger, 1948; Karplus, 1948; Tolman, 1938; Fano, 1957) descriptions of magnetic resonance can be employed to compute saturation-transfer EPR spectral line shapes. With each of these formalisms, the conventional approach must be modified in order to consider explicitly the effects of interaction of the spin system with an applied Zeeman modulation field† and the effects of molecular dynamics that bring various resonance lines (spin packets) into communication.† In the discussion that follows, we first shall consider the addition of Zeeman modulation to the equations of motion of a spin system. Next we shall consider *both* applied and molecular modulation effects,† demonstrating first the treatment of isotropic Brownian diffusion and then setting the stage for a discussion of more general types of motion.

---

† Effects of the Zeeman modulation field and the molecular dynamics have been treated separately by the following workers: Zeeman modulation (Hubbard *et al.*, 1957; Halbach, 1960; Macomber and Waugh, 1965); molecular modulation (Itzkowitz, 1967; Fixman, 1968; Sillescu and Kivelson, 1968; Korst and Lazarev, 1969; Kubo, 1969; Norris and Weissman, 1969; Aleksandrov *et al.*, 1970; Freed *et al.*, 1971; Gordon and Messenger, 1972; McCalley *et al.*, 1972; Vega and Fiat, 1974). Consideration of both applied and molecular modulation events concurrently acting on the spin system has been discussed by Dalton (1973), Dalton *et al.* (1975, 1976), Thomas and McConnell (1974), Robinson *et al.* (1974, 1978a), Smigel *et al.* (1974), Hyde *et al.* (1975), Coffey *et al.* (1976), Thomas *et al.* (1976a), Perkins *et al.* (1976), and Galloway and Dalton (1978a,b).

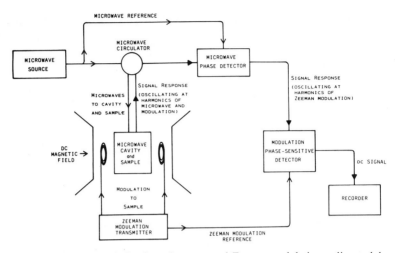

Fig. 8. Block diagram indicating microwave and Zeeman modulation coding and decoding of the signal response of an EPR spectrometer. [From Galloway and Dalton (1978a).]

As is evident from Fig. 8, EPR and ST-EPR spectra are commonly obtained employing phase-sensitive detection with respect to the driving microwave and Zeeman modulation fields. Also, due to the variety and strengths of the various applied and molecular magnetic fields that influence the time evolution of the spins, the spin response will in general be nonlinear, with components existing at various harmonic frequencies and phases with respect to the applied driving fields. We speak of the four unique signal components (in-phase dispersion, in-phase-quadrature dispersion, in-phase absorption, in-phase-quadrature absorption) detected at a given harmonic of the Zeeman modulation.

The computation of continuous-wave magnetic resonance signals (see Fig. 6b) thus involves adding a term describing the sinusoidal modulation to the Bloch or density matrix equations and solving for the signal components that are stationary with respect to the microwave and modulation fields and are in phase or in phase quadrature with respect to these fields. In terms of the "rotating-frame" description commonly employed in the solution of the Bloch and density matrix equations, only one rotating component at the microwave frequency need be considered since the other will be off resonance at twice the microwave frequency. However, the spin system is not resonant with respect to the Zeeman modulation so that components rotating at $\pm m\omega_m$ will contribute to the detected signals. We are able to write the master equation for the magnetic resonance response of an isolated spin system (packet) as follows (see Definitions of Symbols):

$$-T_2^{-1}U_m + m\omega_m U_m' + \Delta V_m = d_m(V_{m-1} + V_{m+1}) \qquad (17a)$$

$$m\omega_m U_m + T_2^{-1}U_m' + \Delta V_m' = d_m(V_{m-1}' + V_{m+1}') \qquad (17b)$$

$$\Delta U_m + T_2^{-1}V_m + m\omega_m V_m' + d_0 Y_m^+ = d_m(U_{m-1} + U_{m+1})$$
$$+ q\omega_\lambda d_0 \, \delta_{m,0} \qquad (17c)$$

$$\Delta U_m' + m\omega_m V_m - T_2^{-1}V_m' + d_0 Y_m^- = d_m(U_{m-1}' + U_{m+1}') \qquad (17d)$$

$$m\omega_m Y_m^- - T_1^{-1}Y_m^+ + 4d_0 V_m = 0 \qquad (17e)$$

$$T_1^{-1}Y_m^- + m\omega_m Y_m^+ + 4d_0 V_m' = 0 \qquad (17f)$$

or in matrix form $(\mathbf{AX} = \mathbf{Z})$ as

$$
\begin{pmatrix}
-T_2^{-1} & m\omega_m & \Delta & 0 & 0 & 0 \\
m\omega_m & T_2^{-1} & 0 & \Delta & 0 & 0 \\
\Delta & 0 & T_2^{-1} & m\omega_m & d_0 & 0 \\
0 & \Delta & m\omega_m & -T_2^{-1} & 0 & d_0 \\
0 & 0 & 4d_0 & 0 & -T_1^{-1} & m\omega_m \\
0 & 0 & 0 & 4d_0 & m\omega_m & T_1^{-1}
\end{pmatrix}
\begin{pmatrix}
U_m \\ U_m' \\ V_m \\ V_m' \\ Y_m^+ \\ Y_m^-
\end{pmatrix}
$$

$$
=
\begin{pmatrix}
d_m(V_{m-1} + V_{m+1}) \\
d_m(V_{m-1}' + V_{m+1}') \\
d_m(U_{m-1} + U_{m+1}) + q\omega_\lambda d_0 \, \delta_{m,0} \\
d_m(U_{m-1}' + U_{m+1}') \\
0 \\
0
\end{pmatrix}
\qquad (18)
$$

It also is common practice to backsubstitute (and thus eliminate) Eqs. (17e) and (17f) into Eqs. (17c) and (17d). With this operation, the matrix equation can be rewritten as

$$
\begin{pmatrix}
-T_2^{-1} & m\omega_m & \Delta & 0 \\
m\omega_m & T_2^{-1} & 0 & \Delta \\
\Delta & 0 & \Lambda_1 & \Lambda_2 \\
0 & \Delta & \Lambda_2 & -\Lambda_1
\end{pmatrix}
\begin{pmatrix}
U_m \\ U_m' \\ V_m \\ V_m'
\end{pmatrix}
=
\begin{pmatrix}
d_m(V_{m-1} + V_{m+1}) \\
d_m(V_{m-1}' + V_{m+1}') \\
d_m(U_{m-1} + U_{m+1}) + q\omega_\lambda d_0 \, \delta_{m,0} \\
d_m(U_{m-1}' + U_{m+1}')
\end{pmatrix}
$$

$$(19)$$

where $\Lambda_1 = T_2^{-1} + T_1^{-1}\zeta$, $\Lambda_2 = m\omega_m(1 - \zeta)$, and $\zeta = 4d_0^2[(T_1^{-1})^2 + (m\omega_m)^2]^{-1}$. The reader will note that, in the absence of modulation, Eqs. (17)–(19) reduce to the familiar forms of the Bloch and density matrix equations. Formally, solution of Eqs. (18) and (19) involves finding the inverse of the $\mathbf{A}$ matrix so that the signal vector $\mathbf{X}$ can be computed by the matrix multiplication $X = A^{-1}Z$. However, the addition of Zeeman modulation results in considerable complexity, and analytical solution of the general equations is no longer practical. The equations, however, can be solved conveniently for a number of special cases, reproducing the results of Portis (1955) and Weger (1960). For example, for the conditions of small Zeeman modulation amplitudes (the coupling to states higher than $m = 1$ is neglected), $4d_0^2 T_1 T_2 \gg 1$ (saturating microwave power) and $T_1^{-1} \gg \omega_m$, one obtains

$$U_1 \simeq K_1(\Delta^2 - 4d_0^2 T_1 T_2^{-1}) \tag{20a}$$

$$U_1' \simeq K_2(4d_0^2 \omega_m T_1)[\Delta^2(2T_1 T_2^{-1} - 1) + 4d_0^2 T_1 T_2^{-1}] \tag{20b}$$

$$V_1 \simeq 2K_1 \Delta T_2^{-1} \tag{20c}$$

$$V_1' \simeq -K_2 \omega_m \Delta[\Delta^2 + 4d_0^2 T_1 T_2^{-1}(2T_1 T_2^{-1} - 1)] \tag{20d}$$

where

$$K_1 = 2q\omega_\lambda d_0 d_m/[\Delta^2 + 4d_0^2 T_1 T_2^{-1}(2T_1 T_2^{-1} - 1)]^2$$

and

$$K_2 = K_1/(\Delta^2 + 4d_0^2 T_1 T_2^{-1})$$

This result shows that the in-phase-quadrature signals go to zero as the modulation frequency goes to zero, and the signal intensity of the in-phase-quadrature dispersion signal maximizes at $\omega_m = T_1^{-1}$, as can be seen from the form of the equation derived under the more relaxed condition of arbitrary Zeeman modulation frequency:

$$U_1' \propto \frac{\omega_m T_1}{1 + (\omega_m T_1)^2} \tag{21}$$

Indeed the relation expressed in Eq. (21) has been employed by several workers (Mailer and Taylor, 1973; Ammerlaan and van der Wiel, 1976) to measure electron spin–lattice relaxation times. From the standpoint of this chapter, the above relationships are important in demonstrating that signals such as the in-phase-quadrature dispersion signal detected at the first harmonic of the Zeeman modulation will be sensitive to events that modulate microwave-induced saturation of the resonance.

Rotational diffusion modulating anisotropic magnetic interactions constitutes this type of event, and we now direct our attention to the simultaneous consideration of both applied and molecular modulation effects. For simplicity, we first consider modulation of the resonance condition arising from isotropic Brownian diffusion modulating axially symmetric magnetic (electron–Zeeman and electron–nuclear hyperfine) interactions. In this case, the resonance frequency of EPR transitions depends on only one angular variable, e.g., θ. A convenient approximation to the continuous reorientation through θ space is to quantize θ space into discrete angular grids, θ_n, and to couple these discrete zones by appropriate transition or jump rates (Norris and Weissman, 1969; Gordon and Messenger, 1972; McCalley et al., 1972; Thomas and McConnell, 1974; Thomas et al., 1976a; Perkins et al., 1976). The precise values of the various transition rates will be determined by the particular motional model governing the time evolution in θ space. For example, the isotropic Brownian rotational diffusion model leads to a tridiagonal supermatrix upon quantization of θ into discrete angular zones. Inversion of such a matrix conveniently is effected by either the Gauss–Jordan or Gauss pivot method. Typical computer-simulated spectra are shown in Fig. 6b, and these can be compared with representative experimental spectra for maleimide-spin-labeled hemoglobin shown in Fig. 6a. It is also useful to define the spectral parameters H''/H, L''/L, and C'/C as shown in Fig. 9. From the comparison of the experimental and theoretical spectra shown in Figs. 6a,b and 9, it is clear that, although the trends observed as a function of changing rotational correlation time are reproduced moderately well, the quantitative agreement between experimental and theoretical spectra is poor. In particular, the agreement is poorer at the centers of the spectra than in the spectral wings or extremities, and poorer at the longer rotational correlation times. In computing the theoretical data reproduced in Figs. 6b and 9, three somewhat questionable approximations were made: (1) The magnetic interactions of the maleimide spin label II (3-maleimido-N-1-oxyl-2,2,6,6-tetramethyl-1-piperidinyloxyl) were assumed to be axially symmetric. This is a good approximation for the nitrogen hyperfine interaction but is a rather poor approximation for the electron–Zeeman interaction, which is characterized by the tensor elements $g_{xx} = 2.00840$, $g_{yy} = 2.00600$, and $g_{zz} = 2.00241$ (McCalley et al., 1972). (2) The effects of nitrogen nuclear relaxation (or the higher-order effects associated with the anisotropic nitrogen hyperfine interaction and the microwave field intensity) were ignored. (3) The effects of proton hyperfine interactions and the effects of dynamic processes coupling proton hyperfine lines were neglected. The first two approximations are largely responsible for the poorer agreement observed for the center portion of the spectrum, and the latter approximation appears to be responsible for the poorer agreement at longer rotational correlation times. When these approximations are abandoned, good quantitative agreement is observed between theoretical and experimental spectra (Robinson and Dalton, 1978). However, such calculations are so time-consuming as to be of little practical value in the analysis of spectra. Thus, the problem of computational efficiency must be considered.

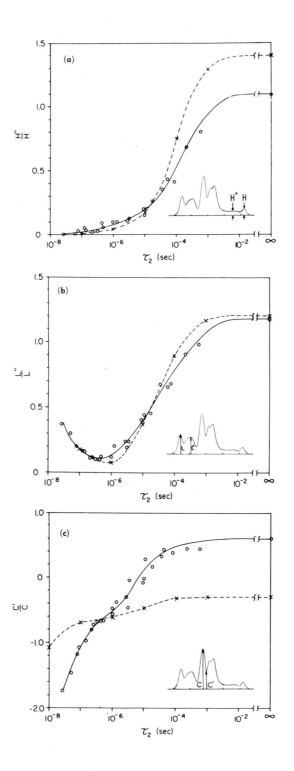

V. COMPUTATIONAL EFFICIENCY

Three approximate procedures have been developed that significantly improve computational speed.

A. Approximation of Inhomogeneous Broadening by Gaussian Convolution

The precise effect of proton hyperfine interactions on resonance spectra is dictated by the coupling of the proton hyperfine lines by Heisenberg spin exchange and by proton–nuclear relaxation. For example, motion characterized by correlation times in the range 10^{-8}–10^{-6} sec is of the proper time scale to couple proton hyperfine lines by the modulation of the anisotropic proton hyperfine interactions. The strong proton relaxation causes the nitrogen hyperfine envelopes to behave as homogeneously (or pseudohomogeneously) broadened lines, and spectra can be accurately simulated by ignoring proton interactions and determining T_2 from the width of the hyperfine envelope (e.g., $T_2 = 2.4 \times 10^{-8}$ sec). For slower molecular motion (e.g., correlation times of the order of 10^{-3} sec), the proton hyperfine lines behave as isolated, discrete spin packets, and consequently the hyperfine envelopes are inhomogeneously broadened. For such a case, a considerably longer T_2 (e.g., 2.0×10^{-7} sec) must be employed in the calculation, and the spectra must be computed by summing over the various proton hyperfine lines. Thomas et al. (1976a) showed that this summation over proton hyperfine lines can be approximated by integration over a Gaussian function. Explicitly, the signal components are computed according to

$$\mathbf{X} = \mathbf{A}^{-1}\mathbf{Z} \tag{22}$$

and this solution for $\mathbf{X}(\omega)$ is followed by convolution to obtain the final (broadened) spectrum $\bar{\mathbf{X}}(\omega)$ from

$$\bar{\mathbf{X}}(\omega) = \int_{\omega'} \mathbf{X}(\omega') \exp[-(\omega - \omega')^2/2\sigma^2] \, d\omega' \tag{23}$$

where σ is the Gaussian broadening parameter. This procedure yields quite good reproduction of spectra for correlation times from 5×10^{-5} sec to the rigid lattice limit. In Fig. 10 experimental rigid lattice spectra are shown, and in Fig. 11 the computer simulations of spectra are shown with and without the Gaussian postbroadening approximation for proton hyperfine interactions.

Fig. 9. Dependence on τ_R (τ_2 in the figure) of parameters derived from second-harmonic absorption out-of-phase spectra. Circles (solid curves) are from maleimide-spin-labeled hemoglobin and crosses (dashed curves) are from computer-simulated spectra. The assumption of axial symmetry in the simulation is responsible for the discrepancy between theory and experiment in (c). [After Thomas, et al. (1976a).]

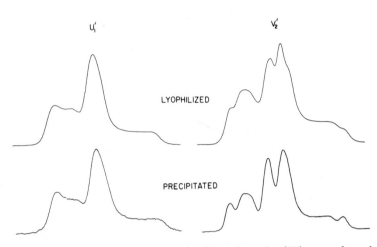

Fig. 10. Saturation-transfer dispersion (U_1') and absorption (V_2') spectra for maleimide-spin-labeled (II) hemoglobin immobilized by freeze-drying and by precipitation with ammonium sulfate. Spectra were recorded employing the customary saturation-transfer conditions: $H_1 = 0.25$ G; $H_m = 5$ G; $\omega_m/2\pi = 100$ kHz (V_1'), 50 kHz (V_2'). Scan width is 100 G. [From Thomas et al. (1976a).]

B. Gaussian Preconvolution Followed by Runge–Kutta Solution of the Master Supermatrix Equation

The preceding treatment of proton hyperfine interactions suggests an alternate scheme for the computation of spectra. Since $X(\omega)$ is the statistically weighted sum of signal responses from all orientations of the spin label in the magnetic field, it is an equivalent operation mathematically to broaden $X(\omega)$ before summation by virtue of the commutation of addition and integration operations. Applying Eq. (23) before inversion of A yields

$$A' \int_{\omega'} X(\omega') \exp[-(\omega - \omega')^2/2\sigma^2] \, d\omega' + B \int_{\omega'} (\omega' - \omega_0)X(\omega')$$

$$\times \exp[-(\omega - \omega')^2/2\sigma^2] \, d\omega' = \int_{\omega'} Z(\omega') \exp[-(\omega - \omega')^2/2\sigma^2] \, d\omega' \quad (24)$$

where the matrix A has been decomposed into a part B and a part A', neither of which depends on ω'. Matrix B, a matrix of ones, is multiplied by $(\omega' - \omega_0)$. By defining

$$\bar{Z}(\omega) \equiv \int_{\omega'} Z(\omega') \exp[-(\omega - \omega')^2/2\sigma^2] \, d\omega' \quad (25)$$

and

$$\bar{X}(\omega) \equiv \int_{\omega'} X(\omega') \exp[-(\omega - \omega')^2/2\sigma^2] \, d\omega' \quad (26)$$

Eq. (23) becomes

$$A'\bar{X}(\omega) + B \int_{\omega'} (\omega' - \omega_0)X(\omega') \exp[-(\omega - \omega')^2/2\sigma^2] \, d\omega' = \bar{Z}(\omega) \quad (27)$$

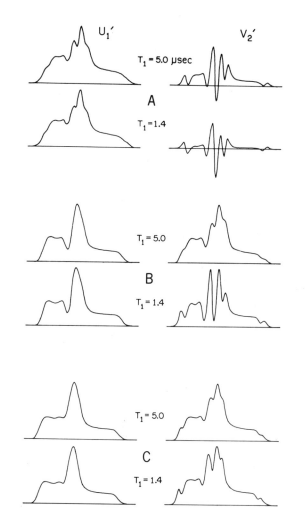

Fig. 11. Effect of inhomogeneous broadening and nonaxial symmetry on simulated saturation-transfer spectra in the rigid lattice limit. Input parameters were chosen to model the maleimide-spin-labeled Hb systems shown in Fig. 10. Values for T_1 of 5.0 μsec (lyophilized) and 1.4 μsec (precipitated) were determined from saturation-recovery and CW saturation measurements as discussed in Thomas *et al.* (1976a). Spectral displays are 100 G wide and centered at 3400 G. (A) Axial symmetry, homogeneous; $A_\parallel = 35$ G, $A_\perp = 7$ G, $g_\parallel = 2.00241$, $g_\perp = 2.00741$, $H_1 = 0.25$ G, $H_m = 5$ G, $T_2 = 2.4 \times 10^{-8}$ sec. (B) Axial symmetry, inhomogeneous; same input parameters as in (A) except that $T_2 = 2.7 \times 10^{-7}$ sec and the inhomogeneity was taken into account by broadening each point in the spectrum with a Gaussian convolution function $g(H_0) = A \exp[\frac{1}{2}(H_0 - H^{res})/\Gamma_G^{*2}]$ where $\Gamma_G^* \equiv 1/(|\gamma| T_2^*) = 4.52$ G. (C) Nonaxial symmetry, inhomogeneous; $g_{xx} = 2.00840$, $g_{yy} = 2.00600$, $g_{zz} = 2.00241$, $A_{xx} = 5.5$ G, $A_{yy} = 5.7$ G, $A_{zz} = 35.0$ G. Other input parameters are as in part (B). [Reproduced from Thomas, *et al.* (1976a)].

The remaining integral in the above equation can be replaced by an equivalent expression derived by differentiating $\bar{X}(\omega)$ with respect to ω,

$$\frac{d\bar{X}(\omega)}{d\omega} = \frac{1}{\sigma^2} \int_{\omega'} (\omega' - \omega) X(\omega') \exp[-(\omega - \omega')^2/2\sigma^2] \, d\omega' \qquad (28)$$

or equivalently

$$\frac{d\bar{X}(\omega)}{d\omega} = \frac{1}{\sigma^2} \int_{\omega'} (\omega_0 - \omega) X(\omega') \exp[-(\omega - \omega')^2/2\sigma^2] \, d\omega'$$

$$- \frac{1}{\sigma^2} \int_{\omega'} (\omega_0 - \omega') X(\omega') \exp[-(\omega - \omega')^2/2\sigma^2] \, d\omega' \qquad (29)$$

Solving Eq. (29) for the integral in Eq. (27) yields

$$\int_{\omega'} (\omega_0 - \omega') X(\omega') \exp[-(\omega - \omega')^2/2\sigma^2] \, d\omega' = -\sigma^2 \frac{d\bar{X}(\omega)}{d\omega} + (\omega_0 - \omega)\bar{X}(\omega)$$

$$(30)$$

which upon substitution into Eq. (27) gives for $\bar{X}(\omega)$

$$\frac{d\bar{X}(\omega)}{d\omega} = \frac{1}{\sigma^2} [\bar{Z}(\omega) - A'\bar{X}(\omega) - (\omega - \omega_0)\bar{X}(\omega)] \qquad (31)$$

Advantage has been taken of the fact that the problem can be written in such a manner that B is the identity matrix.

Examination of Eq. (31) shows that this is a set of coupled first-order differential equations in the variable ω. The numerical techniques developed by Runge and Kutta easily are adapted to solving this set of equations. The only quantities to be determined at the outset are the initial values of the components of $X(\omega)$; then the Runge–Kutta formulas are directly applicable to the initial-value problem of determining $\bar{X}(\omega)$. The Runge–Kutta methods evaluate $\bar{X}(\omega_1 + h)$ from the value of $\bar{X}(\omega_1)$, where h is the step parameter in frequency space. The parameter h may be fixed or varied during the calculation to control convergence. The Runge–Kutta scheme affords a reduction in core requirement and CPU time compared to calculations based on inversion of the full supermatrix A. There appears to be no loss in accuracy.

C. A Perturbation Scheme in ω_m and the Differential Approximation

It has been common practice to compute conventional EPR spectra neglecting Zeeman modulation and approximating the in-phase absorption signal at the first harmonic of the Zeeman modulation as the derivative (with respect to Δ) of the signal at the zeroth harmonic of the modulation (the signal computed ignoring Zeeman modulation). The equations developed here [see Eq. (17)] permit convenient examination of this approximation and a more general understanding of the relationship between signal components at various modulation harmonics. From Eq. (17) we write the zeroth-harmonic signal equations as

$$-T_2^{-1}U_0 + \Delta V_0 = 2d_m V_1 \tag{32a}$$

$$T_2^{-1}U_0' + \Delta V_0' = 0 \tag{32b}$$

$$\Delta U_0 + T_2^{-1}V_0 + d_0 Y_0^+ = 2d_m U_1 + q\omega_\lambda d_0 \tag{32c}$$

$$\Delta U_0' - T_2^{-1}V_0' + d_0 Y_0^- = 0 \tag{32d}$$

$$d_0 V_0 - (T_1^{-1}/4)Y_0^+ = 0 \tag{32e}$$

$$d_0 V_0 + (T_1^{-1}/4)Y_0^- = 0 \tag{32f}$$

Next let us compute the derivatives with respect to Δ of these zeroth-harmonic equations:

$$-T_2^{-1}\frac{dU_0}{d\Delta} + \Delta\frac{dV_0}{d\Delta} = 2d_m\frac{dV_1}{d\Delta} - V_0 \tag{33a}$$

$$T_2^{-1}\frac{dU_0'}{d\Delta} + \Delta\frac{dV_0'}{d\Delta} = -V_0' \tag{33b}$$

$$\Delta\frac{dU_0}{d\Delta} + T_2^{-1}\frac{dV_0}{d\Delta} + d_0\frac{dY_0^+}{d\Delta} = 2d_m\frac{dU_1}{d\Delta} - U_0 \tag{33c}$$

$$\Delta\frac{dU_0'}{d\Delta} - T_2^{-1}\frac{dV_0'}{d\Delta} + d_0\frac{dY_0^-}{d\Delta} = -U_0' \tag{33d}$$

$$\frac{dY_0^+}{d\Delta} = 4d_0 T_1(dV_0/d\Delta) \tag{33e}$$

$$\frac{dY_0^-}{d\Delta} = -4d_0 T_1(dV_0'/d\Delta) \tag{33f}$$

Upon backsubstituting Eqs. (33e) and (33f), we obtain a matrix equation for the four unique signals,

$$\begin{pmatrix} -T_2^{-1} & \Delta & 0 & 0 \\ \Delta & \hat{T}_2^{-1} & 0 & 0 \\ 0 & 0 & T_2^{-1} & \Delta \\ 0 & 0 & \Delta & -\hat{T}_2^{-1} \end{pmatrix} \begin{pmatrix} \dfrac{dU_0}{d\Delta} \\ \dfrac{dV_0}{d\Delta} \\ \dfrac{dU_0'}{d\Delta} \\ \dfrac{dV_0'}{d\Delta} \end{pmatrix} = \begin{pmatrix} 2d_m\dfrac{dV_1}{d\Delta} - V_0 \\ 2d_m\dfrac{dU_1}{d\Delta} - U_0 \\ -V_0' \\ -U_0' \end{pmatrix} \tag{34}$$

where $\hat{T}_2^{-1} = T_2^{-1} + 4d_0^2 T_1$. Now writing out the first-harmonic equations,

$$-T_2^{-1}U_1 + \omega_m U_1' + \Delta V_1 = d_m V_0 \tag{35a}$$

$$\omega_m U_1 + T_2^{-1}U_1' + \Delta V_1' = d_m V_0' \tag{35b}$$

$$\Delta U_1 + T_2^{-1}V_1 + \omega_m V_1' + d_0 Y_1^{+} = d_m U_0 \tag{35c}$$

$$\Delta U_1' + \omega_m V_1 - T_2^{-1}V_1' + d_0 Y_1^{-} = d_m U_0' \tag{35d}$$

$$4d_0 V_1' + \omega_m Y_1^{+} + T_1^{-1}Y_1^{-} = 0 \tag{35e}$$

$$4d_0 V_1 - T_1^{-1}Y_1^{+} + \omega_m Y_1^{-} = 0 \tag{35f}$$

we note that Eqs. (35e) and (35f) can be written as

$$Y_1^{+} = \frac{4d_0}{\omega_m^2 + T_1^{-2}}(T_1^{-1}V_1 - \omega_m V_1') \tag{36a}$$

and

$$Y_1^{-} = \frac{-4d_0}{\omega_m^2 + T_1^{-2}}(T_1^{-1}V_1' + \omega_m V_1) \tag{36b}$$

If the Zeeman modulation frequency ω_m is small relative to the electron spin–lattice relaxation time T_1, then Eqs. (36a) and (36b) can be rewritten as

$$Y_1^{+} = 4d_0 T_1 V_1 \qquad \text{and} \qquad Y_1^{-} = -4d_0 T_1 V_1' \tag{37}$$

Upon backsubstituting these equations into the first-harmonic signal equations, we obtain

$$-T_2^{-1}U_1 + \omega_m U_1' + \Delta V_1 = d_m V_0 \tag{38a}$$

$$\omega_m U_1 + T_2^{-1}U_1' + \Delta V_1' = d_m V_0' \tag{38b}$$

$$\Delta U_1 + \hat{T}_2^{-1}V_1 + \omega_m V_1' = d_m U_0 \tag{38c}$$

$$\Delta U_1' + \omega_m V_1 - \hat{T}_2^{-1}V_1' = d_m U_0' \tag{38d}$$

or in matrix form as

$$\begin{pmatrix} -T_2^{-1} & \Delta & 0 & 0 \\ \Delta & \hat{T}_2^{-1} & 0 & 0 \\ 0 & 0 & T_2^{-1} & \Delta \\ 0 & 0 & \Delta & -\hat{T}_2^{-1} \end{pmatrix} \begin{pmatrix} U_1 \\ V_1 \\ U_1' \\ V_1' \end{pmatrix} = \begin{pmatrix} d_m V_0 - \omega_m U_1' \\ d_m U_0 - \omega_m V_1' \\ d_m V_0' - \omega_m U_1 \\ d_m U_0' - \omega_m V_1 \end{pmatrix} \tag{39}$$

In matrix notation, the derivative zeroth-harmonic equation [Eq. (34)] and the first-harmonic equation [Eq. (39)] can be summarized as

$$\mathbf{A}\frac{d\mathbf{X}(0)}{d\Delta} = 2d_m \frac{d\hat{\mathbf{X}}(1)}{d\Delta} - \bar{\mathbf{X}}(0) \tag{40}$$

and

$$\mathbf{A}\mathbf{X}(1) = d_m \bar{\mathbf{X}}(0) - \omega_m \mathbf{X}(1) \tag{41}$$

where

$$\frac{dX(0)}{d\Delta} = \begin{pmatrix} \dfrac{dU_0}{d\Delta} \\[2mm] \dfrac{dV_0}{d\Delta} \\[2mm] \dfrac{dU_0'}{d\Delta} \\[2mm] \dfrac{dV_0'}{d\Delta} \end{pmatrix} \qquad X(1) = \begin{pmatrix} U_1 \\[2mm] V_1 \\[2mm] U_1' \\[2mm] V_1' \end{pmatrix}$$

$$\frac{d\hat{X}(1)}{d\Delta} = \begin{pmatrix} \dfrac{dV_1}{d\Delta} \\[2mm] \dfrac{dU_1}{d\Delta} \\[2mm] 0 \\[2mm] 0 \end{pmatrix} \qquad \bar{X}(0) = \begin{pmatrix} V_0 \\[2mm] U_0 \\[2mm] 0 \\[2mm] 0 \end{pmatrix}$$

$$d_m = d_m \begin{pmatrix} 1 & 0 & 0 & 0 \\ 0 & 1 & 0 & 0 \\ 0 & 0 & 0 & 0 \\ 0 & 0 & 0 & 0 \end{pmatrix} \qquad \omega_m = \omega_m \begin{pmatrix} 0 & 0 & 1 & 0 \\ 0 & 0 & 0 & 1 \\ 1 & 0 & 0 & 0 \\ 0 & 1 & 0 & 0 \end{pmatrix}$$

We observe that Eqs. (40) and (41) have the term $\bar{X}(0)$ in common. Solving for $\bar{X}(0)$ utilizing Eq. (40) and then substituting this result into Eq. (41), we obtain after some algebraic manipulation

$$X(1) = -(\mathbb{1} + A^{-1}\omega_m)^{-1}A^{-1}d_m A[dX(0)/d\Delta] \tag{42}$$

If ω_m is small enough to allow $A^{-1}\omega_m$ to be considered as a perturbation to the unit matrix, we can expand the matrix inverse in a power series in ω_m. To first order in ω_m we write

$$X(1) = -(\mathbb{1} + A^{-1}\omega_m)A^{-1}d_m A[dX(0)/d\Delta] \tag{43}$$

It is instructive to examine the matrix characteristics of this first-order equation. Matrices in the above equation can be written in block form

$$A = \begin{pmatrix} A_I & 0 \\ 0 & A_O \end{pmatrix} \qquad A^{-1} = \begin{pmatrix} A_I^{-1} & 0 \\ 0 & A_O^{-1} \end{pmatrix} \qquad \omega_m = \begin{pmatrix} 0 & \hat{\omega}_m \\ \hat{\omega}_m & 0 \end{pmatrix}$$

$$d_m = \begin{pmatrix} \hat{d}_m & 0 \\ 0 & 0 \end{pmatrix} \qquad \mathbb{1} = \begin{pmatrix} 1 & 0 \\ 0 & 1 \end{pmatrix}$$

Eq. (43) can be rewritten in terms of these components as

$$
\begin{pmatrix} X_1(1) \\ X_0(1) \end{pmatrix} = \begin{pmatrix} -\hat{d}_m & 0 \\ A_0^{-1}\hat{\omega}_m\hat{d}_m & 0 \end{pmatrix} \begin{pmatrix} \dfrac{dX_1(0)}{d\Delta} \\ \dfrac{dX_0(0)}{d\Delta} \end{pmatrix} \tag{44}
$$

where $X_1(1)$ and $X_0(1)$ denote, respectively, the in-phase and out-of-phase (in-phase-quadrature) signal components at the first harmonic of the Zeeman modulation, and $dX_1(0)/d\Delta$ and $dX_0(0)/d\Delta$ denote, respectively, the derivatives of the in-phase and in-phase-quadrature signal components at the zeroth harmonic of the modulation. For the conditions of the preceding derivation (d_m and ω_m small relative to molecular relaxation rates), the in-phase first-harmonic signals are proportional to derivatives of the zeroth-harmonic signals, namely,

$$
X_1(1) = -\hat{d}_m \frac{dX_1(0)}{d\Delta} \tag{45}
$$

Thus, we see the mathematical justification for the derivative approximation routinely employed in the computation of conventional EPR spectra. The validity of this approximation can be readily ascertained by examining the magnitude of the higher-order terms in the perturbation expansion in ω_m.

The in-phase-quadrature signals are obtained from the derivatives of the in-phase signal at the zeroth harmonic of the Zeeman modulation by the relationship

$$
X_0(1) = A_0^{-1}\hat{\omega}_m\hat{d}_m \frac{dX_1(0)}{d\Delta} \tag{46}
$$

This result suggests a perturbation approximation to the theoretical results of Thomas et al. (1976a). Treating molecular reorientation by the transition rate matrix description, we write the master supermatrix equation as

$$
A(m, n, v)X(m, n, v) = Z(m, n, v) \tag{47}
$$

where the indices m, n, v denote, respectively, Zeeman modulation harmonic, orientational grid point, and nitrogen nuclear spin state. Also,

$$
X(m, n, v) = \begin{pmatrix} X_1(0, n, v) \\ X_1(1, n, v) \\ X_0(1, n, v) \end{pmatrix} \tag{48}
$$

where

$$
X_1(m, n, v) = \begin{pmatrix} U(m, n, v) \\ V(m, n, v) \\ Y^+(m, n, v) \end{pmatrix} \quad \text{and} \quad X_0(m, n, v) = \begin{pmatrix} U'(m, n, v) \\ V'(m, n, v) \\ Y^-(m, n, v) \end{pmatrix}
$$

We can define $Z(m, n, v)$, which contains the equilibrium part of the magnetization, in an analogous manner. Also,

$$
A(m, n, v) = \begin{pmatrix} B_1(n, v) & 0 & 0 \\ \hat{d}_m & B_1(n, v) & \omega_m \\ 0 & \omega_m & B_0(n, v) \end{pmatrix}
\tag{49}
$$

and

$$
B_{\binom{1}{0}}(n, v) = \begin{pmatrix} A_{\binom{1}{0}}(1, v) & C_{\binom{1}{0}}_{1,2} & 0 & \cdots \\ C_{\binom{1}{0}}_{2,1} & A_{\binom{1}{0}}(2, v) & C_{\binom{1}{0}}_{2,3} & \cdots \\ 0 & C_{\binom{1}{0}}_{3,2} & A_{\binom{1}{0}}(3, v) & \cdots \\ 0 & 0 & C_{\binom{1}{0}}_{4,3} & \cdots \\ & & & \\ & & & \\ & & & \end{pmatrix}
\tag{50}
$$

$$
C_{In, n'} = t_{n, n'}^{-1} \begin{pmatrix} 1 & 0 & 0 \\ 0 & -1 & 0 \\ 0 & 0 & 1 \end{pmatrix} \qquad C_{On, n'} = t_{n, n'}^{-1} \begin{pmatrix} -1 & 0 & 0 \\ 0 & 1 & 0 \\ 0 & 0 & -1 \end{pmatrix}
\tag{51a,b}
$$

where $t_{n, n'}^{-1}$ for the various angular zones are defined in terms of the rotational correlation time for isotropic Brownian diffusion (McCalley *et al.*, 1972). The $A_{\binom{1}{0}}(n, v)$ matrices are given explicitly as

$$
A_1(n, v) = \begin{pmatrix} -T_2^{-1}(n, v) & \Delta(n, v) & 0 \\ \Delta(n, v) & T_2^{-1}(n, v) & d_0 \\ 0 & d_0 & -T_1^{-1}(n, v)/4 \end{pmatrix}
\tag{52}
$$

and

$$
A_0(n, v) = \begin{pmatrix} T_2^{-1}(n, v) & \Delta(n, v) & 0 \\ \Delta(n, v) & -T_2^{-1}(n, v) & d_0 \\ 0 & d_0 & T_1^{-1}(n, v)/4 \end{pmatrix}
\tag{53}
$$

where $T_2^{-1}(n, v) = T_2^{-1}(v) + t_n^{-1}$, and $T_1^{-1}(n, v) = T_1^{-1}(v) + t_n^{-1}$, with t_n defined by the rotational correlation time; ω is an $n \times n$ matrix defined as

$$\omega = \omega_m \begin{pmatrix} 1 & 0 & 0 & \cdot & \cdot & \cdot \\ 0 & 1 & 0 & \cdot & \cdot & \cdot \\ 0 & 0 & 1 & \cdot & \cdot & \cdot \\ \cdot & & \cdot & \cdot & & \\ \cdot & & & \cdot & \cdot & \\ & \cdot & & \cdot & \cdot & \end{pmatrix} \qquad \text{where} \qquad 1 = \begin{pmatrix} 1 & 0 & 0 \\ 0 & 1 & 0 \\ 0 & 0 & \frac{1}{4} \end{pmatrix} \qquad (54)$$

We define $\mathbb{D}$ as the $n \times n$ matrix

$$\mathbb{D} = d_m \begin{pmatrix} \hat{1} & 0 & 0 & \cdot & \cdot & \cdot \\ 0 & \hat{1} & 0 & \cdot & \cdot & \cdot \\ 0 & 0 & \hat{1} & \cdot & \cdot & \cdot \\ \cdot & & \cdot & \cdot & & \\ \cdot & & & \cdot & \cdot & \\ & \cdot & & \cdot & \cdot & \end{pmatrix} \qquad \text{where} \qquad \hat{1} = \begin{pmatrix} 0 & 1 & 0 \\ 1 & 0 & 0 \\ 0 & 0 & 0 \end{pmatrix} \qquad (55)$$

We see that this form [Eq. (47)] also is adapted readily to the perturbation

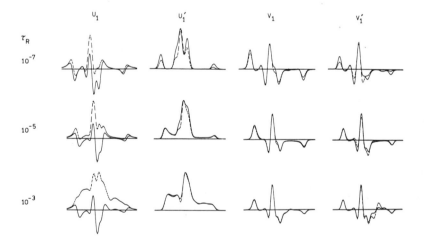

Fig. 12. Comparison of first-harmonic signal responses U_1, U_1', V_1, and V_1' calculated employing the ω_m derivative approximation (solid line) with the signals calculated using the full matrix inversion (dashed line) in the limit of low modulation amplitude. Input parameters: $H_1 = 0.25$ G, H_m (peak to peak) = 0.002 G, $\omega_m/2\pi = 100$ kHz, $A_\parallel = 35$ G, $A_\perp = 7$ G, $g_\parallel = 2.00241$, $g_\perp = 2.00741$, $T_1 = 6.6 \times 10^{-6}$ sec, $T_2 = 2.0 \times 10^{-7}$ sec. The width of the spectral display is 100 G. [From Galloway and Dalton (1978a).]

treatment discussed earlier; for example, the master equation can be solved to yield

$$X_I(0, n, v) = B_I^{-1}(n, v)Z(0, n, v) \tag{56a}$$

$$X_I(1, n, v) = -\bar{B}_I^{-1} \mathbb{D} X_I(0, n, v) \tag{56b}$$

$$X_0(1, n, v) = -B_0^{-1}(n, v)\omega X_I(1, n, v) \tag{56c}$$

where $\bar{B}_I = B_I(n, v) - \omega B_0^{-1}(n, v)\omega$. In Fig. 12 we compare the results of computation of the several first-harmonic signals by full matrix inversion as discussed by Thomas *et al.* (1976a) and by the perturbation scheme discussed above. The validity of the approximate scheme for computing 100-kHz signals for the various relaxation rates is evident from a consideration of the spectra shown in Fig. 13. The perturbation approach is useful for the computation of first-harmonic signals even for the condition of finite Zeeman modulation amplitudes. However, the perturbation approach is unsatisfactory for the computation of the second-harmonic in-phase-quadrature absorption signal as is seen from Fig. 14.

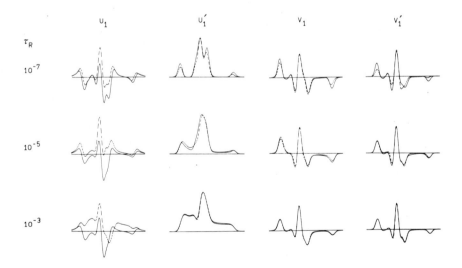

Fig. 13. Comparison of first-harmonic signal responses U_1, U_1', V_1, and V_1' calculated employing the ω_m derivative approximation (solid line) with signals calculated using the full matrix inversion (dashed line) for finite modulation amplitude. Input parameters are the same as in Fig. 12, except H_m (peak to peak) = 5 G. [Taken from Galloway and Dalton (1978b).]

U_2 U_2' V_2 V_2'

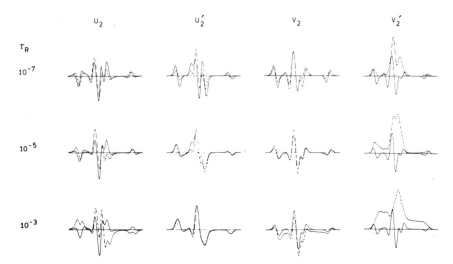

Fig. 14. Comparison of second-harmonic signal responses U_2, U_2', V_2, and V_2' calculated employing the derivative approximation (solid line) with signals calculated by full matrix inversion (dashed line). Magnetic and molecular relaxation parameters are the same as employed in Fig. 13 (note that $\omega_m/2\pi$ is 100 kHz). [Taken from Galloway and Dalton (1978b).]

VI. APPLICATIONS OF SATURATION-TRANSFER ELECTRON PARAMAGNETIC RESONANCE

The first application of ST-EPR to a biomolecular system was in model studies employing the soluble protein, maleimide-spin-labeled (4-maleimido-2,2,6,6-tetramethyl-1-piperidinyloxyl) **(II)** human oxyhemoglobin (Hyde and Thomas, 1973; Thomas *et al.*, 1976a). Supramolecular aggregations of soluble proteins seemed suitable for ST-EPR techniques, and early applications included the study of the interaction of the fundamental muscle proteins actin and myosin (Thomas *et al.*, 1975a,b, 1976b, 1978a,b; Thomas, 1978) and their polymers and of the self-association of human deoxyhemoglobin S (sickle cell hemoglobin) (Robinson *et al.*, 1977; Johnson *et al.*, 1978).

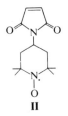

II

A natural extension of the method was to the study of membrane-bound proteins, since the interaction between membrane proteins and lipids is likely to result in protein motions in the very slow tumbling region. Recent applications include the measurement of the rotational diffusion of the visual receptor protein rhodopsin in the rod outer segment membrane (Baroin *et al.*, 1977; Devaux *et al.*, 1977), the cholinergic receptor protein in *Torpedo marmorata* membrane fragments (Devaux *et al.*, 1977), and the Ca^{2+}-ATPase of sarcoplasmic reticulum (Thomas and Hidalgo, 1978; Hidalgo *et al.*, 1978).

The motion of lipid hydrocarbon chains near the surface of membrane bilayers in the gel phase is sufficiently restricted to be characterized by elements of a diffusion tensor with values that fall in the very slow motion region ($10^7 < \tau^{-1} < 10^3$ rad/sec). The study of phospholipid motion in gel-phase lipid bilayers by ST-EPR has demonstrated that hydrocarbon chain motion is highly anisotropic, with diffusion tensor elements in the very slow motion region (D. Marsh, unpublished). D. D. Thomas, J. S. Hyde, and H. M. McConnell (unpublished) have investigated the rotational motion of palmitic acid spin labels in highly ordered multilamellar arrays of egg lecithin in an effort to measure the slow (collective) motions of hydrocarbon chains in phospholipid bilayers. ST-EPR has been applied to the study of the modification of membrane structure by myopathies (muscle diseases) (Wilkerson *et al.*, 1978).

Long-chain-polymer motion is another application of ST-EPR, and studies of synthetic polymers (Dorio and Chien, 1974; Brown, 1976, 1978; Brown and Lind, 1977) and the biological polymer DNA (Robinson *et al.*, 1979) have been executed. A number of biological supramolecular assemblies such as antibodies and viral particles (Hemminga *et al.*, 1977) are logical candidates for study by ST-EPR.

Although the orientation of this chapter is toward consideration of biological systems employing nitroxide spin labels, many other applications of ST-EPR can be envisioned, including the binding of paramagnetic substrates to surfaces, dynamic Jahn–Teller distortion (e.g., in copper complexes), exciton motion, and chemical reaction kinetics. The investigation of the "hopping" or two-dimensional diffusion processes of O_2^- adsorbed on silver (Clarkson and Kooser, 1978) is a good example.

A comprehensive review of ST-EPR applications lies outside the objectives of this chapter; rather, we choose to review applications that demonstrate important features of ST-EPR methodology and seem likely to stimulate further efforts.

A. Actin–Myosin Interactions

1. ROTATIONAL MOTION OF MYOSIN CROSSBRIDGES ("HEADS") IN THE ABSENCE OF ATP

The most thorough application of ST-EPR has been in characterizing the mobility of myosin "heads" (subfragment-1, see Fig. 15) in the absence

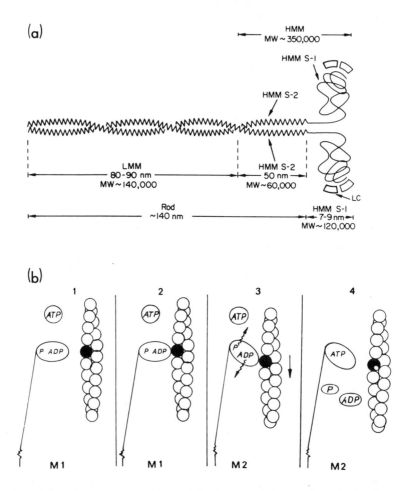

Fig. 15. (a) Schematic representation of the structure of the myosin molecule. The rod portion of the molecule has a coiled-coil α-helical structure. Hinge regions postulated in the mechanism of contraction are at the junction of HMM (heavy meromyosin) S-1 and HMM S-2 and of HMM S-2 and LMM (light meromyosin). It should be noted that HMM S-1 has one chief polypeptide chain, whereas the other fragments have two (LC, light chain associated with the S-1 region). (b) Schematic representation of the crossbridge cycle coupled to the hydrolysis of ATP. Here M1 and M2 denote two possible orientations of the myosin head region (showing only one head) resulting from rotation between the S-1 and S-2 portions of HMM. The other point at which a hinge has been postulated is the junction of the connecting lines and the body of the thick filament, indicated by the break at the bottom left of each panel. [Parts (a) and (b) reproduced with permission from Thomas *et al.* (1976b).]

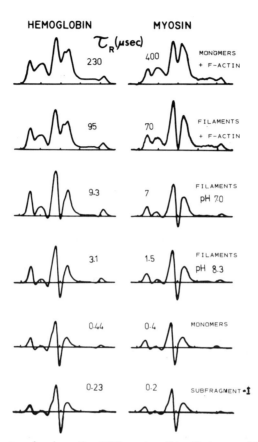

Fig. 16. Saturation-transfer absorption EPR spectra, V_2', of iodoacetamide-spin-labeled (III) myosin and its derivatives (right). The rotational correlation time τ_R is in microseconds. The spectra on the left are for comparison and were obtained from maleimide-spin-labeled (II) hemoglobin in glycerol–water solutions. Their τ_R values were calculated from the solvent viscosity and known size of hemoglobin. The τ_R values for the myosin preparations were determined by comparing their spectra with hemoglobin spectra and with theoretically simulated spectra (Thomas *et al.*, 1976a). Preparative and spin-labeling procedures are given in Thomas *et al.* (1975a,b, 1976b). [Reproduced from Thomas *et al.* (1976b).]

of ATP (Thomas *et al.*, 1975a,b, 1976b). As can be seen from Fig. 16, the spectra obtained for iodoacetamide-spin-labeled (N-1-oxyl-2,2,6,6-tetra-methyl-4-piperidinyliodoacetamide) (III) myosin subfragment-1 under various conditions are quite comparable to those obtained for the model system of maleimide-spin-labeled human oxyhemoglobin. As discussed elsewhere in this chapter (p. 58), this observation is consistent with the existence of coincident principal axes of the diffusion and magnetic tensors. Thus, the rotational correlation times shown for subfragment-1 in Fig. 16 correspond to rotation about an axis perpendicular to the long axis, $\tau_\perp$.

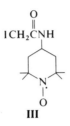

III

Comparison of the ST-EPR spectrum of the free subfragment-1 with that of intact myosin monomers (Fig. 16) indicates the existence of considerable segmental flexibility of the S-1 "heads" relative to the α-helical rod portion of the molecule. The remaining spectra shown in Fig. 16 demonstrate the motion of the myosin heads in various self-assembled systems, including myosin filaments and actomyosin. The rotational correlation times observed for myosin filaments indicate that, although filament formation restricts crossbridge motion significantly relative to that observed for myosin monomers in solution, there is sufficient crossbridge mobility for consistency with a model of the swiveling crossbridge as the site of force generation in muscle contraction (Huxley, 1969). It also is clear that there is less crossbridge motion in filaments prepared at pH 7, where the filaments are larger and form visible aggregates, than at pH 8.3.

The top two spectra of Fig. 16 show that F-actin has a large immobilizing effect on myosin heads when slightly more than one F-actin molecule is present per myosin head. The greatest immobilization occurs upon addition of F-actin to monomeric myosin (at high ionic strength), demonstrating that both heads of each myosin molecule are immobilized upon combination with F-actin in the absence of ATP. Thomas *et al.* (1975a,b) also showed that about the same degree of immobilization occurs when F-actin is added to isolated subfragment-1. The spectra in Fig. 16 also demonstrate that the effect of F-actin on the heads in myosin filaments (at low ionic strength) is not as great, probably because steric effects make it difficult for all of the crossbridges in the bulky, thick filaments to bind to F-actin filaments.

2. ROTATIONAL MOTION OF THE MYOSIN CROSSBRIDGES DURING ATP HYDROLYSIS

The spectra discussed in the preceding section permit characterization of F-actin and myosin interactions in the state of rigor (associated with death). However, the crossbridge motion that is proposed as the force-generating step in muscle contraction occurs during hydrolysis of ATP. The iodoacetamide spin label cannot be employed to provide information about these motions, since it does not remain rigidly bound to the myosin head in the presence of ATP. Recently, Thomas *et al.* (1978a) utilized a maleimide spin label (4-maleimido-2,2,6,6-tetramethyl-1-piperidinyloxyl) (**II**) that was observed to bind specifically to the heads of the myosin and to be strongly immobilized in both the presence and absence of ATP. They therefore were able to measure the rotational motion of myosin heads during ATP hydrolysis. Although the motion is sensitive to the concentrations of KCl, phosphate, and other ions, the rotational motion of heads in myosin filaments is no different in the presence of ATP (during ATP hydrolysis) than in its absence. That is, even in the absence of ATP, the rotational correlation time is already in the microsecond range, and the ATP has no additional effect.

3. Rotational Motion of Crossbridges in Myofibrils (The Minimum Contractile Unit)

By blocking some of the fast-reacting sulfhydryl groups with *N*-ethylmaleimide, reacting with the maleimide spin label, and then treating with ferricyanide, Thomas *et al.* (1978a) succeeded in specifically spin labeling the head region of myosin in intact myofibrils. The label remained rigidly attached during ATP hydrolysis, and thus they were able to measure crossbridge motion in the three characteristic states of muscle: rigor (no ATP, crossbridges presumed to be in contact with actin), relaxation (ATP but no Ca^{2+}, crossbridges presumed to be detached from actin), and contraction (ATP and Ca^{2+}, crossbridges presumably attaching and detaching cyclically). Preliminary results of Thomas *et al.* (1978a) indicate that in rigor the crossbridges are rigidly immobilized on the millisecond time scale of ST-EPR, but addition of ATP, either in the presence or absence of Ca^{2+}, results in approximately the same rotational motion ($\tau_R \sim 10^{-5}$ sec) as observed for myosin filaments in the absence of ATP.

4. Rotational Dynamics of F-Actin

Although attention has been focused on the crossbridges formed by myosin heads in thick filaments in order to understand the molecular dynamics of muscle contraction, there has also been interest in the motions occurring within the actin-containing thin filament. Moreover, evidence is accumulating that actin, not myosin, is the main component in contractile systems in nonmuscle cells.

Thomas *et al.* (1978b) utilized ST-EPR to investigate the effect of myosin on motions within actin filaments employing maleimide-spin-labeled actin. They detected rotational motions of 10^{-4} sec, which are at least ten times faster than the large-scale motions observed with light-scattering techniques (Fujime and Ishiwata, 1971) and are probably due to more localized bending

or flexing motions within the actin filaments. When myosin, heavy meromy-
sin, or isolated subfragment-1 is added to spin-labeled F-actin, the
saturation-transfer spectra indicate that the actin motion is greatly reduced.
As shown in Fig. 17, which is taken from the work of Thomas *et al.* (1976b),
a maximal immobilization of actin is observed at a low heavy meromyosin–
actin ratio. This suggests that the immobilization is propagated "cooperatively" along the actin filament. Thomas *et al.* also observed essentially the
same immobilization in the rigor complex with a spin label on the myosin
head as with the label on actin, adding to the evidence that this complex
between the actin molecules and the myosin head is a rigid unit. Thomas

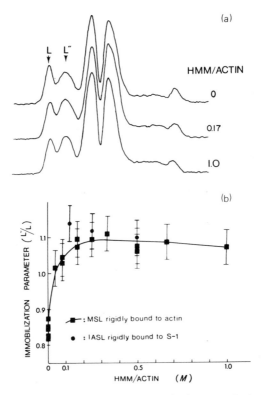

Fig. 17. (a) Saturation-transfer absorption EPR (V_2') spectra obtained from maleimide-
spin-labeled (**II**) F-actin combined with various amounts of heavy meromyosin (HMM) as
indicated by the molar ratios given at the right. The actin concentration was 2.8 mg/ml in
buffer. Temperature was 20°C. (b) The dependence of the L''/L parameter on the ratio
HMM/F-actin. For the case of isotropic Brownian rotational diffusion, a change in L''/L from
0.85 to 1.1 indicates a change in τ_R from 100 to 400 μsec (see Fig. 9) (MSL, maleimide spin
label; IASL, iodoacetamide spin label). [Taken from Thomas *et al.* (1976b).]

et al. (1976b, 1978b) obtained results similar to those shown in Fig. 17 upon adding subfragment-1 to actin.

Tropomyosin and troponin are known to be important in thin-filament regulation. Thomas *et al.* (1976b) found, using ST-EPR, that the effect of myosin heads on the slow motions in F-actin is present even if tropomyosin is combined with actin.

5. ACTIN–MYOSIN INTERACTIONS IN NONMUSCLE SYSTEMS

The work of Thomas and co-workers has focused on actin–myosin interactions in skeletal muscle. The extension of ST-EPR methodology to the investigation of these interactions in smooth muscle is obvious. Perhaps even more exciting is the possibility of employing ST-EPR to characterize actin–myosin interaction in nonmuscle systems, such as developing neurons. For example, for this system it is not yet clear whether the Ca^{2+} influence on fetal brain actomyosin Mg^{2+}-ATPase is mediated through actin or through myosin.

B. Hemoglobin S Self-Association

Johnson, Lionel, and Dalton (Robinson *et al.*, 1977; Lionel *et al.*, 1978) examined supramolecular aggregates of sickle cell hemoglobin (Hb-S) both from the standpoint of utilizing this system as a model system for anisotropic motion and to determine the effect of various substances on the structure of the aggregates and on the kinetics of "polymerization." Typical results are shown in Fig. 18. When compared with model system Hb-A spectra (see, for example, Fig. 16), it is clear that the Hb-S spectra are characterized by anisotropic motion. A detailed consideration of the general theory of anisotropic motion (Robinson *et al.*, 1977) indicates that the spectra are consistent with the existence of nearly orthogonal magnetic and diffusion tensor axes and with the model for supramolecular aggregate growth proposed by Hofrichter *et al.* (1973) in which fiber growth occurs along the X molecular axis of hemoglobin. Perhaps even more interesting than the information on aggregate structure provided by ST-EPR is the information obtained concerning the kinetics of self-association. Johnson and co-workers (Robinson *et al.*, 1977; Johnson *et al.*, 1978) observed the kinetic rates to increase with increasing 2,3-diphosphoglycerate (DPG) concentration, with increasing inositol hexaphosphate (IHP) concentration, and with lowering the pH toward 6.7. The addition of 1 M salt significantly inhibited self-association. The effects of DPG and IHP are probably due to the induction of small structural changes in the deoxy conformation that enhance the stability of the aggregate. The salt and pH effects indicate that polar interactions play a significant role in polymer stability. Erythrocyte membrane fragments also

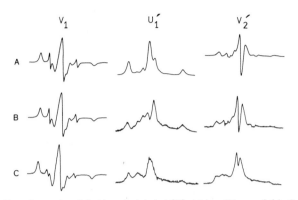

Fig. 18. Experimental maleimide-spin-labeled (II) sickle cell hemoglobin S (Hb-S) spectra are shown at various stages in the self-association process. The spectra were recorded at 25°C on a deoxy Hb-S sample at a hemoglobin concentration of 29.3 gm/dl. The conventional absorption EPR (V_1) and saturation-transfer dispersion (U_1') and absorption (V_2') spectra are shown. The ST-EPR spectra were recorded at $H_1 = 0.25$ G, H_m (peak to peak) $= 5$ G. Display width for all spectra is 100 G. (A) Spectra recorded immediately after the temperature has been raised from 5° to 25°C. Note that the V_1 EPR spectrum is comprised of overlapping spectra from spin label free in buffer and from label bound to the β-93 position of Hb-S. (B) Spectra recorded 60 min after the temperature jump. The EPR spectrum has changed little from that shown in (A); on the other hand, the ST-EPR spectra are characterized by considerably slower and highly anisotropic motion. (C) Spectra recorded 120 min after the temperature jump.

appear to play a role in the kinetics of polymerization, although the exact dependence is obviously difficult to quantitate. The action of membrane fragments could be through an effect either on the deoxy conformation or on the molecular contact regions of the associating subunits.

Two criticisms can be leveled at the study of Hb-S self-association employing spin-labeled Hb-S and ST-EPR methodology: (1) the presence of the spin label may significantly perturb the structure of supramolecular aggregates and the kinetics of polymerization; and (2) Hb-S gels represent a heterogeneous system, since aggregates of more than one size will be present. Neither of these criticisms can be totally refuted, but they should be considered in the light of the following comments. Nuclear magnetic resonance measurements following the methods of Fung and Ho (1975) were used to compare resonances arising from intersubunit hydrogen bonds of labeled Hb with those of unlabeled Hb, with the result that the resonances were unaffected by labeling. This suggests that the labeling process does not induce significant distortions in the quaternary conformations. Fully spin-labeled Hb-S does exhibit gelation under standard conditions, indicating that modification of the label site (β-93) is not very important to the polymerization process. Any perturbation by the label is reflected at most by an effect on the kinetics of self-association.

Regarding the second criticism, we note from a consideration of the spectra of unassociated Hb-S that unaggregated Hb-S molecules do not contribute to the spectral regions denoted by L' and H''. Kinetic experiments that monitor these points yield results that reflect only the contribution of supramolecular aggregates. The problem of the heterogeneity of supramolecular aggregates is more difficult. Ideally, one would take advantage of the fact that the spectral contribution from each aggregate depends on the Zeeman modulation frequency and the rotational frequencies characteristic of the aggregate. Thus, by varying Zeeman modulation frequencies, one could hope to gain insight into the distribution of rotational frequencies (hence aggregates) present. Unfortunately, such variable modulation frequency measurements are not practical with present instrumentation.

There are two goals in the study of Hb-S self-association. These are to identify in structural terms the factors that influence polymerization and to understand and optimize promising desickling agents. Realization of the first objective requires the study of the effect of Hb structure (Hb hybrids) and of the other components of the red cell. In this regard, Johnson and Danyluk (1978) investigated aggregation of carbonmonoxyHb-S (Hb-S-CO) and observed a slight aggregation of Hb-S-CO in the presence of organic phosphate.

An interesting spin-label study of the effect of potential desickling agents is suggested by the work of Votano et al. (1977), who showed that polypeptides of the general structure Phe-R-X, where X = Arg or Lys, substantially inhibit the gelation of deoxy Hb-S, with the tetrapeptide being significantly more effective than the tripeptide. Since these polypeptides do not bind covalently, do not alter the oxygen-binding properties of Hb, and do not appear to alter Hb structure, it is almost certain that they block critical intermolecular polymerization sites. It would be interesting to attempt to fix the location of these sites and to probe the detailed effect of polypeptides on Hb-S self-association by employing spin-labeled analogues of these materials.

C. Protein–Lipid Interactions

In Volume I of "Spin Labeling: Theory and Application" (Berliner, 1976), Griffith and Jost (1976) discuss the investigation of membrane lipids and protein–lipid interactions employing conventional EPR spectroscopy. Membrane interactions with both intrinsic proteins (that penetrate the lipid bilayer) and extrinsic proteins (that interact at the surface with the polar ends of the lipids) are discussed. ST-EPR obviously is important in studying the slower motions characteristic of constrained proteins. The following questions concerning membrane proteins can be answered: (1) Are given membrane proteins free to diffuse (float) in the bilayer structure, or are they localized to given portions (e.g., the surface of the bilayer) of the membrane? (2) Are the measured protein rotational rates comparable to the rates of biological events, such as the rate of calcium pumping? (3) What is the extent and role of protein–protein interactions? The examples discussed below illustrate the present state of knowledge concerning these questions.

1. ROTATIONAL DIFFUSION OF THE Ca^{2+}-ATPASE OF SARCOPLASMIC RETICULUM

Active transport of Ca^{2+} in sarcoplasmic reticulum (SR) must involve molecular motions. Thomas and co-workers (Thomas and Hidalgo, 1978; Hidalgo et al., 1978) and Kirino et al. (1978) have employed ST-EPR in an effort to elucidate the role of large-scale rotational motions of proteins in enzyme action.

1. Protein Motion in Native SR Membranes. By blocking the fast-reacting sulfhydryl groups with *N*-ethylmaleimide and then adding a maleimide spin label (**II**), Thomas and Hidalgo (1978) succeeded in specifically labeling the Ca^{2+}-ATPase, despite the presence of other sulfhydryl-containing proteins. The conventional EPR spectrum indicates that the label is strongly immobilized relative to the Ca^{2+}-ATPase. Enzymatic assays show that the enzyme is still active. The ST-EPR spectrum recorded by Hidalgo et al. (1978) at 4°C is reproduced in Fig. 19, and the measured ratio of $L''/L = 0.8$ corresponds to an "effective" time of 10^{-4} sec, since the Ca^{2+}-ATPase is almost certainly undergoing anisotropic diffusion. However, it is interesting that this effective time is in the same range as that observed for maximal enzyme activity ($\sim 10^4$ sec^{-1}). Several experiments were carried out to establish that the observed motion was that of the protein within the membrane and not that of the overall tumbling of the membrane vesicles. The membranes were centrifuged into a pellet, and its ST-EPR spectrum was recorded. Only slightly slower motion resulted, indicating that rotation of vesicles was not a major contributor to label motion. Second, the proteins were cross-linked with glutaraldehyde, which stopped the label motion ($L''/L = 1.2$). Since the main effect of glutaraldehyde cross-linking is to immobilize the proteins on the membrane surface, the motion observed before cross-linking is probably the motion of the protein relative to the membrane in which it is embedded.

2. Protein Motion in Synthetic Membranes: Correlating the Lipid Environment, Enzyme Activity, and Protein Motion. Hidalgo et al. (1978) purified the Ca^{2+}-ATPase and inserted it into well-defined lipid environments. Their objective was to correlate the effects of various perturbations of enzyme activity with the effects on motion. If a certain rate of rotational motion is

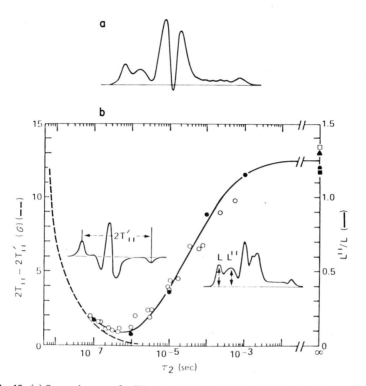

Fig. 19. (a) Saturation-transfer V_2' spectrum of the sarcoplasmic reticulum Ca^{2+}-ATPase protein spin labeled with maleimide (II) by the procedure of Thomas and Hidalgo (1978). The measured ratio of $L''/L = 0.8$ corresponds to an effective τ_R of 10^{-4} sec, as discussed in (b). (b) Immobilization parameters from conventional EPR V_1 and ST-EPR V_2' spectra used to characterize τ_R of spin-labeled biomolecules undergoing isotropic rotational diffusion and to estimate effective rotational times for molecules undergoing anisotropic motion. The dashed curve (scale at left) is from computer simulations of V_1 spectra by McCalley et al. (1972) and is frequently referred to as the "ΔS method." The solid curve (scale at right) is from computer-simulated (closed circles) and experimental maleimide-spin-labeled Hb (open circles) V_2' ST-EPR spectra (Thomas and McConnell, 1974; Thomas, et al., 1976a). At τ_R (τ_2 in figure) $= \infty$, the solid square is from crystallized Hb, the open square from ammonium sulphate precipitated hemoglobin, and the solid triangle from a centrifuged (100,000 g) pellet of maleimide-spin-labeled actin. [Reproduced from Hidalgo et al. (1978).]

required for a given level of enzyme activity, the activity should correspondingly decrease as the protein motion decreases. To explore this hypothesis, Hidalgo et al. (1978) performed the following five experiments:

(1) Motion in native lipids. A preparation of "purified ATPase" containing the native SR lipids but only the one protein was examined. The spectrum of this preparation (Fig. 20) was very similar to that of native SR (Fig. 19).

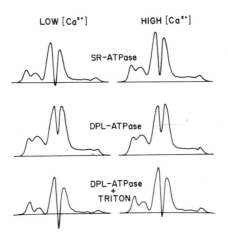

Fig. 20. Saturation-transfer EPR V_2' spectra of maleimide-spin-labeled (**II**) ATPase enzyme of rabbit sarcoplasmic reticulum, showing the effects of Ca^{2+} concentration and the nonionic detergent Triton. Spectra at left: $[Ca^{2+}] \leq 10^{-5}\ M$; right: $[Ca^{2+}] = 10^{-2}\ M$; top row: purified ATPase in buffer; middle: ATPase + dipalmitoyl lecithin (DPL), [DPL]/gm protein $= 3 \times 10^{-4}\ M$/gm; bottom: addition of 1.5 gm Triton X-100 per gram of protein to the DPL–ATPase. All spectra were recorded at 4°C; scan width is 100 G. [Reproduced from Hidalgo *et al.* (1978).]

(2) Effect of dipalmitoylphosphatidylcholine (DPPC). Substituting DPPC for native lipids produced a rigid lipid environment at 4°C, stopped the enzyme activity, and stopped the motion of the protein (Fig. 20).

(3) Effect of Triton after DPPC substitution. The nonionic detergent Triton X-100 (Sigma Chemical Company) dissolved the membrane, fluidized the lipids, activated the enzyme, and increased protein motion (Fig. 20).

(4) The effect of high concentrations of Ca^{2+}. High Ca^{2+} concentrations were observed to partially reverse the effect of Triton (Fig. 20) without any apparent effect on lipids. Similar effects were observed on the intact membrane preparation (purified ATPase).

(5) Effect of temperature. Increasing the temperature resulted in both greater enzymatic activity and more motion of the protein.

In summary, the work of Hidalgo *et al.* (1978) shows that when the motion of the Ca^{2+}-ATPase is slowed to an effective rotational correlation time greater than 10^{-4} sec, the enzyme activity decreases, and this observation is valid regardless of whether the motional slowing is accomplished by lipid substitution, by high Ca^{2+} concentrations, or by lowering the temperature. Thus, a direct relationship between the rotational motion of the Ca^{2+}-ATPase and enzyme activity appears to have been demonstrated.

2. ROTATIONAL DIFFUSION OF RHODOPSIN IN THE
   VISUAL RECEPTOR MEMBRANE

Baroin *et al.* (1977), Devaux *et al.* (1977), Kusumi *et al.* (1978), and Rousselet and Devaux (1978) investigated the rotational diffusion of rhodopsin by ST-EPR employing spin labels II, IV, V and VI.

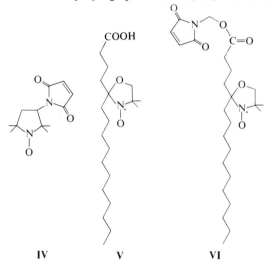

A typical ST-EPR spectrum of **IV** attached to membrane-bound rhodopsin is shown in Fig. 21 and corresponds to a rotational correlation time at 4°C of 50 μsec. This number is very close to the value of 20 μsec at 20°C measured by Cone (1972), who monitored transient photodichroism. Also shown in Fig. 21 is the effect of addition of glutaraldehyde (5%), which can be seen from the ST-EPR spectrum to result in complete immobilization of the protein. In a similar way, delipidation (50% decrease of the phospholipid–protein ratio by treatment with phospholipase A_2) completely stops the motion of the maleimide-labeled rhodopsin. Temperature, and hence viscosity, dependence of the rotational diffusion of rhodopsin is given in Fig. 21. By a comparison of the H''/H and L''/L ratios with those of systems undergoing isotropic diffusion, it is clear that rhodopsin is undergoing anisotropic diffusion. Of substantial interest is the observation by Baroin *et al.* (1977) that bleaching does not appear to influence the motion of the protein.

The results of experiments with labels **V** and **VI** are summarized in Fig. 22. The ST-EPR spectrum shown in Fig. 22A is typical of the fatty acid spin label dissolved in the lipid bilayer of the membrane. The maleimide moiety of **VI** binds to the SH groups of rhodopsin and is thus more restricted than **V**. One striking feature of the ST-EPR spectra of **VI** shown in

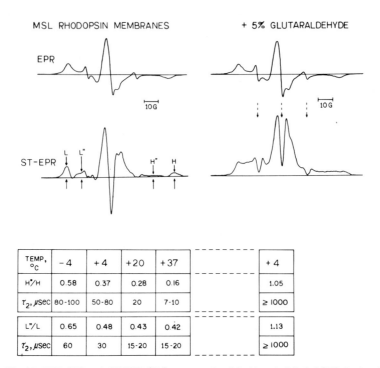

MSL RHODOPSIN MEMBRANES · + 5% GLUTARALDEHYDE

EPR · ST-EPR

TEMP, °C	− 4	+ 4	+20	+ 37		+ 4
H''/H	0.58	0.37	0.28	0.16		1.05
τ_2, μsec	80-100	50-80	20	7-10		≥ 1000
L''/L	0.65	0.48	0.43	0.42		1.13
τ_2, μsec	60	30	15-20	15-20		≥ 1000

Fig. 21. EPR (V_1) and ST-EPR (V_2') spectra of maleimide-spin-labeled (**IV**) rhodopsin in bovine rod outer segment membranes suspended in buffer (left) and cross-linked with glutaraldehyde (right). Immobilization parameters H''/H, L''/L, and the corresponding rotational correlation times are compared in the table below the spectra. Both dark-adapted and illuminated rhodopsin membranes gave the same immobilization parameter profile with temperature. For the cross-linked rhodopsin, the arrows indicate points of overlap of weakly immobilized label on the signal from the strongly immobilized label. Note the suppression in amplitude of the contributions from the weakly immobilized label in the ST-EPR V_2' spectrum in comparison to the EPR display. [Taken from Baroin *et al.* (1977).]

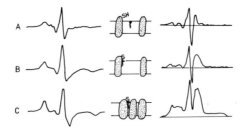

Fig. 22. EPR (V_1, left) and ST-EPR (V_2', right) spectra referable to rotational diffusion of rhodopsin and fluidity of the lipid annulus around rhodopsin. (A) Rod outer segment membrane fragments labeled with palmitic acid (**V**). (B) Membrane-bound rhodopsin labeled with palmitic acid ester (**VI**). (C) Membrane fragments as in (B) from which the bulk lipid phase has been removed, leaving only the lipid annulus around the protein. [Reproduced from Devaux *et al.* (1977).]

Fig. 22B,C is the difference between normal membranes and delipidated membranes, which indicates that the lipid annulus has a viscosity that depends on the presence of the bulk lipids. This suggests a continuity between the lipid annulus and the bilayer.

3. ROTATIONAL DIFFUSION OF THE CHOLINERGIC RECEPTOR IN *Torpedo marmorata* MEMBRANE FRAGMENTS

Rousselet and Devaux (1977) investigated the rotational diffusion of the cholinergic receptor protein in its membranous state, employing the maleimide spin label **IV** and the following spin-labeled choline derivative (**VII**):

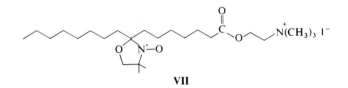

VII

Both labels **IV** and **VII** show by ST-EPR that the cholinergic receptor proteins are completely immobilized (see Fig. 23). When label **VII** is displaced from the protein by addition of 5×10^{-4} M acetylcholine, an ST-EPR spectrum that is similar to **VII** in a lipid bilayer environment is obtained.

The work of Rousselet and Devaux (1977) demonstrates that the cholinergic receptor proteins are completely immobilized ($\tau_R > 10^{-3}$ sec) despite

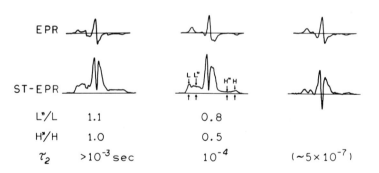

Fig. 23. EPR (V_1, top row) and ST-EPR (V_2', second row) spectra of the cholinergic receptor protein in *Torpedo marmorata* membrane fragments. Left column: maleimide-spin-labeled (**IV**) membrane fragments; center: membrane fragments labeled with 8-doxylpalmitoylcholine (**VII**); right: displacement of label **VII** from receptor sites upon addition of acetylcholine is demonstrated by increase in mobility of label **VII**. Correlation times τ_R (τ_2 in figure) were derived from the L''/L and H''/H parameters of Thomas, *et al.* (1976a). For the lower right ST-EPR spectrum, τ_R was estimated from the C'/C parameter (Thomas *et al.*, 1976a; see also Fig. 9 of this chapter). [Adapted from Rousselet and Devaux (1977).]

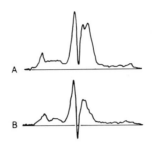

Fig. 24. ST-EPR spectra of rat heart mitochondria labeled with the doxylpalmitate **VII** showing (A) label bound to the ADP carrier (atractyloside-binding protein, MW ~ 50,000) and (B) label displaced by addition of excess atractyloside. [Reproduced from Devaux *et al.* (1977).]

the proximity of a fluid hydrophobic phase. The observed immobilization suggests that strong protein–protein interactions must exist in the post-synaptic membrane.

4. Rotational Diffusion of the ADP Carrier in Mitochondria

Devaux and co-workers (1977) employed label **VIII** to investigate the rotational diffusion of the ADP carrier. Typical results are shown in Fig. 24. The spectrum in Fig. 24A is characteristic of an "effective" rotational correlation time of the order of 10^{-4} sec. However, in **VIII** the nitroxide spin label is rather far from the protein binding site and may reflect more motion that what actually is experienced by the protein. When the molecular weight (50,000) of the atractyloside-binding protein is considered, the results of Devaux and co-workers (1977) suggest that the atractyloside-binding protein (the ADP carrier) is associated with some other component that slows down its motion.

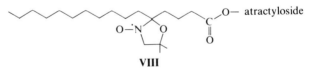

VIII

D. Phospholipid Motion in Gel-Phase Lipid Bilayers

D. Marsh (unpublished) used a phospholipid derivative of label **V** to study phase transitions in lipid bilayer membranes composed of a single type of lipid. In this model study, he compared two membrane systems, DPPC and dimyristoylphosphatidylethanolamine (DMPE). A phase transition from the gel to liquid–crystalline state occurs in DPPC at 41°C and in DMPE at 48°C, which can be monitored by either ST-EPR or conventional EPR. It was observed by ST-EPR, however, that DPPC has a pretransition at 25°C,

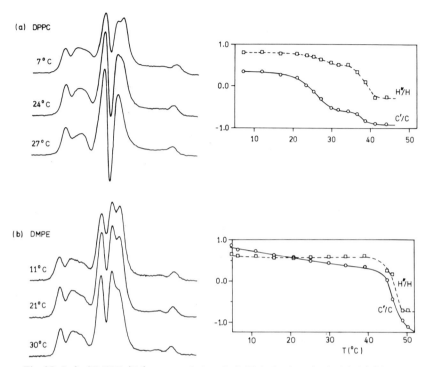

Fig. 25. Left: ST-EPR (V_2') spectra of phospholipid derivative of spin label (**V**) in multi-bilayer dispersions of (a) dipalmitoylphosphatidylcholine (DPPC) and (b) dimyristoylphosphatidylethanolamine (DMPE) at selected temperatures. Microwave power level, 63 mW (0.25 G); modulation amplitude, 5 G (peak to peak); spectral display width, 100 G. Right: Temperature profile of the Thomas *et al.* (1976a) immobilization parameters C'/C and H''/H from the (a) DPPC and (b) DMPE spectra. [Reproduced from D. Marsh (unpublished) and Marsh *et al.* (1977).]

accompanied by a change in C'/C (see Fig. 25 and Table I) by two orders of magnitude from 10^{-4} to 10^{-6} sec, although L''/L and H''/H showed no change. This phenomenon was not observed in DMPE. The pretransition step in DPPC in the plot of C'/C versus temperature was interpreted by Marsh as the onset of motion about the long axis of label **V**.

Since specific involvement of the lipid in membrane transport processes must involve slow lipid motions, studies of slow motion, such as those conducted by Marsh, are important. Marsh's work also is noteworthy for introducing a calibration technique for studying anisotropic motion. The idea is perhaps best understood by studying the experimental data of Marsh presented in Table I, which lists the rotational correlation times derived from the parameters L''/L, H''/H, and C'/C measured as a function of temperature. All three parameters should yield the same rotational correlation time

TABLE I

ROTATIONAL CORRELATION TIMES IN GEL-PHASE DPPC AND DMPE DEDUCED FROM CALIBRA-
TION OF PEAK HEIGHT RATIOS OF HEMOGLOBIN REFERENCE SPECTRA[a]

Membrane system	τ_2	12°C	25°C	30°C	45°C	50°C
DPPC	L''/L	6×10^{-4}	1×10^{-4}	0.8×10^{-4}	$\sim 10^{-9}$	
	H''/H	4×10^{-4}	2×10^{-4}	1×10^{-4}	$\sim 10^{-9}$	
	C'/C	0.8×10^{-4}	5×10^{-6}	0.9×10^{-6}	$\sim 10^{-9}$	
DMPE	L''/L	2×10^{-4}	2×10^{-4}	2×10^{-4}	8×10^{-6}	$\sim 10^{-9}$
	H''/H	1×10^{-4}	1×10^{-4}	1×10^{-4}	1×10^{-5}	$\sim 10^{-9}$
	C'/C	$\leq 10^{-3}$	10^{-4}	10^{-4}	10^{-6}	$\sim 10^{-9}$

[a] Reference spectra taken from Thomas *et al.* (1976a).

if the motion is isotropic Brownian diffusion. They are used by Marsh to calibrate a system in which the motion is clearly anisotropic. Parameters L''/L and H''/H are redundant and should yield the same rotational correlation time corresponding to motion about an axis (axes) perpendicular to the axis of the probe. These ratios are, in fact, nearly always the same at a particular temperature in Table I. It is at 25°C that C'/C departs sharply from the other two ratios for DPPC, showing a transition from isotropic to anisotropic motion that correlates with known thermodynamic data on the phase transition in DPPC.

E. Molecular Motion on Surfaces

There are nonbiological systems and paramagnetic materials other than nitroxides that are suited for investigation by ST-EPR techniques. The study of the diffusion of O_2^- adsorbed on a silver surface is an example. Clarkson and Kooser (1978) investigated the "hopping" or two-dimensional diffusion processes of O_2^- adsorbed on silver over the temperature range 173°–293°K by observing changes in the in-phase-quadrature first-harmonic dispersion spectra. Their results are summarized in Fig. 26. They attempted, by carrying out computer simulations, to ascertain whether the rotational diffusion of O_2^- on the surface is characterized by large- or small-angle jump (Brownian) diffusion. The results of their computer calculations are shown at the right in Fig. 26 and seem to indicate a slight preference for the small-angle reorientational model. On the basis of this model, the dwell time, τ, varies from about 10^{-6} sec at 273°K to greater than 10^{-4} sec at 173°K. Clarkson and Kooser indicate that such numbers are in good agreement with Langmuir-type adsorption and with what is known about the chemisorption behavior.

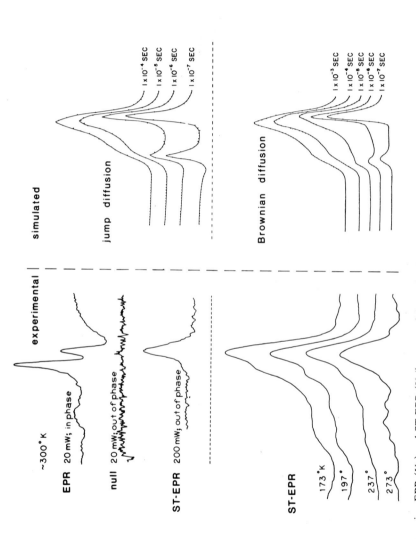

Fig. 26. Dispersion EPR (U_1) and ST-EPR (U_1') spectra of O_2^- adsorbed on a silver surface. Computer simulations of the ST-EPR spectra as a function of temperature indicate that the O_2^- diffusion process is best described as Brownian diffusion. Input parameters for the computations: $g_\parallel = 2.0279$, $g_\perp = 2.0067$, $H_1 = 0.1$ G, $H_m = 1.0$ G, $\omega_m/2\pi = 100$ kHz; Brownian diffusion, $\Delta\theta = 2.5°$; jump diffusion, $\Delta\theta = 9°$, where $\Delta\theta$ is the width of angular zones on an octant of a spherical surface. [Adapted from Clarkson and Kooser (1978).]

DeBoer (1953) demonstrated that the average immobilized dwell time τ for a molecule is given by

$$\tau = \frac{\delta}{\eta(1 - \delta/\delta_0)} \tag{57}$$

where δ is the number of molecules adsorbed per (centimeter)^2, δ_0 is the number of molecules adsorbed per (centimeter)^2 at monolayer coverage, and η is the number of molecules competing for an adsorption site per (centimeter)^2 second. Using values of δ and δ_0 determined by EPR (Clarkson and Cirillo, 1974) and assuming $\eta = 7 \times 10^{14}$ molecules per (centimeter)^2 second after the work of Kilty et al. (1973), Clarkson and Kooser calculated $\tau = 2 \times 10^{-4}$ sec at 173°K. The excellent agreement between the preceding Langmuir calculation and the value of τ obtained by comparison of theoretical and experimental ST-EPR spectra lends strong support to a surface mobility model of the Langmuir type for the $O_2{}^-$–silver system.

VII. ANISOTROPIC OR SPECIAL MOTIONS

It has become clear that a very wide range of useful biological studies using saturation-transfer spectroscopy are possible in which empirical comparisons can be made between spectra of the sample under investigation and reference spectra derived from maleimide-spin-labeled hemoglobin undergoing isotropic rotational Brownian diffusion. The detailed theoretical simulations fit the observed spectra with an accuracy limited only by computer costs.

Most biological systems are not moving isotropically. In the remainder of this chapter we discuss the use of saturation-transfer spectroscopy to study anisotropic motions. The discussion is of a rather speculative nature. We believe that many of the most important contributions of the technique in the future will be in this area.

A. Rotational Diffusion of a Spin-Labeled Ellipsoid of Revolution in an Isotropic Medium

The situation discussed in this section is perhaps the simplest case of anisotropic motion. We introduce the diffusion coefficients $D_\parallel$ and $D_\perp$ corresponding to rotational diffusion about the symmetry axis and about an axis perpendicular to the symmetry axis, respectively.

We further assume that the magnetic interactions can be approximated by an axial Hamiltonian with the p_z axis coincident with the $D_\parallel$ axis (Fig. 27). It is obvious that rotational diffusion about the z axis can have no

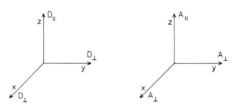

Fig. 27. The special-case situation in which parallelism of the symmetry axis of a figure of revolution with the z axis of an axial Hamiltonian renders detection of anisotropic motion impossible.

effect on the spectrum, since the tensorial components of A along x and y are identical. Thus, in this special case there is no way to obtain information about $D_{\parallel}$. Saturation-transfer spectra will be identical in shape to those obtained from spherical molecules and will permit the determination of $D_{\perp}$.

One can relate $D_{\parallel}$ and $D_{\perp}$ to D, the diffusion coefficient calculated for a sphere of the same volume, at the same temperature and viscosity (Tao, 1969, and references therein) by the equations

$$D_{\parallel} = \frac{3}{2} \frac{\rho(\rho - S')}{\rho^2 - 1} D \tag{58a}$$

$$D_{\perp} = \frac{3}{2} \frac{\rho[(2\rho^2 - 1)S' - \rho]}{\rho^4 - 1} D \tag{58b}$$

where ρ is the ratio of the longitudinal semiaxis to the equatorial semiaxis of the ellipsoid, and

$$S' = (\rho^2 - 1)^{-1/2} \ln[\rho + (\rho^2 - 1)^{1/2} \qquad \text{for } \rho > 1 \tag{59a}$$

$$S' = (1 - \rho^2)^{-1/2} \tan^{-1}[(1 - \rho^2)^{-1/2}/\rho] \qquad \text{for } \rho < 1 \tag{59b}$$

Thomas *et al.* (1975a,b) encountered just this situation in their study of subfragment-1 of myosin. The molecular weight, viscosity, and ρ were known. Saturation-transfer spectroscopy gave a spectrum that looked like isotropic rotational diffusion with a rotational diffusion coefficient that was the same as that calculated for $D_{\perp}$. It therefore was concluded that colinearity of $D_{\parallel}$ and $A_{\parallel}$ had been established.

Another possible geometry is illustrated in Fig. 28. Rotational diffusion about z gives rise to spectral diffusion that depends on both $A_{\parallel}$ and $A_{\perp}$. The resulting saturation-transfer spectrum cannot be the same as any that can be obtained from systems undergoing isotropic diffusion.

For arbitrary orientation of the diffusion tensor with respect to the magnetic tensor, one could, assuming the orientation were known, reasonably hope to determine both $D_{\parallel}$ and $D_{\perp}$. It is not at all clear whether one could

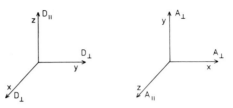

Fig. 28. With axial symmetries of the diffusion and magnetic tensors but noncoincidence of the principal axes, as here, information on anisotropic motion can be obtained.

determine uniquely $D_{\parallel}$, $D_{\perp}$, and also the relative orientation of the two tensors. It is at this point that our discussion has encountered the leading edge of current theoretical and experimental research.

B. Experimental Methods

Effects of anisotropic motions were evident in many of the applications of saturation-transfer spectroscopy previously discussed. If special types of motions are to be analyzed successfully, the experimentalist must address himself to the task of designing experiments that yield optimal information.

One can expect to obtain more detailed information on motion when $T_1/\tau_R \leq 1$. Fast motion causes rapid averaging over all angles, and the details of individual random walk processes are lost, whereas for motion near the rigid lattice limit the particle does not walk far enough to give much insight into the random walk process. Examination of the transition rate and eigenfunction expansion formalisms gives a mathematical statement of this physical observation (*vide infra*).

Microwave frequency is a little used but potentially very important experimental variable. In Fig. 29 simulations of pure absorption spectra of a typical nitroxide radical are shown as a function of microwave frequency. It is instructive to study this figure from the point of view of analysis of anisotropic motions.

At low microwave frequencies (e.g., 3 GHz), the approximation of axially symmetric magnetic interactions is appropriate, and the effects of anisotropic diffusion will be detectable only if the magnetic and diffusion tensor axes are noncoincident. As an aside, we note that this observation also applies to fluorescence depolarization measurements made employing planar dyes.

It is evident, then, that the anisotropic Zeeman interaction affords an additional sensitivity to anisotropic rotational motion compared to that which can be realized with optical techniques and that this sensitivity will be enhanced by performing experiments at higher microwave frequencies.

Special importance is attributed to the use of 35 GHz, where it appears that the similar magnitudes of the hyperfine and electron–Zeeman anisotro-

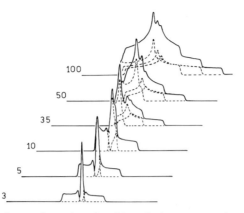

Fig. 29. Simulated pure absorption nitroxide radical spectra as a function of microwave frequency (in gigahertz). The g values were 2.0096, 2.0063, and 2.0022; the hyperfine values were 4.0, 6.0, and 35.0 G. Linewidths that were introduced into the calculation were selected to give best agreement with typical X-band experimental spectra.

pies make the spectrum unusually sensitive to motions of different types. At this frequency the ratio of the width of the spectral fragment associated with the high-field (-1) nuclear spin configuration to the width of the low-field $(+1)$ fragment is at a maximum. Thus, at this frequency there is sensitivity to the greatest range of motions. It has been established that the spin–lattice relaxation characteristics of very slowly tumbling spin labels at 35 GHz are favorable for saturation-transfer experiments (Hyde, 1978), although more data on a wider variety of systems would be desirable. In these preliminary experiments, it appears that 35-GHz spectra are unusually sensitive to very slow motions relative to X-band spectra.

Table II lists a number of saturation-transfer methods, some of which are

TABLE II

SATURATION-TRANSFER TECHNIQUES

1. Rapid passage under sinusoidal field modulation
   a. Absorption second harmonic out of phase
   b. Dispersion first harmonic out of phase
2. Rapid passage under repetitive trapezoidal field modulation
3. ELDOR
   a. Stationary
   b. Pulsed
4. Microwave power saturation
   a. CW (or progressive) saturation
   b. Saturation recovery
   c. Spin lock

well developed and some of which are speculative. These methods may be expected to have different efficacies in the analysis of anisotropic motion. For example, it already has been argued that anisotropic motions are better studied if $T_1/\tau_R \leq 1$ and that under this condition second-harmonic absorption out-of-phase spectra are more sensitive to motion than are the first-harmonic dispersion out-of-phase spectra.

The field modulation methods had the greatest utility at the time this chapter was being written. This is largely because of the ease in performing experiments and considerable experience in interpreting results. It should be recognized that the signals are the result of a rather complicated integration of spectral responses over the interval of the spectrum that corresponds to the field modulation amplitude, which seems likely to make investigation of special motions difficult. One interesting possibility in order to improve the total information content of a field modulation ST-EPR experiment is to obtain spectra at various field modulation frequencies. Thus far, however, little information is available on the utility of this approach.

ELDOR permits the preparation of an ensemble of saturated spins corresponding to a subset of orientations and the monitoring of a second set of spins corresponding to a second subset of orientations. This selection process provides considerable enhancement of the sensitivity to the details of molecular motional processes. Pulsed ELDOR permits the further measurement of the time evolution of the spectral diffusion profile.

One should evaluate the merits of variation of the relevant experimental parameters associated with each of the techniques listed in Table II with the goal of obtaining added information on spectral diffusion. The saturation-recovery technique contains an interesting additional instrumental parameter—the duration of the saturating microwave pulse. Because the *disappearance* of the x component of magnetization at the pumping point is observed rather than the *appearance* somewhere else, it may not in general be easy to obtain a clear demonstration of spectral diffusion of saturation. The condition $T_1/\tau_R \sim 1$ is most favorable for obtaining motional information.

The as yet untried experiments of spin locking and rapid passage under trapezoidal field modulation are similar in nature. Let us consider the trapezoidal modulation in more detail (Fig. 30). Period A is for relaxation of the spin system and is, for example, $3T_1$ long. Period B is the controlled field sweep dH_0/dt. This might vary from 10^7 to 10^3 G/sec and is an experimental variable. Slow sweeps are used to study the slow motions, and fast sweeps are used to study fast motions. Region C is the examining time, during which one would try to measure 10–100 points using a swept boxcar integrator or a fast analogue-to-digital converter. Region D is for fast return to the starting point. Since regions C and B are always of the order of T_1 or shorter (no

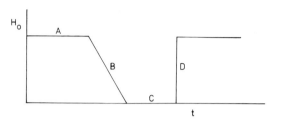

Fig. 30. The trapezoidal field modulation cycle.

transient signals can be detected at times longer than T_1), the overall repetition rate of the experiment is of the order of $1/(4T_1) \sim 10^5$. In the rotating frame, the various magnetization vectors after an adiabatic sweep from one point to another yield a resultant vector along H_1 that is spin locked and will decay because of spectral diffusion. With the field position during the resting period A, the rate of sweep during the interval B, and the field position during the observing period C as experimental variables, one has the potential in this method of extracting very detailed information on molecular dynamics.

In summary, there often may be a question of uniqueness in determination of a motional model for a particular experimental observation. In all cases, theoretical simulations are desirable, and every effort should be made to increase the intrinsic information content in the experiment. Studying the same system as a function of microwave frequency is obvious. Three-, four-, or even five-dimensional displays of data should remove most ambiguities. (An example of a five-dimensional display is pulsed ELDOR signal height as a function of microwave frequency, pumping resonant condition, observing resonant condition, and time.)

C. Theoretical Comments

Earlier theoretical discussion in this chapter has been limited to molecular dynamics describable by the isotropic Brownian diffusion model. From the review of experimental applications of ST-EPR, it is clear that most recent applications have involved more complicated motional effects and, in particular, anisotropic diffusion. Generalization to other motional models is straightforward, and in terms of the transition rate matrix approach discussed earlier, it simply implies the appropriate definition of the elements of the transition rate matrix. For example, the model of isotropic jump diffusion by finite angular jumps also yields a tridiagonal supermatrix; however, the off-diagonal elements are now removed from the diagonal row according to the magnitude of the jump angle. Of particular interest are the effects of anisotropic motion. In the most general sense (arbitrary rotational diffusion tensor, arbitrary magnetic tensor, and arbitrary spatial relationship between the diffusion and magnetic tensors), the equation describing the time evolution of the orientational variables can be written as

$$\Gamma_\Omega = d_+ M^2 + (d_0 - d_+)M_0^2 + \frac{1}{2}d_-(M_+^2 + M_-^2) + \frac{1}{\sqrt{2}}$$

$$\times \delta_+(2M_0 + 1)M_- + \frac{1}{\sqrt{2}}\delta_-(2M_0 - 1)M_+ - \frac{1}{2}\delta_0(M_+^2 - M_-^2) \quad (60)$$

where

$$d_+ \equiv \frac{1}{2}(d_{11} + d_{22}) \qquad \delta_+ \equiv \frac{1}{\sqrt{2}}(d_{31} + id_{23})$$

$$d_0 \equiv d_{33} \qquad\qquad \delta_0 \equiv d_{12}$$

$$d_- \equiv \frac{1}{2}(d_{11} - d_{22}) \qquad \delta_- \equiv \frac{1}{\sqrt{2}}(d_{31} - id_{23})$$

with d_{11}, d_{22}, etc., the elements of the rotational diffusion tensors in the frame of the magnetic tensors. In terms of Euler angles (α, β, γ), $\mathbf{M}$ can be written as

$$\mathbf{M} = \frac{i}{\sin \beta}
\begin{vmatrix}
-\cos \gamma & \sin \beta \sin \gamma & \cos \beta \cos \gamma \\
\sin \gamma & \sin \beta \cos \gamma & -\cos \beta \sin \gamma \\
0 & 0 & \sin \beta
\end{vmatrix}
\begin{vmatrix}
\dfrac{\partial}{\partial \alpha} \\[4pt]
\dfrac{\partial}{\partial \beta} \\[4pt]
\dfrac{\partial}{\partial \gamma}
\end{vmatrix} \quad (61)$$

In the above form, the equations are readily adapted to solution either by the orthogonal eigenfunction expansion or by the transition rate matrix algorithm. However, rather than discuss the details of particular solution schemes, it is perhaps of greater interest to examine the form of Eq. (60) under various conditions. If the magnetic and diffusion tensors are coincident, Eq. (60) reduces to

$$\Gamma_\Omega = d_+ M^2 + (d_0 - d_+)M_0^2 + \tfrac{1}{2}d_-(M_+^2 + M_-^2) \quad (62)$$

and if axially symmetric magnetic tensors are assumed, the equation further reduces to

$$\Gamma_\Omega = d_+ M^2 + (d_0 - d_+)M_0^2 \quad (63)$$

For such a case, the resonance spectrum no longer will be sensitive to anisotropic diffusion, as already discussed. If the diffusion is isotropic, the following expression obtains:

$$\Gamma_\Omega = dM^2 \tag{64}$$

Here, $d = d_{11} = d_{22} = d_{33}$, and M^2 is now the familiar angular Laplacian operator.

Equation (60) describes the arbitrary anisotropic diffusion of a particle (arbitrary inertial tensors) in an isotropic potential field. The effects of a nonisotropic environment also can be considered with the most general form of the equation being

$$\Gamma_\Omega = \mathbf{M} \cdot [\mathbf{D} \cdot \mathbf{M} + D(\mathbf{M}U(\Omega)/kT)] \tag{65}$$

where the restoring potential can be written in the most general form as

$$U(\Omega) = \sum_{\mathscr{L}, \mathscr{K}, \mathscr{M}} \varepsilon_{\mathscr{K}\mathscr{M}}^{\mathscr{L}} \mathscr{D}_{\mathscr{K}\mathscr{M}}^{\mathscr{L}}(\alpha, \beta, \gamma)$$

with the $\mathscr{D}_{\mathscr{K}\mathscr{M}}^{\mathscr{L}}(\alpha, \beta, \gamma)$ being Wigner rotation matrices.

Of course, one can envision many types of special motions that will lead to very special forms of the motional operator. Two examples that merit special mention are (1) Heisenberg spin exchange arising from the collision of spin labels associated with translational diffusion and (2) fluxional motion such as the inversion of strained ring systems or the rotation of functional groups. Heisenberg spin exchange acts to effect equal coupling of all orientations and nuclear spin states and as such leads to a particularly simple form of the transition rate matrix. Fluxional motion is often describable by two- or three-site jump models.

The introduction of special motions naturally leads to the idea of superimposed motions, since special motions such as Heisenberg spin exchange and fluxional modes can occur simultaneously with rotational diffusion of the spin label. A single spectral measurement is not necessarily sufficient to permit discrimination of superimposed modes. For example, the superposition of isotropic Brownian diffusion with a dynamic ring inversion that modulates the magnetic interactions yields a spectral result that is indistinguishable from that for an isotropic free diffusion model. However, the dependence of the individual superimposed motions on temperature and viscosity may differ, so that the variation of these parameters can be employed to discriminate among the superimposed motions. In like manner, the investigation of the pressure dependence of spectra can be utilized to discriminate between inter- and intramolecular modes.

ACKNOWLEDGMENTS

We are grateful to Dr. P. F. Devaux, Dr. M. Hemminga, Dr. M. E. Johnson, Dr. D. Marsh, and Dr. D. D. Thomas for communicating to us results of their research in advance of publication.

Preparation of this chapter was assisted by grants GM 22923 and 1 P41 RR01008 from the National Institutes of Health, GP-42998X and CHE-7701018 from the National Science Foundation, 9830-AC6 from the Petroleum Research Fund, by a Teacher-Scholar grant to L.R.D. from the Camille and Henry Dreyfus Foundation, and by an Alfred P. Sloan Foundation fellowship.

L. A. Dalton provided invaluable scientific, technical, and editorial assistance in the preparation of the manuscript.

REFERENCES

Aleksandrov, I. V., A. N. Ivanova, N. N. Korst, A. V. Lazarev, A. I. Prikhozhenko, and V. B. Stryukov (1970). The E.S.R. line shape for the iminoxyl radical in high viscosity media. *Mol. Phys.* **18**, 681–691.
Ammerlaan, C. A. J., and A. van der Wiel (1976). The divacancy in silicon: Spin-lattice relaxation and passage effects in electron paramagnetic resonance. *J. Magn. Reson.* **21**, 387–396.
Barlow, A. J., J. Lamb, and A. J. Matheson (1966). Viscous behaviour of supercooled liquids. *Proc. R. Soc. London, Ser. A* **292**, 322.
Baroin, A., D. D. Thomas, B. Osborne, and P. F. Devaux (1977). Saturation transfer electron paramagnetic resonance on membrane-bound proteins. I. Rotational diffusion of rhodopsin in the visual receptor membrane. *Biochem. Biophys. Res. Commun.* **78**, 442–447.
Berliner, L. J. (1970). Refinement and location of the hydrogen atoms in the nitroxide 2,2,6,6-tetramethyl-4-piperidinol-1-oxyl. *Acta Crystallogr., Sect. B* **26**, 1198–1202.
Berliner, L. J., ed. (1976). "Spin Labeling: Theory and Applications," Vol. 1. Academic Press. New York.
Bloch, F. (1946). Nuclear induction. *Phys. Rev.* **70**, 460–474.
Brière, R., H. Lemaire, A. Rassat, P. Rey, and A. Rousseau (1967). Nitroxydes. XXIV. Résonance magnétique nucléaire de radicaux libres nitroxydes pipéridiniques. *Bull. Soc. Chim. Fr.* pp. 4479–4484.
Brown, I. M. (1976). Molecular oxygen effects on the electron spin relaxation in a spin-labeled polystyrene. *J. Chem. Phys.* **65**, 630–638.
Brown, I. M. (1978). EPR of spin probes in block copolymers of dimethylsiloxane and bisphenol A carbonate. *Macromolecules* (to be published).
Brown, I. M., and A. C. Lind (1977). The study of the glass transition in crosslinked polymers using nuclear and electron magnetic resonance. *U.S. Nav. Air Syst. Command, Rep. No.* MDC QO633.
Clarkson, R. B., and A. C. Cirillo, Jr. (1974). The formation and reactivity of oxygen as O_2^- on a supported silver surface. *J. Catal.* **33**, 392–401.
Clarkson, R. B., and R. G. Kooser (1978). Molecular motion on surface as investigated by saturation transfer EPR, *Surf. Sci.* **74**, 325–332.
Coffey, P., B. H. Robinson, and L. R. Dalton (1976). Rapid computer simulation of ESR spectra. II. Saturation transfer spectroscopy of axially symmetric ^{14}N-nitroxide spin labels. *Mol. Phys.* **31**, 1703–1715.
Cone, R. A. (1972). Rotational diffusion of rhodopsin in the visual receptor membrane. *Nature (London), New Biol.* **236**, 39–43.

Dalton, L. R. (1973). Theory of nonlinear spin response: Rapid passage for very slow molecular reorientation. *Bull. Am. Phys. Soc.* **18**, 1571.

Dalton, L. R., P. Coffey, L. A. Dalton, B. H. Robinson, and A. D. Keith (1975). Theory of nonlinear spin response: Rapid passage for very slow molecular reorientation. *Phys. Rev. A* **11**, 488–498.

Dalton, L. R., B. H. Robinson, L. A. Dalton, and P. Coffey (1976). Saturation transfer spectroscopy. *Adv. Mag. Reson.* **8**, 149–259.

deBoer, J. H. (1953). "The Dynamical Character of Adsorption" p. 55, Oxford Univ. Press, London and New York.

Devaux, P. F., A. Baroin, A. Bienvenue, E. Favre, A. Rousselet, and D. D. Thomas (1977). Rotational diffusion of various membrane bound proteins as determined by saturation transfer EPR spectroscopy. *In* "Membrane Bioenergetics" (L. Packer, G. Papageorgiou, and A. Trebst, eds.), pp. 47–54. Elsevier, Amsterdam.

Dorio, M. M., and J. C. W. Chien (1974). A study of molecular motion in polymeric solids by electron-electron double resonance. *Macromolecules* **8**, 734–739.

Fano, V. (1957). Description of states in quantum mechanics by density matrix and operator techniques. *Rev. Mod. Phys.* **29**, 74–93.

Feher, G. (1959). Electron spin resonance experiments on donors in silicon. I. Electronic structure of donors by the electron nuclear double resonance technique. *Phys. Rev.* **114**, 1219–1244.

Fixman, M. (1968). NMR line shapes and path averages. *J. Chem. Phys.* **48**, 223–226.

Freed, J. H., G. V. Bruno, and C. F. Polnaszek (1971). Electron spin resonance lineshapes and saturation in the slow motional region. *J. Phys. Chem.* **75**, 3385–3399.

Fujime, S., and S. Ishiwata (1971). Dynamic study of F-actin by quasielastic scattering of laser light. *J. Mol. Biol.* **62**, 251–265.

Fung, L. W.-M., and C. Ho (1975). A proton nuclear magnetic resonance study of the quaternary structure of human hemoglobins in water. *Biochemistry* **14**, 2526–2535.

Galloway, N. B., and L. R. Dalton (1978a). Approximate methods for the fast computation of EPR and ST-EPR spectra. I. A perturbation approach. *Chem. Phys.* **30**, 445–459.

Galloway, N. B., and L. R. Dalton (1978b). Approximate methods for the fast computation of EPR and ST-EPR spectra. III. Extension of the perturbation approximation to conditions of overmodulation. *Chem. Phys.* **32**, 189–200.

Goldman, S. A., G. V. Bruno, and J. S. Freed (1973). ESR studies of anisotropic rotational reorientation and slow tumbling in liquid and frozen media. II. Saturation and nonsecular effects. *J. Chem. Phys.* **59**, 3071–3091.

Gordon, R. G., and T. Messenger (1972). Ch. XIII. Magnetic resonance line shapes in slowly tumbling molecules. *In* "Electron Spin Relaxation in Liquids" (L. T. Muus and P. W. Atkins, eds.), pp. 341–381. Plenum, New York.

Griffith, O. H., and P. C. Jost (1976). Lipid spin labels in biological membranes. *In* "Spin Labeling: Theory and Applications" (L. J. Berliner, ed.), Vol. 1, pp. 453–523. Academic Press, New York.

Halbach, K. (1954). Über eine neue methode zur messung von relaxationszeiten und über den spin von Cr^{53}. *Helv. Phys. Acta* **27**, 259–282.

Halbach, K. (1960). Modulation-effect corrections for moments of magnetic resonance line shapes. *Phys. Rev.* **119**, 1230–1233.

Hemminga, M. A., and J. L. de Wit (1977). Spin-label probes of virus protein. *Abstr., Int. Symp. Magn. Reson. 6th, 1977*, p. 347.

Hemminga, M. A., P. A. de Jager, and J. L. de Wit (1977). Saturation transfer electron paramagnetic resonance spectroscopy of spin-labeled tobacco mosaic virus protein. *Biochem. Biophys. Res. Commun.* **79**, 635–639.

Hidalgo, C., D. D. Thomas, and N. Ikemoto (1978). Effect of the lipid environment on protein motion and enzymatic activity of the sarcoplasmic reticulum calcium ATPase. *J. Biol. Chem.* **253**, 6879–6887.

Hofrichter, J., D. G. Hendriker, and W. A. Eaton (1973). Structure of hemoglobin S fibers: Optical determination of the molecular orientation in sickled erythrocytes. *Proc. Natl. Acad. Sci. U.S.A.* **70**, 3604–3608.

Hubbard, P. S., Jr., and T. J. Rowland (1957). Solution of the Bloch equations for determination of relaxation times in liquids. *J. Appl. Phys.* **28**, 1275–1281.

Huisjen, M., and J. S. Hyde (1974). A pulsed EPR spectrometer. *Rev. Sci. Instrum.* **45**, 669–675.

Huxley, H. E. (1969). The mechanism of muscular contraction. *Science* **164**, 1356–1366.

Hyde, J. S. (1960). Magnetic resonance and rapid passage in irradiated LiF. *Phys. Rev.* **119**, 1483–1492.

Hyde, J. S. (1978). Saturation-transfer spectroscopy. *In* "Methods in Enzymology" (C. H. W. Hirs and S. N. Timasheff, eds.), Vol. 49G, No. 19, pp. 480–511. Academic Press, New York.

Hyde, J. S., and L. Dalton (1972). Very slowly tumbling spin labels: Adiabatic rapid passage. *Chem. Phys. Lett.* **16**, 568–572.

Hyde, J. S., and D. D. Thomas (1973). New EPR methods for the study of very slow motion: Application to spin-labeled hemoglobin. *Ann. N.Y. Acad. Sci.* **222**, 680–692.

Hyde, J. S., L. E. G. Eriksson, and A. Ehrenberg (1970). EPR relaxation of slowly moving flavin radicals: "Anomalous" saturation. *Biochim. Biophys. Acta* **222**, 688–692.

Hyde, J. S., M. D. Smigel, L. R. Dalton, and L. A. Dalton (1975). Molecular and applied modulation effects in electron electron double resonance. IV. Stationary ELDOR of very slowly tumbling spin labels. *J. Chem. Phys.* **62**, 1655–1667.

Itzkowitz, M. S. (1967). Monte Carlo simulation of the effects of molecular motion on the EPR spectrum of nitroxide free radicals. *J. Chem. Phys.* **46**, 3048–3056.

Johnson, M. E. (1978). Librational motion of an "immobilized" spin label: Hemoglobin spin labeled by a maleimide derivative. *Biochemistry* **17**, 1223–1228.

Johnson, M. E., and S. S. Danyluk, (1978). Spin label detection of intermolecular interactions in carbonmonoxy sickle hemoglobin. *Biophys. J.* (in press).

Johnson, M. E., T. Lionel, and L. R. Dalton (1978). Organic phosphate effects upon sickle hemoglobin aggregation. *Biophys. J.* **21**, 50a.

Karplus, R. (1948). Frequency modulation in microwave spectroscopy. *Phys. Rev.* **73**, 1027–1034.

Karplus, R., and J. Schwinger (1948). A note on saturation in microwave spectroscopy. *Phys. Rev.* **73**, 1020–1026.

Kevan, L., and L. D. Kispert (1976). "Electron Spin Double Resonance Spectroscopy." Wiley, New York.

Kilty, P. A., M. C. Rol, and W. M. H. Sachtler (1973). Identification of oxygen complexes adsorbed on silver and their function in the catalytic oxidation of ethylene. *Proc. Int. Congr. Catal., 5th, 1972,* pp. 929–943.

Kirino, Y., T. Ohkuma, and H. Shimizu (1978). Saturation transfer electron spin resonance study on the rotational diffusion of calcium- and magnesium-dependent adenosine triphosphatase in sarcoplasmic reticulum membranes. *J. Biochem.* (Tokyo) **84**, 111–115.

Korst, N. N., and A. V. Lazarev (1969). Calculation of the E.S.R. line shape in highly viscous media. *Mol. Phys.* **17**, 481–487.

Kubo, R. (1969). Stochastic theories of randomly modulated systems. *J. Phys. Soc. Jpn., Suppl.* **26**, 1–5.

Kusumi, A., S. Ohnishi, T. Ito, and T. Yoshizawa (1978). Rotational motion of rhodopsin in the visual receptor membrane as studied by saturation-transfer spectroscopy. *Biochim. Biophys. Acta* **507**, 539–543.

McCalley, R. C., E. J. Shimshick, and H. M. McConnell (1972). The effect of slow rotational motion on paramagnetic resonance spectra. *Chem. Phys. Lett.* **13**, 115–119.

Macomber, J. D., and J. S. Waugh (1965). Theory of sideband production in spectroscopic experiments. *Phys. Rev. A* **140**, 1494–1497.

Mailer, C., and C. P. S. Taylor (1973). Rapid adiabatic passage EPR of ferricytochrome *c*: Signal enhancement and determination of the spin-lattice relaxation time. *Biochim. Biophys. Acta* **322**, 195–203.

Marsh, D., A. Watts, and P. F. Knowles (1977). Cooperativity of the phase transition in single- and multibilayer lipid vesicles. *Biochim. Biophys. Acta* **465**, 500–514.

Muromtsev, V. I., N. Ya. Shteinshneider, S. N. Safronov, V. P. Golikov, A. I. Kuznetsov, and G. M. Zhidomirov (1975). Spin-lattice relaxation of iminoxyl radicals in frozen solution. *Sov. Phys.—Solid State (Engl. Transl.)* **17**, 517–519.

Norris, J. R., and S. I. Weissman (1969). Studies of rotational diffusion through electron-electron dipolar interaction. *J. Phys. Chem.* **73**, 3119–3124.

Percival, P. W., and J. S. Hyde (1976). Saturation-recovery measurements of the spin-lattice relaxation times of some nitroxides in solution. *J. Magn. Reson.* **23**, 249–257.

Percival, P. W., J. S. Hyde, L. A. Dalton, and L. R. Dalton (1975). Molecular and applied modulation effects in electron electron double resonance. V. Passage effects in high resolution frequency and field swept ELDOR. *J. Chem. Phys.* **62**, 4332–4342.

Perkins, R. C., Jr., T. Lionel, B. H. Robinson, L. A. Dalton, and L. R. Dalton (1976). Saturation transfer spectroscopy: Signals sensitive to very slow molecular reorientation. *Chem. Phys.* **16**, 393–404.

Portis, A. M. (1955). Rapid passage in electron spin resonance. *Phys. Rev.* **100**, 1219–1221.

Robinson, B. H., and L. R. Dalton (1979). EPR and saturation-transfer EPR spectra at high microwave field intensities. *Chem. Phys.* (in press).

Robinson, B. H., L. R. Dalton, L. A. Dalton, and A. L. Kwiram (1974). Fast computer calculation of ESR and nonlinear spin response spectra from the fast motion to the rigid lattice limits. *Chem. Phys. Lett.* **29**, 56–64.

Robinson, B. H., M. E. Johnson, R. C. Perkins, Jr., and L. R. Dalton (1977). ST-EPR of anisotropically diffusing biomolecules. *Abstr., Int. Symp. Magn. Reson., 6th, 1977*, p. 354.

Robinson, B. H., A. H. Beth, P. S. Crooke, and L. R. Dalton (1978). Approximate methods for the fast computation of EPR and ST-EPR spectra.II. Gaussian preconvolution followed by Runge–Kutta solution of the master supermatrix equation. *Chem. Phys.* **30**, 461–468.

Robinson, B. H., L. S. Lerman, A. H. Beth, H. L. Frisch, L. R. Dalton, and C. Auer (1979). Torsional flexibility in DNA. *J. Mol. Biol.* (submitted).

Rousselet, A., and P. F. Devaux (1977). Saturation transfer electron paramagnetic resonance on membrane bound proteins. II. Absence of rotational diffusion of the cholinergic receptor protein in *torpedo marmorata* membrane fragments. *Biochem. Biophys. Res. Commun.* **78**, 448–454.

Rousselet, A., and P. F. Devaux (1978). Interaction between spin-labeled rhodopsin and spin-labeled phospholipids in the retinal outer segment disc membranes. *FEBS Lett.* **93**, 161–164.

Sillescu, H., and D. Kivelson (1968). Theory of spin-lattice relaxations in classical liquids. *J. Chem. Phys.* **48**, 3493–3505.

Smigel, M. D., L. R. Dalton, J. S. Hyde, and L. A. Dalton (1974). Investigation of very slowly tumbling spin labels by nonlinear spin response techniques: Theory and experiment for stationary electron electron double resonance. *Proc. Natl. Acad. Sci. U.S.A.* **71**, 1925–1929.

Tao, T. (1969). Time-dependent fluorescence depolarization and Brownian rotational diffusion coefficients of macromolecules. *Biopolymers* **8**, 600–632.

Thomas, D. D. (1978). Large-scale rotational motions of proteins as detected by EPR and fluorescence. *Biophys. J.* (in press).

Thomas, D. D., and C. Hidalgo (1978). Rotational motion of the calcium-ATPase in sarcoplasmic reticulum membranes. *Proc. Natl. Acad. Sci. U.S.A.* (in press).

Thomas, D. D., and H. M. McConnell (1974). Calculation of paramagnetic resonance spectra sensitive to very slow rotational motion. *Chem. Phys. Lett.* **25**, 470–475.

Thomas, D. D., J. C. Seidel, J. Gergely, and J. S. Hyde (1975a). The quantitative measurement of rotational motion of the subfragment-1 region of myosin by saturation transfer EPR spectroscopy. *J. Supramol. Struct.* **3**, 376–390.

Thomas, D. D., J. C. Seidel, J. S. Hyde, and J. Gergely (1975b). Motion of subfragment-1 in myosin and its supramolecular complexes: Saturation transfer electron paramagnetic resonance. *Proc. Natl. Acad. Sci. U.S.A.* **72**, 1729–1733.

Thomas, D. D., L. R. Dalton, and J. S. Hyde (1976a). Rotational diffusion studied by passage saturation transfer electron paramagnetic resonance. *J. Chem. Phys.* **65**, 3006–3024.

Thomas, D. D., J. C. Seidel, and J. Gergely (1976b). Molecular motions and the mechanism of muscle contraction and its regulation. *In* "Contractile Systems in Non-Muscle Tissues" (S. V. Perry, A. Margreth, and R. S. Adelstein, eds.), pp. 13–21. Elsevier, Amsterdam.

Thomas, D. D., J. C. Seidel, and J. Gergely (1978a). Rotational motion of crossbridges in contracting myofibrils. (to be published).

Thomas, D. D., J. C. Seidel, and J. Gergely (1978b). Rotational dynamics of spin-labeled F-actin in the sub-millisecond time range. *J. Mol. Biol.* (submitted for publication).

Tolman, R. C. (1938). "Principles of Statistical Methanics." Oxford Univ. Press, London and New York.

Vega, A. J., and D. Fiat (1974). Orientation dependent spin density matrix of tumbling molecules in thermal equilibrium. *J. Chem. Phys.* **60**, 579–583.

Votano, J. R., J. Gorecki, and A. Rich (1977). Sickle hemoglobin aggregation: A new class of inhibitors. *Science* **196**, 1216–1219.

Weger, M. (1960). Passage effects in paramagnetic resonance experiments. *Bell Syst. Tech. J.* **39**, 1013–1112.

Wilkerson, L. S., R. C. Perkins, Jr., R. Roelofs, L. Swift, L. R. Dalton, and J. H. Park (1978). Erythrocyte membrane abnormalities in Duchenne muscular dystrophy monitored by saturation transfer EPR spectroscopy. *Proc. Natl. Acad. Sci. U.S.A.* **75**, 838–841.

2

The Spin-Probe–Spin-Label Method

JAMES S. HYDE, HAROLD M. SWARTZ,
and WILLIAM E. ANTHOLINE

NATIONAL BIOMEDICAL ESR CENTER,
DEPARTMENT OF RADIOLOGY,
MEDICAL COLLEGE OF WISCONSIN,
MILWAUKEE, WISCONSIN

I. Introduction 71
II. Theory 74
  A. Dipolar Interactions in Solids 74
  B. Heisenberg Exchange in Liquids 83
  C. Dipole–Dipole Interactions in Liquids 87
III. Survey of Chemical and Magnetic Properties of Paramagnetic Metal-Ion
  Probes 90
  A. Chemical and Magnetic Properties of Lanthanide Ions 90
  B. Lanthanides as Probes in Biochemistry 91
  C. Chemical and Magnetic Properties of 3d Transition-Metal Ions 93
  D. Complementary Probes 94
IV. Applications of the Method 101
  A. Nature of the Paramagnetic Species 102
  B. Special Experimental Precautions 102
  C. Experimental Parameters 103
  D. Use of the Metal Ion as the "Label" 103
  E. Types of Biochemical and Biological Applications of Spin-Probe–Spin-
  Label Methods 103
  References 109

I. INTRODUCTION

The title of this chapter was borrowed from Chapter 4 of Likhtenshtein's book (1976) on spin labeling. It is generalized here, however, to include a wider range of interactions and biological systems. The spin probe is a paramagnetic metal ion, almost always of the transition or lanthanide series. The spin label is usually a nitroxide but may also be an intrinsic free radical (natural label). The purpose of this chapter is to discuss the various ways in which the simultaneous use of probe and label can yield useful information over the entire range of systems of interest in biochemistry and biophysics. Kulikov and Likhtenshtein (1977) and Eaton and Eaton (1978) have provided reviews that are useful supplements to the "textbook" approach used here.

The important role of metal ions in biological systems makes the spin-probe–spin-label method especially significant. Several of the naturally occurring ions may be paramagnetic; others may be replaced by homologous paramagnetic ions. The increasing use of elements of the lanthanide series as replacements for Ca^{2+} in a wide variety of biological systems is of special importance and is reviewed in Section III,B.

It is assumed throughout that T_1, T_2, and T_2^* of the metal ion are much shorter than those of the spin label and are not significantly altered by the presence of the spin label. The situation is asymmetric, however. Because the relaxation times of the spin label are much longer, they can be significantly affected by the presence of the metal ion. It is important to recognize that the metal ion not only can alter the appearance of the spectrum of the spin label, but also can alter its spin–lattice relaxation time. The free-radical spectral shape, signal intensity, and spin–lattice relaxation time are of roughly comparable importance as observables in detecting magnetic interaction between spin probe and spin label.

Magnetic interactions between spin probe and spin label are either dipole–dipole or Heisenberg exchange. The interactions may be modulated by the spin–lattice relaxation time of the metal ion, by rotational diffusion of the metal-ion–free-radical assembly, or by translational diffusion of the metal ion with respect to the free radical. In order to use the spin-probe–spin-label method, it is necessary to determine which of the two interactions is dominant and how the modulation of the dominant interaction affects the experimental observables.

The major historical event leading to the method was the publication of the paper by Taylor et al. (1969) in which it was reported that a covalently bound spin label on a protein showed a loss of intensity and no apparent broadening upon specific binding of Mn^{2+}. The subsequent theory of Leigh (1970) showed that this surprising result was a consequence of dipole–dipole interaction and not, as had been conjectured, a manifestation of chemical effects. Leigh's theory assumes a rigid lattice and a single radial distance between spin probe and spin label. There is an implicit assumption that the protein tumbled so slowly in solution on the electron spin resonance (ESR) time scale that a "solid-state" theory was appropriate. The theory is a cornerstone of the method and is discussed at some length in Section II.

In the Soviet Union, an independent body of work leading eventually to the paper by Salikhov *et al.* (1971) was developed in which Heisenberg exchange between metal ions and nitroxide radicals in the *liquid* phase was investigated both theoretically and experimentally. Both the translational correlation time and the metal spin–lattice relaxation time are carried in the theoretical treatment. Thus, the theory is the most complete of the various calculations relevant to the spin-probe–spin-label method. However, there are not many biological systems in which interactions can be considered on the ESR time scale as purely liquidlike. Thus, this theory is applicable only as a limiting situation and may be more useful in the conceptual insights it provides than in immediate applications to biological systems.

Likhtenshtein (1976) bases his theoretical comments on the work of Salikhov *et al.* and favors Heisenberg exchange as the most likely interaction between metal ions in solution and nitroxide radicals covalently bound to proteins. The present authors know of no physicochemical investigations of exchange and dipole–dipole interactions at a liquid–solid interface and believe such basic research is necessary before the approaches outlined by Likhtenshtein can be pursued with confidence.

There are many aspects of the literature on nuclear magnetic resonance (NMR) that are immediately relevant, most of which are lucidly discussed in Abragam's book (1961). Chief of these are nuclear magnetic relaxation induced by paramagnetic ions in single crystals and NMR relaxation in liquids induced by paramagnetic ions. The reader should be cautioned, however, that many of the equations in the literature on NMR are incomplete because of the assumption that $\omega\tau \ll 1$, where ω is the radiofrequency and τ the relevant correlation time. In general in ESR the inequality is reversed.

Our group at the National Biomedical ESR Laboratory at the Medical College of Wisconsin became involved in the spin-probe–spin-label method somewhat indirectly. In the course of a study of the naturally occurring free radical in the biological pigment melanin (Sarna *et al.*, 1976), addition of Cu^{2+} (which binds tightly to the surface of melanin granules) was found to cause loss of ESR intensity in a manner adequately described by the theory of Leigh. Because of our ongoing interest in nitroxide radicals it was natural to generalize our viewpoint, which led to the writing of this chapter. Special emphasis in our publications is given to measurements of spin–lattice relaxation characteristics as well as spectral features and the usefulness of elements of the lanthanide series as spin probes.

With this brief introduction and historical review, we will consider the theory in more detail. We also will provide an elementary review of those aspects of metal-ion chemistry and paramagnetism that are especially relevant to the spin-probe–spin-label method. Finally, we will discuss some of the various types of biologically relevant information that can be obtained by the method.

II. THEORY

The solid state and the liquid state are the two limiting phases of matter in which theoretical analysis of magnetic interactions between spin probe and spin label is reasonably straightforward. In many cases the analysis is directly applicable to biological systems. In other cases, however, the biological system may be better modeled as a mixture of these phases, or perhaps a degree of motion will be present such that the system lies intermediate between liquid and solid phases. The latter situations are to a considerable extent ignored in the present treatment, mainly because so little theoretical work has been done. We discuss here dipolar interactions in solids and dipolar and exchange interactions in liquids. An understanding of the physics of these limiting situations provides a reasonable base for use of the method.

A. Dipolar Interactions in Solids

1. ELECTRON SPIN RESONANCE RIGID-LATTICE LINE SHAPE IN A SYSTEM OF TWO INTERACTING SPINS

Leigh's calculation considers only the effect of the A component (Bloembergen et al., 1948) of the dipolar Hamiltonian. The local magnetic field at the free-radical site is the vector sum of the dipolar field from the metal ion and the applied field. Classically, for a particular orientation of the polarizing magnetic field H_0 with respect to the free-radical coordinate system and a particular orientation of H_0 with respect to the radial vector between probe (of spin 1/2) and label, the free-radical signal will be split in two, corresponding to the two possible orientations of the metal-ion spin. *In the absence of any time-dependent modulation* of the dipolar interactions, in a powder sample, one would integrate over all orientations of the free radical with respect to the laboratory frame in order to simulate the spectrum. The results would depend on the orientation of the radial vector with respect to the free-radical frame. The dipolar interaction depends on the sixth power of the radial distance, so a detailed theoretical fit of an experimental spectrum should yield both the radial distance and the orientation with respect to the free-radical frame. (There is substantial similarity between this discussion and that of Pake doublets in texts on NMR.)

If modulation of the interaction is now introduced because of spin–lattice relaxation of the metal ion at a rate similar to the inverse of the separation of the lines (i.e., reciprocal radians per second), broadening of the split lines—smearing out—and eventual reappearance at the "center of gravity" occur in a manner first described by McConnell (1958) in a discussion of chemical exchange in NMR. The detailed calculation of the powder line shape in the presence of motion is based on a theory due to Redfield (1957) and follows Slichter's treatment (1963). The principal assumptions in Leigh's theory are as follows:

(1) Each free radical interacts with one metal ion at a separation r, and this radial vector has a fixed orientation in the free-radical frame.

(2) All orientations of the free radical with respect to the laboratory frame are equally probable.

(3) The A component of the dipolar Hamiltonian dominates all other components in determining the spin-label spectral shape, i.e., $\omega\tau > 1$.

(4) The spin–lattice relaxation time of the metal ion is comparable to the inverse of the dipolar interaction.

(5) The spin–lattice relaxation time of the metal ion is determined by the lattice (and not the spin label) and has no angular dependence.

(6) There is no motional modulation of the dipolar interaction.

The calculation proceeds in a straightforward way. Let $L(\theta, \varphi, H)$ be the spectrum of the free radical at a particular orientation θ, φ of the molecular coordinate system with respect to the laboratory frame. Let $g(H)$ be the line shape function for the individual lines that make up the spectrum. Each line $g(H)$ is broadened by dipolar interaction with the metal-ion probe,

$$g(H) = \frac{\delta H}{(H - H_0)^2 + \delta H^2} \tag{1}$$

where

$$\delta H = C(1 - 3\cos^2\theta_r')^2 + \delta H_0 \tag{2}$$

and

$$C = \frac{g\beta\mu^2 T_{1k}}{\hbar r^6} \tag{3}$$

Here, δH_0 is the natural linewidth in the absence of the metal-ion probe, T_{1k} is the metal spin–lattice relaxation time, and θ_r' is the angle between the radial vector (of magnitude r) and the applied field. The effective magnetic moment μ is an experimental quantity derived from static susceptibility measurements that was introduced by Van Vleck (1932). It eliminates the necessity of explicit consideration of orbital angular momentum and high spin values of the metal ion.

One determines the composite line shape by integrating over θ, φ [Eq. (4)] using the line shape function given in Eq. (1). Some geometry is involved in relating θ_r' to θ and φ: one must know the orientation of the radial vector (θ_r, φ_r) in the molecular coordinate system.

$$L(H) = \int_0^{2\pi} \int_0^{\pi} L(\theta, \varphi, H) \sin \theta \, d\theta \, d\varphi \qquad (4)$$

Computer-generated line shapes as a function of C, θ_r, and φ_r, taken from Leigh's paper, are shown in Fig. 1. The main qualitative conclusion is that the spectra do not appear to change very much in shape but *appear* to lose intensity. The word *appear* is crucial. There is no loss of integrated intensity or total radiofrequency susceptibility; there is a substantial change in line shape. But when field modulation and phase-sensitive detection are employed, broad components of the spectrum are effectively suppressed relative to narrow components. For those powder fragments where **r** with respect to $\mathbf{H}_0$ is *close* to the magic angle of 54.7°, the dipolar interaction is *near* zero and the free-radical spectrum is unaltered. Away from the magic angle, the dipolar broadening smears spectral intensity into a broad continuum with very small first-derivative intensity. The magnitude of the parameter C defines the words *near* and *close* in the sentence above.

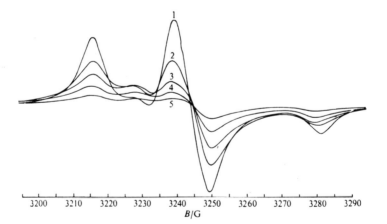

Fig. 1. Computed "nitroxide" ESR spectra. The parameters used were $g_x = 2.0089$, $g_y = 2.0061$, $g_z = 2.0027$, $A_x = 1.25 \times 10^8$ rad/sec, $A_y = 0.986 \times 10^8$ rad/sec, $A_z = 5.63 \times 10^8$ rad/sec, and $\delta H_0 = 4.5$ G, $\theta_r = 0.2\pi$, $\varphi_r = \pi/2$. The dipolar interaction coefficient C is 0, 3, 10, 30, and 100 in curves 1–5, respectively. The magnetic field scale begins at 3200 G, and the marks show 5-G increments (Leigh, 1970).

As the spin–lattice relaxation time gets shorter and shorter, the dipolar interactions become increasingly well averaged, $\delta H \rightarrow \delta H_0$ [Eq. (2)], and the effect of the presence of the paramagnetic metal ion becomes negligible. Thus, relatively slowly relaxing metal ions such as Cu^{2+}, Gd^{3+}, and Mn^{2+} are particularly effective.

As the temperature is decreased, the spin–lattice relaxation time of metal ions increases and the magnitude of the Leigh effect increases. Thus, the free-radical signal will actually decrease in amplitude as the temperature is lowered, in contrast to the situation ordinarily encountered in which the signal shows an inverse linear dependence on absolute temperature because of the Boltzmann difference in populations. As the microwave frequency is changed, say from X band to 35 GHz, one may expect changes in the magnitude of the Leigh effect for at least two reasons: both the spin–lattice relaxation time of metal ions and $L(H, \theta, \varphi)$ will change. The reader should be cautioned that the characteristic apparent loss of intensity without accompanying line shape change may not always occur. In our laboratory broadening and line shape changes have been observed at 35 GHz in some systems. Computer simulation of experimental results is desirable.

Leigh's theory has often been used to determine radial distances. Although knowledge of metal spin–lattice relaxation times is usually poor, estimates correct within an order of magnitude can generally be made. Because of the radial sixth power dependence of C it may often be possible to obtain reliable estimates of r despite the uncertainty of T_{1k}. Workers have not been very successful in learning a great deal about orientation of the radial vector with respect to the free-radical frame. Rao et al. (1979) carried out simulations in which rotational diffusion of the protein was additionally introduced into Leigh's theory. They were able to reproduce the spectrum obtained by Taylor et al. (1969) from creatine kinase quite satisfactorily, independent of their initial choice of θ_r. It is perhaps not surprising that angular information is lost when slow rotational diffusion occurs but that information on the radial separation between spin label and spin probe is preserved.

The Leigh theory has been remarkably successful. It can be refined and extended by consideration of other components of the dipolar Hamiltonian and by summation over a variable r, including the possibility that more than one metal ion interacts with each free radical.

2. DIPOLAR-INDUCED ELECTRON SPIN–LATTICE RELAXATION IN SOLIDS

If the spin–lattice relaxation time is equal to or greater than the reciprocal of the dipolar interaction between spin probe and spin label, then the theory of Leigh, which considers only the A component of the dipolar Hamiltonian, is a reasonably complete description of the interaction. As the metal spin–lattice relaxation time becomes shorter and shorter, the Leigh effect decreases, but when $T_{1k} \simeq \omega^{-1}$, where ω is the microwave frequency, another aspect of the dipolar interaction enters: the fluctuating dipolar field at the free radical from the rapidly relaxing metal ion becomes a powerful spin–lattice relaxation mechanism for the free radical. In this section this effect is analyzed. The effects discussed in this and in the preceding section are of approximately equal importance in spin-probe–spin-label methodology and are to a considerable extent complementary.

Bloembergen (1949) and Abragam (1955) calculated the influence of paramagnetic metals in solids on relaxation of nearby nuclei, deriving the expression

$$T_{1j}^{-1} = J(J + 1)\left[\frac{1}{3} b_{jk}^2\left(\frac{2T_{1k}}{1 + (\omega_{pm} - \omega)^2 T_{1k}^2}\right)\right.$$
$$\left. + \frac{2}{3} c_{jk}^2\left(\frac{2T_{1k}}{1 + \omega^2 T_{1k}^2}\right) + \frac{2}{3} e_{jk}^2\left(\frac{2T_{1k}}{1 + (\omega_{pm} + \omega)^2 T_{1k}^2}\right)\right] \quad (5)$$

where

$$b_{jk}^2 = \tfrac{1}{4}\gamma_j^2\gamma_k^2\hbar^2 r_{jk}^{-6}(1 - 3\cos^2\theta_{jk})^2$$
$$c_{jk}^2 = \tfrac{9}{4}\gamma_j^2\gamma_k^2\hbar^2 r_{jk}^{-6}\sin^2\theta_{jk}\cos^2\theta_{jk}$$
$$e_{jk}^2 = \tfrac{9}{8}\gamma_j^2\gamma_k^2\hbar^2 r_{jk}^{-6}\sin^4\theta_{jk} \quad (6)$$

The subscripts j and k refer to the nucleus (or free radical here) and metal ion; θ is the angle between the radial vector r_{jk} and the magnetic field H_0; J is the total angular momentum of the metal; the γ's are the magnetogyric ratios; ω_{pm} is the resonance frequency of the paramagnetic metal in the magnetic field H_0 used to observe resonance of the free radical; ω is the resonance frequency of the free radical; and T_{1k} is the spin–lattice relaxation time of the metal ion. The first term arises from the B component of the dipolar Hamiltonian, the second from the C and D components, and the third from the E and F components. For nuclear relaxation, $(\omega_{pm} \pm \omega)^2 T_{1k}^2 \gg \omega^2 T_{1k}^2 \sim 1$, and the CD term dominates. For free-radical relaxation, however, all three terms must be considered.

If $(\omega_{pm} - \omega)T_{1k} \geq 1$, the B term dominates. This is cross-relaxation and is discussed in the next section. If $\omega T_{1k} \ll 1$, all three terms are of similar order of magnitude. The differing coefficients and angular dependencies are, however, insignificant compared with the r^6 dependence. Thus, even in the absence of a detailed analysis of these three terms, rather good values of r can be obtained from saturation experiments. Hyde and Rao (1978) showed that the CD term dominates the others by a factor of about 4 for Dy^{3+} and Tm^{3+} at $-150°C$, which permits simplification of the interpretation of experimental results.

Let us assume that r is fixed, only one metal ion interacts with one free radical, and all orientations are equally probable; that is, the physical model is the same as was assumed by Leigh as discussed in the preceding section. In magnetically dilute, unordered solids, the free-radical line shape is a resultant of unresolved superhyperfine interactions and anisotropic Zeeman and hyperfine interactions. Such lines are described as inhomogeneous. For sufficient dilution in the absence of motion, $T_{1j} = T_{2j}$. Portis (1953) first wrote a theory of saturation for these systems. This theory was rewritten by Hyde (1960) using an argument similar to Redfield's (1955) to take into account a breakdown in the Bloch equations that occurs in the solid phase at high microwave powers. Hyde's Eq. (10) is

$$\chi''(\omega)H_1 = \frac{\pi}{2}\chi_0\,\omega h(\omega - \omega_0)H_1\frac{1}{(1 + \gamma_j{}^2 H^2 T_{1j}^2)^2} \tag{7}$$

where $h(\omega - \omega_0)$ is the inhomogeneous line shape. It then becomes a straightforward matter to insert one, two, or all three terms of Eq. (5) into Eq. (7) and integrate over θ, which would predict the shape of the observed experimental progressive-saturation curve.

Similarly one might attempt an integral of the form

$$f(t) = \int_0^{2\pi}\int_0^{\pi}(1 - e^{-t/T_{1j}})\sin\theta\,d\theta\,d\varphi \tag{8}$$

to predict the result from a saturation–recovery experiment. Note that it is incorrect to integrate Eq. (5) over θ to determine an apparent T_{1j}. Such a procedure would lead to the prediction that the saturation–recovery signal should be a simple exponential. One must integrate an experimental observable over θ.

Equations (7) and (8) integrated over θ lead to inhomogeneous spin–lattice relaxation behavior; that is, there is a distribution of spin–lattice relaxation times.

When $\omega T_{1k} < 1$, the relaxation times are so fast that the Leigh-type mechanism is no longer active. That is, the "A" terms are effectively averaged to zero. Yet these ions are optimally effective in inducing free-radical spin–lattice relaxation. Metal ions with these fast relaxation times show no detectable ESR spectra because of their great spectral widths.

Comparison of the method of this section with the Leigh approach indicates some advantages in both methods. The saturation technique intrinsically permits the observation of longer-range interactions. This is because the fluctuating dipolar field need only be of the order of $(\gamma T_{1j})^{-1}$ to induce relaxation, whereas it must be of the order of the free-radical spectral width for there to be much effect through the Leigh mechanism. Saturation techniques on *slowly* tumbling spin labels, however, do not work very well. This is because even though T_{1j} is relatively long, T_{2j} is very short, resulting in great difficulty in saturation. In the solid phase or even the *very slowly* tumbling domain, however, spin-label spectra typically show the onset of microwave power saturation at an incident level of a few milliwatts, and the saturation methods are applicable. An advantage of saturation methods, as found by Sarna *et al.* (1976), is that measurements can be made over a much wider range of dipolar interactions than for the Leigh mechanism. For example, if the signal-to-noise ratio of the spin-label spectrum were 10 : 1 in the absence of metal ion, conditions would have to be just right when a slowly relaxing metal ion was added or no signal at all would be seen. But saturation measurements could be made over a much wider range of fast-relaxing metal-ion concentrations.

Hyde and Rao (1978) extended this type of calculation in order to consider a situation in which a number of metal ions could interact with each free radical. The model is shown in Fig. 2A. The free radical is located at the center. The potential metal-ion sites are indicated by squares, some of which are occupied by metal ions with a probability of occupation *f*. Although the distribution of squares around each free radical is assumed always to be the same (i.e., good short-range order), the occupation of squares by metal ions varies over the full range of statistical probabilities. The distance *a* is

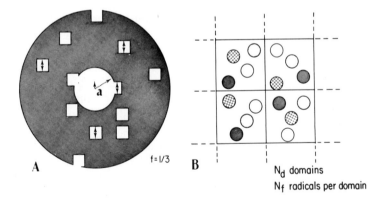

$f = 1/3$

N_d domains

N_f radicals per domain

Fig. 2. The model in the calculation of Hyde and Rao. (A) The free radical is at the center. The local geometry of potential metal binding sites around each free radical is identical. The probability of occupation of a metal binding site is *f*. In the figure $f = 1/3$, for example. The distance of closest approach is *a*. (B) There are N_d domains, each large enough that every combination of metal-ion occupations (including, for example, that of Fig. 2A) is represented. The number of domains is sufficiently large that macroscopic solutions of the Bloch equations for each possible combination can be summed to obtain the total system response (Hyde and Rao, 1978).

the distance of closest approach of the metal ion to the free radical. The sample is further envisaged as being divided into domains (Fig. 2B), each domain being large enough that every possible combination of metal-ion occupations is represented. Finally, the number of domains actually observed by the spectrometer is sufficiently large that one can write macro-scopic solutions to the Bloch equations for each combination and sum over the combinations.

Hyde and Rao observed that in a purely formal sense one could calculate second and higher "moments" of the B, CD, and EF terms following Van Vleck (1948),

$$\langle \Delta \omega^2 \rangle_h = \frac{1}{3} J(J+1) \hbar f \sum_{jk} h_{jk}^2 \qquad (9)$$

where the sum is over every possible metal-ion binding state (the squares of Fig. 2A) whether or not occupied. The subscript h stands for either b, c, or e in Eq. (6). Similar expressions for the fourth "moments" were written. The sums appearing in these expressions for the moments were identical with the sums that appeared when Eq. (5) was inserted into the model of Fig. 2A,B, and therefore the final results of the calculation were expressed in terms of these moments.

In the analysis of saturation–recovery experiments, the initial slope of the recovery signal $S(t)$ is given by Eq. (10)

$$\left[\frac{dS(t)}{dt} \right]_{t=0} = 2t \left[\frac{2T_{1k}}{1 + \omega^2 T_{1k}^2} \right] \langle \Delta \omega^2 \rangle N_j \qquad (10)$$

(there being N_j free radicals observed), assuming a dominance of the CD terms. The second term in the expansion of the expression for the recovery signal was that expected for an exponential recovery except for an additional multiplicative factor of $1/f$. Only if f is equal to 1 (i.e., every site identical) is the recovery purely exponential.

Analysis of the progressive-saturation experiment shows that $P_{1/2}$ (the incident power at which the signal is half as large as it would be in the absence of microwave power saturation) varies as the square of the second moment.

The second moment, that is, the sum in Eq. (9), can be evaluated for any chosen structural model. Assuming a simple cubic lattice of lattice pa-rameter a with some lattice sites occupied by free radicals, most sites vacant, and other sites occupied by metal ions, it was found that $\langle \Delta \omega^2 \rangle \propto N a^{-3}$, where N is the concentration of free radicals. One can immediately draw an important general conclusion: results of saturation experiments in the model of Fig. 2A,B are dominated by the distance of closest possible approach even if the probability of occupation of the close sites is low. A

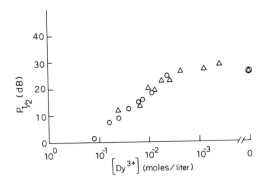

Fig. 3. The power for which the signal intensity is one-half the unsaturated intensity, $P_{1/2}$, for approximately 2×10^{-4} M galvinoxyl (Aldrich) versus the concentration (moles/liter) of dysprosium chloride (Ventron) in alcohol reagent A-962, (Fisher Lab Chem., containing ethyl alcohol 90%, methanol 5%, isopropyl alcohol 5%) at 77°K: $\triangle$, degassed samples; $\bigcirc$, samples with dissolved oxygen (Antholine *et al.*, 1978).

further prediction is that $P_{1/2}$ will vary as N^2. In an effort to test this theory, Antholine *et al.* (1978) collected the data shown in Fig. 3. The predicted N^2 dependence was observed. They also studied systems in which the distance of closest approach was varied. For example, complexing Dy^{3+} with EDTA in 75% alcohol–25% H_2O reduced $P_{1/2}$ of the free radical galvinoxyl by 10 dB, all metal-ion and free-radical concentrations remaining unchanged.

Thus, a main conclusion of the analysis is that paramagnetic metal ions can be added to systems containing spin labels and, even though there is a wide range of radial distances, observed magnetic interactions will be determined primarily by the distance of closest approach. There will be a numerical factor that will depend on specific geometry, but it will tend to be of little significance after the sixth root is taken.

3. CROSS-RELAXATION—THE B TERM

The literature contains two theoretical models for cross-relaxation (the simultaneous change of free-radical and metal-ion quantum numbers). The first is simply that already described by Eq. (5). The physical problem in cross-relaxation arises in accounting for the difference in energy when the simultaneous spin flips correspond to different resonant frequencies. In the model of Eq. (5), the dipolar interaction is modulated so fast that there exists a spectral density at the site of the slowly relaxing spin at a frequency corresponding to its resonance frequency. It has already been mentioned that when $(\omega_{pm} - \omega)T_{1k} \geq 1$, the B term as given by Eq. (5) is the dominant one.

The second theory is that of Bloembergen *et al.* (1959). This theory introduces a cross-relaxation line shape given by

$$g_{\alpha\beta} = \iint g_\alpha(v')g_\beta(v'')\delta(v' - v'') \, dv' \, dv'' \tag{11}$$

where α and β refer to the two spin systems. Cross-relaxation in this theoretical framework depends on the extent of overlap of the resonance lines of the free radical and paramagnetic metal ion. No spectral densities appear in the theory. Since, however, we are considering cross-relaxation between unlike spins, one of which (the metal ion) has a much shorter spin–lattice relaxation time than the other, the net effect of cross-relaxation, following Eq. (11), is to shorten the apparent spin–lattice relaxation time of the free radical.

This theory appears to be consistent in detail with the experimental investigations of Wong and Wan (1971), who studied cross-relaxation between various metal ions and the hyperfine lines of trapped hydrogen atoms. Only when there was significant overlap of the metal-ion spectrum with one of the hyperfine lines was significant cross-relaxation observed. Wong and Wan reported no effects that could be associated with any other mechanisms discussed in this chapter. It is concluded that, when there is significant overlap, the mechanism of Bloembergen *et al.* (1959) is by far the dominant and longest-range magnetic interaction.

This brief review of cross-relaxation is presented here mainly for pedagogical reasons. Nothing exists in the literature concerning spin-probe–spin-label cross-relaxation interactions applied to biological systems. It has occurred to us that cross-relaxation between Gd^{3+} and free radicals (where $\omega T_{1k} \gg 1$ and there is significant overlap of resonances) might be an unusually long-range interaction and that useful experiments could be designed based on this observation.

B. Heisenberg Exchange in Liquids

In this section and in the next, interactions between metal ions and free radicals in pure liquids are discussed. Straightforward application to biological systems is not at all obvious. We believe, however, that an understanding of these interactions is desirable even when one is using one of the solid-state models already treated. Furthermore, we view the spin-probe–spin-label method as being in an early stage of development. Because some biological systems may have both liquid- and solidlike properties, accurate application of this method to such systems will require an understanding of the liquid-state mechanisms.

The subject of this section is Heisenberg exchange in solution between two different paramagnetic species. One of them, the paramagnetic metal ion, is at relatively high concentration and has a relatively short spin–lattice relaxation time. This time is assumed to be unaltered to any significant degree by either inter- or intraspecies magnetic interactions. The other species is a free radical. It is at a relatively low concentration, has relatively narrow lines, and has a rather long spin–lattice relaxation time. The linewidth and spin–lattice relaxation time are the experimental observables and are investigated in model studies as functions of macroscopic viscosity and metal-ion concentration.

The theory of liquid-phase Heisenberg exchange began with the paper of Pake and Tuttle (1959). A number of authors have contributed to the subject since then. The papers by Kivelson (1960) and Eastman *et al.* (1969) have been particularly useful to the present authors. *Intraspecies* Heisenberg exchange is characterized by the following physical ideas:

(1) The theory assumes an exchange interaction J, which is abruptly turned on during a collision.

(2) Although early theories considered the possibilities of both weak exchange (many collisions per event) and strong exchange (each collision results in exchange) experimental work on free radicals showed, without exception, to our knowledge, that strong exchange always occurred.

(3) The exchange process is a lifetime-limiting event contributing directly to the observed linewidth. Incremental changes in the observed linewidths $\Delta(\gamma T_{2j})^{-1}$ obey the following proportionalities:

$$\Delta\omega_{ex}/\gamma = \Delta(\gamma T_{2j})^{-1} \propto \text{collision frequency} \propto (\eta/T)^{-1} \times [M] \qquad (12)$$

Here η is the viscosity and $[M]$ the concentration.

(4) Exchange is a spectral diffusion mechanism coupling free radicals that have different nuclear spin configurations. It thus leads to a change in "apparent" spin–lattice relaxation as observed using progressive-saturation methods, in which all relaxation paths must be included. It does not affect the relaxation probability between pairs of levels. The true spin–lattice relaxation time can be observed directly by saturation–recovery techniques and is unaffected by Heisenberg exchange.

(5) Since exchange depends on the collision frequency, which in turn depends on translational diffusion, parallel study of Heisenberg exchange and line shape analysis (which yields rotational diffusion information) permits investigation of microscopic models of liquids.

Salikhov and co-workers (1971, and earlier works cited therein) provided a definitive theoretical and experimental investigation of Heisenberg exchange between free radicals and metal ions. The 1971 paper is particularly recommended as background for the spin-probe–spin-label method.

The key theoretical result is Eq. (13), where Z is the frequency of collisions of a given free radical with all metal ions, S_k is the spin of the metal ion,

τ_c is the translational correlation time [see Eq. (15)], T_{1k} is the spin–lattice relaxation time of the metal ion, and J is the exchange integral.

$$\Delta\omega_{ex} = Z\tfrac{2}{3}J^2 S_k(S_k + 1)T_{1k}\tau_c[1 + \tfrac{2}{3}J^2 S_k(S_k + 1)\tau_c T_{1k}]^{-1} \qquad (13)$$

Thus, if $\tfrac{2}{3}J^2 S_k(S_k + 1)\tau_c T_{1k} \gg 1$, then $\Delta\omega_{ex} = Z$, just as in intraspecies Heisenberg exchange. It is clear from the experiments that Heisenberg exchange with relatively slowly relaxing metal-ion probes (Cu^{2+}, Mn^{2+}, VO^{2+}, etc.) is in this limit. The experimentalist can verify that he is in the strong-exchange limit by checking that the linewidths exhibit an $(\eta/T)^{-1} \times [M]$ dependence. We note that the collision frequency Z is predicted on rather classic grounds to be independent of molecular size. The translational diffusion slows with increasing size but the collisional cross section increases, resulting in a cancellation. Thus, varying the size of the interacting species is not useful in testing the dominance of strong Heisenberg exchange.

If T_{1k} is sufficiently short, as for Co^{2+} in solution, the weak-exchange limit is reached. Hyde and Sarna (1978) also showed that Gd^{3+} is in the weak-exchange limit because J is small. In this limit, $\Delta\omega_{ex} \propto ZT_{1k}\tau_c$. The viscosity dependencies of Z and τ_c cancel in the usual classic model. Since T_{1k} generally decreases as the viscosity decreases, $\Delta\omega_{ex}$ is expected to decrease when the viscosity decreases rather than increase linearly with viscosity as is the case when strong exchange dominates. This predicted viscosity dependence of exchange in the weak-exchange limit was verified by Skubnevskaya and Molin (1972).

The work of Skubnevskaya and Molin (1967, 1972), Keith et al. (1977), and Hyde and Sarna (1978) provides the experimental basis for the theory of spin exchange between metal ions and free radicals. Although the broad outlines of the theory were amply verified, there were difficulties with detail. Both the Soviet workers and Keith et al. found it necessary to introduce an empirical constant into the theory. One might combine their constants in an expression of the form

$$\omega_{ex} = Z\frac{f}{k} \qquad (14)$$

where the f factor is due to Skubnevskaya and Molin (1967) and is interpreted as arising from the solute molecules. It is a steric factor; not all collisions are imagined to be effective for exchange, depending on the geometry of the colliding molecules. In Eq. (14) the k term reflects properties of the solvent. Equation (14) implies that, even though the kinetics of exchange exhibit the usual strong exchange $(\eta/T)^{-1}$ dependence, the magnitude of the exchange broadening depends on the solute molecules and the solvent.

Hyde and Sarna and Keith *et al.* all found that complexing of copper to EDTA makes very little change in the exchange frequency. However, Keith *et al.* observed a factor of 3 difference in the exchange frequency between free radical and $NiCl_2$ and between free radical and nickel–EDTA. Skubnevskaya and Molin (1972) studied nickel acetylacetonate, which showed weak exchange in pyridine and strong exchange in aniline. This was attributed to a difference in delocalization of spin density onto the solvent molecules occupying the axial position.

We conclude that, if one wants to investigate microviscosity in the strong-exchange limit, use of a paramagnetic ion with relatively long relaxation time is desirable, Cu^{2+} being the most promising candidate. If one wants to learn about details of a probe–radical encounter, use of an ion such as Ni^{2+} is an interesting possibility. It is on the borderline between strong and weak exchange because of a short T_{1k}. Steric factors and spin density delocalization can affect J and shift the interaction from one domain to the other.

Coulomb repulsion between like-charged labels and probes might be expected to reduce Heisenberg exchange. Figure 4, from Keith *et al.* (1977), shows the difference between uncharged TEMPOL, negatively charged TEMPO sulfate, and still more negatively charged TEMPO phosphate, all exchanging with minus-three-charged ferricyanide. The conclusion is that Coulomb forces profoundly affect Heisenberg exchange.

Hyde and Sarna (1978) investigated the effect of Heisenberg exchange on

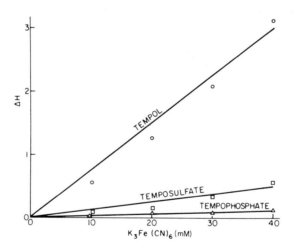

Fig. 4. Charge dependence for line broadening. Spin labels were maintained at $10^{-3}\ M$; $K_3Fe(CN)_6$ was used as the broadening agent (Keith *et al.*, 1977).

spin–lattice relaxation of the free radical. Since the metal has a very short relaxation time, every exchange event provides a direct coupling of the free-radical spin system with the lattice. Thus, it is predicted that if Heisenberg exchange is the dominant magnetic interaction, the changes in transverse and in longitudinal relaxation times because of the presence of the metal ion will be identical. This was verified experimentally. Since the intrinsic information content is the same, it is easier and preferable to study exchange by observing linewidths. However, the spin–lattice relaxation measurement is useful in establishing the dominance of Heisenberg exchange over dipolar interactions.

Heisenberg exchange is generally the dominant magnetic interaction in liquids between metal ions and free radicals. It is usually "strong," and the observed line broadening yields immediately the collision frequency and information on translational diffusion at a molecular level. Extensive efforts to extract more detailed information on the nature of collisions have been thus far only marginally successful. Except for effects of charge, concepts such as the steric factor f, spin density delocalization onto ligands, and the k factor of Keith et $al.$ should be viewed as higher-order refinements and not strongly supported by experiments.

C. Dipole–Dipole Interactions in Liquids

Dipole–dipole interactions between spin probe and spin label are generally dominant in solids, and Heisenberg exchange tends to be dominant in liquids. In this section the magnitude of dipolar interactions in liquids is considered, and the relative roles of the two kinds of magnetic interactions in biological systems (which on the ESR time scale are neither liquid nor solid) are discussed.

Eastman et $al.$ (1969), Skubnevskaya and Molin (1972), and Keith et $al.$ (1977) make qualitative arguments that dipolar interactions are negligible in ordinary liquids of ordinary viscosity. These rough arguments were amply justified by the good agreement between their experiments and the theory of Heisenberg exchange. Hyde and Sarna (1978), however, in a study of magnetic interactions between ions of the lanthanide series and nitroxide–radical spin labels, concluded that dipolar interactions dominate over Heisenberg exchange under all conditions using these ions. The reason is that the paramagnetism resides in inner 4f orbitals that are well shielded by the outer 5s and 5p electrons. Thus, the exchange integral J is small. In addition, the effective magnetic moments of the elements of the lanthanide series are quite high. Thus, for example, everything else being equal, dipolar interactions will be 21 times greater for Gd^{3+} than for Cu^{2+}. [This is the ratio of $S(S + 1)$ for Gd^{3+} to $S(S + 1)$ for Cu^{2+}.]

In the liquid phase, the correlation time for dipolar interactions in the most elementary form is

$$\tau_c = 12\pi\langle a\rangle^3 \frac{\eta}{kT} \tag{15}$$

This is the characteristic time that the two spins interact as they diffuse past each other. The interaction distance $\langle a\rangle$ is an average radial separation at the distance of closest approach; τ_c is of the order of 10^{-10}–10^{-11} sec. K. V. S. Rao (unpublished) derived an expression for the spin–lattice relaxation time of a free radical induced by a paramagnetic ion:

$$
\begin{aligned}
T_{1j}^{-1} = \frac{8\pi N}{15\langle a\rangle^3}\, \gamma^2\langle\mu^2\rangle \Bigg[& \frac{T_{1k}\tau_c(T_{1k}+\tau_c)}{(T_{1k}+\tau_c)^2 + (\omega_{\mathrm{pm}}-\omega)^2 T_{1k}^2\tau_c^2} \\
+ & \frac{3T_{1k}\tau_c(T_{1k}+\tau_c)}{(T_{1k}+\tau_c)^2 + \omega^2 T_{1k}^2\tau_c^2} \\
+ & \frac{6T_{1k}\tau_c(T_{1k}+\tau_c)}{(T_{1k}+\tau_c)^2 + (\omega_{\mathrm{pm}}+\omega)^2 T_{1k}^2\tau_c^2} \Bigg]
\end{aligned}
\tag{16}
$$

Here N is the concentration of metal ions, and the other quantities have already been defined. The derivation follows the method on p. 302 of Abragam (1961). In deriving Eq. (16) it is assumed that the main contribution to T_{1j}^{-1} comes when the ion and free radical are at the minimal distance of approach. Hence, the factor u on p. 302 of Abragam has been set equal to 1, which makes the integral

$$\int_0^\infty J_{3/2}(u)\,\frac{du}{u} = \frac{1}{3}$$

Even though this approach is not rigorous, Eq. (16) has the advantage of being in a closed form.

A more sophisticated theory can be constructed on the basis of the theory of random flights. Hexem et al. (1976) have done this, carrying along in their calculation the spin–lattice relaxation time of the metal ion. In this model the form of the spectral density function varies according to the relative magnitudes of the mean jump length squared $\langle r^2\rangle$ and $\langle a\rangle$. In the limit of large jumps, $T_{1j}^{-1} \propto \eta$, just as predicted by Rao's equation, whereas in the diffusive limit $T_{1j}^{-1} \propto \eta^{1/2}$. These dependencies are to be contrasted with the Heisenberg exchange dependence, $\omega_{\mathrm{ex}} \propto \eta^{-1}$. Experimentally, Hyde and Sarna (1978) observed that the contributions to the spin-label linewidths because of dipolar interactions with Gd^{3+} were nearly independent of viscosity. (See also Freed, 1978.)

The Mn^{2+} ion is very commonly used in spin-probe–spin-label studies. It has a high effective moment, and there is good evidence that Heisenberg exchange with the spin label and dipolar interactions between probe and label will make approximately comparable contributions to the spin-label linewidth.

As the viscosity increases, Heisenberg exchange decreases and dipolar interactions increase. Hyde and Sarna (1978) attempted to follow this process over four orders of magnitude using water–glycerol mixtures as a function of temperature (Fig. 5). The data show a break near 35 cP (1.0 on the abscissa). At higher viscosities it may be assumed that rigid-lattice type of dipolar broadening of the free-radical lines by the paramagnetic metal ion is the dominant interaction. This viscosity, of course, is relevant only when one is considering ordinary low molecular weight nitroxides (MW ca. 300). For a spin-labeled protein we can extrapolate from Fig. 5. The break can be estimated to occur at 35(protein MW/300) cP. Experimentally one can readily establish the magnitude of these incipient solid-state type of contributions to line broadening, since the difference between contributions of metal-ion interactions to T_{1j}^{-1} and to T_{2j}^{-1} is attributed to this source. Heisenberg exchange and liquid-phase dipolar-induced spin–lattice relaxation contribute to each equally.

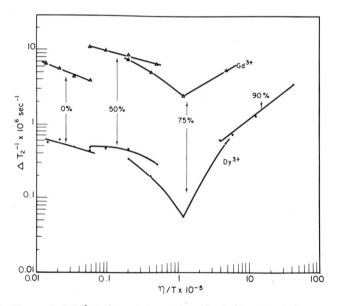

Fig. 5. Changes in ΔT_{2j}^{-1} of tanone that are induced by Gd^{3+} and by Cu^{2+} in 0, 50, 75, and 90% glycerol-water solutions. The experimental data were obtained with 10^{-2} M Gd^{3+} and 0.1 M Dy^{3+}, and the latter values were divided by 10 to make them comparable to the Gd^{3+} data. The four experimental points in each segment were obtained at temperatures of 2°, 20°, 40°, and 60°C. The viscosity is in Poises (Hyde and Sarna, 1978).

Finally, we call attention to the preliminary experiments of Keith *et al.* (1977) designed to study magnetic interactions in a heterogeneous environment (polymer beads). These experiments are a first effort to model biological systems. Sorting out the magnetic interactions in complex biological samples should be a goal of persons using the spin-probe–spin-label methods. It is our opinion that dipolar and exchange interactions may often turn out to be of comparable significance.

III. SURVEY OF CHEMICAL AND MAGNETIC PROPERTIES OF PARAMAGNETIC METAL-ION PROBES

The chemistry and magnetic resonance theory of transition-metal complexes may not be familiar to many who would like to use them as probes. Lanthanide chemistry and resonance theory are even less well known. We felt, therefore, that an elementary survey would be valuable in the initial considerations of double-probe experiments. More complete descriptions of the chemistry are given in basic inorganic chemistry texts, such as Cotton and Wilkinson's "Advanced Inorganic Chemistry" (1972) or Eichhorn's "Inorganic Biochemistry," Vol. 1 (1973). Lanthanide probes are well covered in Moeller's books and review articles (Moeller, 1963, 1972). Abragam and Bleaney's "Electron Paramagnetic Resonance of Transition Ions" (1970) covers the appropriate magnetic resonance theory in depth.

A. Chemical and Magnetic Properties of Lanthanide Ions

Lanthanides (Ln) are usually in the $+3$ oxidation state. The complexes that can be formed are usually with oxygen donors. The variety of complexes that can be formed in biological systems is expected to be rather limited. Following Moeller (1972), the reasons for this are as follows: (1) the partially filled 4f orbitals are buried and unavailable for bond formation, and the 5s and 5p orbitals are filled; (2) ionic bond strengths are minimized because of the large ionic size; (3) ligand exchange reactions for lanthanides are rapid; and (4) hydroxide ion and water are strong ligands for lanthanides. Unless strong chelating agents are available, lanthanides are expected to form soluble aqueous and hydrated complexes in biological systems or precipitate as hydroxide and phosphate complexes. Common coordination numbers for lanthanide complexes (and uncommon for transition metals) are 8, 9, 10, and 12. Complexes with coordination number 9, $[Ln(H_2O)_9]^{3+}$, $[Ln(OH)_9]^{6-}$, $[Ln(H_2O)_6(SO_4)_3]^{3-}$, and $[Ln(EDTA)(H_2O)_3]^-$, may be typical for lanthanide complexes formed in biological systems. Coordination numbers in solution are usually not definite and can change if a sample is frozen.

Although the chemical properties of the lanthanide compounds are similar due to the buried f electrons, the magnetic properties vary greatly (Moeller, 1963; Abragam and Bleaney, 1970). Here J is a good quantum number, and the effective moment can be approximated by $g[J(J + 1)]^{1/2}$. At room temperature ESR signals are observed for the S-state ions Gd^{3+} and Eu^{2+}, where the 4f shell is half-filled (Bielski and Gebicki, 1967; Abragam and Bleaney, 1970; Sarna et al., 1976). The non-S-ground-state lanthanides are fast relaxers; low-temperature ($4°K$) experiments are necessary to observe their ESR signals. The ions La^{3+} and Lu^{3+} at the ends of the series are diamagnetic. They are useful as controls to separate chemical and magnetic effects.

Because the paramagnetism arises from the inner electrons that are not involved in complex formation, the magnetic properties tend to resemble those of the free ion and are relatively unaffected by the environment. This relative constancy of magnetic properties with respect to changes in ligands is one of the virtues of the lanthanides.

The ligand field potential lifts the degeneracy of the J manifold. The magnitude of the zero-field splitting is about 10^2 cm^{-1} for non-S states (Abragam and Bleaney, 1970). Below about $77°K$, the J manifold begins to empty, which results in a temperature-dependent magnetic moment. Non-S-state ions have very anisotropic g values (see Table 5.9 in Abragam and Bleaney, 1970) because the ligand field potential is so small.

The ion Gd^{3+}, which is important in spin-probe–spin-label methodology, is an exception to the general rule that lanthanides exhibit a weak dependence of magnetic properties on the crystal field. The ground state is $^8S_{7/2}$ and is split in zero field into four closely spaced, thermally accessible levels. In solution the zero-field splitting is modulated, giving rise to solvent- and temperature-dependent Gd^{3+} ESR linewidths (and also, presumably, spin–lattice relaxation times). This property can be used to study different local environments of the spin probe.

Little information is available on spin–lattice relaxation times of lanthanides. A time of 10^{-9}–10^{-10} sec can be estimated for Gd^{3+} in solution on the basis of the facts that the ESR spectrum can be observed and no microwave power saturation is evident. Relaxation times for the other lanthanides in the liquid phase are reported to be between 10^{-13} and 5×10^{-13} sec on the basis of NMR experiments (Reuben and Fiat, 1969; Lenkenski and Reuben, 1976). Sarna et al. (1976) estimated that T_1 for Dy^{3+} and Ho^{3+} in an ice matrix at $77°K$ is 10^{-11} sec, i.e., very close to ω^{-1}, where ω is the X-band microwave frequency.

B. Lanthanides as Probes in Biochemistry

There are 15 lanthanides with similar chemical properties; the fine differences in chemical properties, such as ionic radius, can be used to generate some specificity. The widely varying magnetic properties permit the design of many interesting magnetic resonance double-probe experiments. The following examples illustrate the special magnetic and chemical properties of lanthanides when used as probes in biochemical experiments.

The most extensive application of these probes in biochemistry is as NMR shift reagents (Horrocks, 1973). The shifting ability of the lanthanides varies widely. There are large upfield shifts with Dy > Tb > Ho, smaller upfield shifts for Pr, negligible shifts for Nd and Sm, small downfield shifts for Eu, and large downfield shifts with Tm > Er ~ Yb (Horrocks, 1973). Gadolium is a slow relaxer and broadens NMR signals, making them undetectable (Dwek, 1973).

The work of Campbell et al. (1973a,b) on lysozyme exemplifies the application of shift reagents in biochemistry. The shift and broadening of NMR resonances due to lanthanide ions bound to lysozyme showed that the structure around the binding site in solution mimicked the X-ray crystallographic structure.

Barry et al. (1974) took advantage of both the chemical and magnetic properties of the lanthanide ions to determine the conformational structure of Ln^{3+}–cyclic AMP. Proton peaks were shifted by Dy^{3+} and Pr^{3+} reagents, and changes in proton T_1 and T_2 were produced by addition of Gd^{3+}.

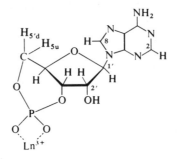

Dwek (1973, p. 68)

Specific lanthanide binding to enzymes and proteins has been a particularly useful tool. Europium($+2$) may be a better substitute than Mn^{2+} for Ca^{2+} due to the affinity of Eu^{2+} for oxygen over nitrogen ligands (Williams, 1970). Lanthanum($+3$) behaves like "supercalcium" when substituted for calcium in the lobster axon (Taketa et al., 1966; Blanstein and Goldman, 1968; Williams, 1970). Trypsin has a binding site at which lanthanides compete with calcium (Epstein et al., 1974). Lanthanides can compete for calcium sites, inhibit at the calcium sites, or block calcium transport (Williams, 1970). Lanthanide binding, however, does not always parallel calcium binding. Bovine serum albumin has four Gd^{3+} binding sites that seem to be nonspecific for calcium (Reuben, 1971).

The lanthanide binding constant for porcine trypsin increases as the lanthanide ionic radius decreases (Epstein *et al.*, 1974). Transferrin (Luk, 1971) has two binding sites for Tb^{3+}, Eu^{3+}, Er^{3+}, and Ho^{3+} as well as binding sites for the natural iron and copper ions. In transferrin there is only a single binding site for Nd^{3+} and Pr^{3+} due to the increase in ionic radius. Also, very few water molecules are bound to the Tb^{3+}–transferrin complex.

C. Chemical and Magnetic Properties of 3d Transition-Metal Ions

The crystal-field parameters are larger, the atomic distances smaller, and the spin-orbit coupling coefficients smaller for 3d transition-metal ions than for 4f lanthanide ions. The d orbitals are outermost, not shielded, and used in bond formation. Transition-metal ions form strong bonds. They are often natural constituents of proteins and enzyme structures. These ions may be added by chelating directly to the protein or by binding them as inert low molecular weight complexes to probe the protein environment.

Because the angular momentum is quenched, the effective magnetic moment is dependent on the ligands. Molecular geometries for d-orbital transition-metal complexes include octahedral, tetragonal, and square planar. The five d orbitals are split by an octahedral field into a threefold degenerate t_{2g} orbital and a twofold degenerate e_g orbital. Distortion to tetragonal, square planar, or rhombic symmetry removes the degeneracy of the t_{2g} and e_g orbitals. A trade-off between the energy needed for the pairing of spins and the crystal-field splitting determines whether high-spin (minimal pairing of spin) or low-spin (maximal pairing of spin) complexes are formed.

The usual oxidation states in aqueous solution for 3d transition metals are Cr^{3+}, Mn^{3+} and Mn^{2+}, Fe^{3+} and Fe^{2+}, Co^{2+} and Co^{3+}, Ni^{2+}, and Cu^{2+} and Cu^{1+} (Buckingham, 1973). Although most of these ions have 3d unpaired electrons, the ESR spectra are rarely observed at room temperature except for Mn^{2+} and Cu^{2+}. Even after sufficient dilution in a diamagnetic host, liquid helium temperatures may be necessary to observe the ESR signals. One use of spin labels in double-probe experiments is to observe the interaction with metal complexes that are not directly detectable with ESR because of very short metal spin–lattice relaxation times.

For 3d transition-metal complexes (M^{2+}) bound to the same ligands, the stability of the complexes increases across the first-row transition-metal ions from Mn^{2+} to Cu^{2+}, which is known as the Irving–Williams order. Factors that influence the stability of these complexes are the following: (1) pK_a values of the ligands; (2) formation of five-membered rings with the metal ion and the ligands; (3) the number of such rings, i.e., bidentate, tridentate, or tetradentate chelation; and (4) formation of mixed ligand complexes (Angelici, 1973). Complexes exist that are completely water soluble and completely lipid soluble. In aqueous solution, the 3d transition-metal ions are hydrated, and only strong ligands can substitute for H_2O in the $M(OH_2)_6$ complex.

D. Complementary Probes

If the ESR spectrum of the paramagnetic metal can be detected, the power of the spin-probe–spin-label method is increased. We term these metal ions *complementary probes*. Spectra are obtained of the probe in the absence of the label and of the label in the absence of the probe, and then magnetic interactions when both probe and label are present are analyzed. The complementary probes that exhibit characteristic room-temperature

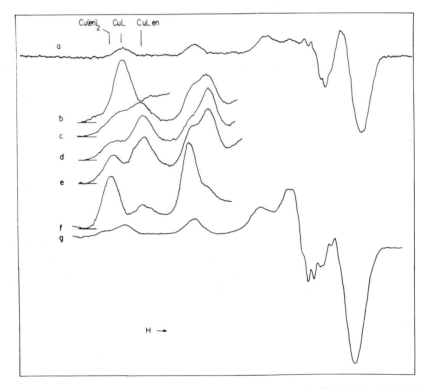

Fig. 6. The ESR spectra of CuL^+ at 77°K. Titration of 6.13×10^{-4} M CuL^+ at (a–f) pH 8.2 ± 0.1 and (g) pH 1.85 in 0.10 M KCl and 30% Me_2SO. Microwave power, 2 mW; microwave frequency, 9.147 GHz; modulation amplitude, 5 G. (b)–(f) × 10 gain: (b) CuL^+, (c) $CuL^+ + 1$ en; (d) $CuL^+ + 2$ en; (e) $CuL^+ + 10$ en; (f) $CuL^+ + 100$ en (Antholine *et al.*, 1977a).

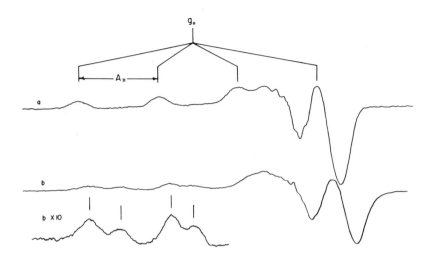

Fig. 7. The ESR spectra of CuL$^+$: (a) CuL(acetate) in water; (b) CuL$^+$ in human plasma; (b × 10) tenfold expansion of the $g_\parallel$ region of (b). All spectra were taken at 77°K; microwave power, 5 mW; modulation amplitude, 5 G; modulation frequency, 100 kHz; microwave frequency, 9.089 GHz. The concentration of CuL$^+$ was 60 μM (Antholine *et al.*, 1977b).

spectra include VO^{2+}, Mn^{2+}, Cu^{2+}, and Gd^{3+}. These four probes are of particular significance because their spectra can be readily observed in a wide variety of circumstances.

1. COPPER (Cu^{2+})

More ESR studies have probably been done on Cu^{2+} complexes than on the complexes of any other paramagnetic metal ion. The literature of copper chemistry is vast indeed. This ESR and chemical background makes copper especially favorable as a probe of biological systems. Copper has a number of properties of special interest in spin-probe–spin-label studies. Among the transition-metal ions, Cu^{2+} gives the biggest "Leigh effect," i.e., the greatest loss of apparent free-radical intensity. It forms tightly bound complexes and binds well to amino acids, proteins, and DNA (Angelici, 1973). Vänngård (1972) reviewed the ESR spectroscopy of copper from a biological viewpoint.

Copper complexes are usually square planar or slightly distorted square planar. The immobilized powder spectrum contains a low-field or axial g value, $g_\parallel$, and a high-field or in-plane g value, $g_\perp$ (Figs. 6 and 7). The nuclear spin, $I = 3/2$ for ^{63}Cu and ^{65}Cu, splits both the parallel and perpendicular lines into four hyperfine lines. Ordinarily, at least two and sometimes three

TABLE I

PARAMETERS FOR COPPER COMPLEXES

Compound[a]	$g_\parallel$	$A_\parallel$ (G)	Donor atoms
Cu(dtc)$_2$	2.090	130	S–S–S–S
Cu(cysteine)$_2$	2.13	190	S–S–N–N
CuKTS	2.14	187	S–S–N–N
Cu(fPTS) + ascites cells	2.13	178	S–S–N–N
Cu(fPTS)(en)	2.188	169	S–N–N–N
Cu(fPTS)(His)	2.188	183	S–N–N–N
Cu(fPTS)(OH$_2$)	2.206	185	S–N–N–O
Cu(en)$_2$	2.213	190	N–N–N–N
CuBLM	2.21	175	N–N–N–N
Cu^{2+} + ascites cells	2.25	180	N–N–N–N (possibly)
Cu^{2+} + melanin	2.3	160	N–O–O–O (possibly)
CuEDTA (pH 6)	2.29	155	N–N–O–O
Cu(OH)$_4$	2.40	113	O–O–O–O
CuSO$_4$ in 50% DMF	2.40	128	O–O–O–O

[a] Abbreviations: dtc, dithiocarbamate; KTS, oxobutyraldehyde bis-(thiosemicarbazone); fPTS, 2-formylpyridine monothiosemicarbazone; en, ethylenediamine; His, histidine; BLM, bleomycin; EDTA, ethylenediaminetetraacetic acid; DMF, dimethylformamide.

or four of the hyperfine lines around $g_\parallel$ are shifted far enough from $g_\perp$ to be readily assigned. To a first approximation, the $g_\parallel$ value is obtained in a powder spectrum from the center of gravity of these four hyperfine lines. Often the in-plane ligands are nitrogens $(I = 1)$, which gives rise to a superhyperfine splitting of the copper hyperfine lines. Nitrogen couplings are occasionally seen in powders and frozen solution.

1. Variation of Experimental Observables. Common donor atoms are sulfur, nitrogen, and oxygen. As a general rule, $g_\parallel$ decreases when S replaces N and decreases further when N replaces O. [For a discussion of this with respect to proteins, see Vänngård (1972) or Peisach and Blumberg (1969 and 1974).] When S replaces nitrogen, $A_\parallel$ tends to decrease; it increases when N replaces O. Table I contains a list of $g_\parallel$ and $A_\parallel$ values for copper complexes that we have studied. The last column gives the probable in-plane donor atoms. The trend for $g_\parallel$ holds remarkably well, and that for $A_\parallel$ holds reasonably well. Values for other copper complexes can be found in the papers of Laurie *et al.* (1975), Crawford and Dalton (1969), and Walker *et al.* (1972).

Observation of superhyperfine structure helps identify in-plane donor atom(s). Axial (out-of-plane) ligands, even though they possess nuclear moments, do not give rise to superhyperfine structure (Wayland and Kapur,

1974). In the study of Wayland and Kapur, the presence of in-plane or out-of-plane phosphorus could be determined. Representative spectra showing nitrogen superhyperfine structure can be found in the following references: Peisach *et al.* (1967), Rist and Hyde (1969), and Vänngård (1972). Nonaka *et al.* (1974) determined the conditions that result in resolved ^{14}N superhyperfine structure of Cu^{2+} complexes. Two prominent factors, besides temperature and dilution, were minimal deviation from square planar configuration and the absence of protons that can couple with the unpaired electron. Spectral simulation is often necessary for interpretation of the superhyperfine structure.

2. Example of the Use of a Designed Copper Probe. The ligand 2-formylpyridine monothiosemicarbazone (L) forms a tight tridentate complex with Cu^{2+} (Fig. 6) (Antholine *et al.*, 1977a). The tridentate coordination leaves a single coordination site where interaction with another ligand X can occur (in plane as well as out of plane). In general three situations can occur when this probe is introduced into a biological system: (1) reduction of the complex to Cu^{+1} by, for example, dithiothreitol, (2) no reaction, and (3) adduct formation.

Addition of ethylenediamine(en) to CuL(acetate) results in adduct formation and presumably a five-coordinate complex. Formation of this complex was suggested by the ESR spectra (Fig. 6). The shifts in the g and hyperfine values were sufficient to resolve three complexes: CuL(acetate), CuL(en), and $Cu(en)_2$. Inclusion of the mixed ligand complex CuL(en) into the expressions for competitive binding of L and en with Cu^{2+} gave a good fit to the ultraviolet–visible spectroscopic data (Antholine *et al.*, 1977a).

Addition of CuL(acetate) to blood plasma resulted in two complexes that could be detected by examination of the low-field $g_\parallel$ region of the ESR spectra (Fig. 7). One set of $g_\parallel$ and $A_\parallel$ values could be modeled by addition of histidine to CuL (Table II).

TABLE II

ESR PARAMETERS FOR CuLX IN FROZEN SOLUTIONS

Species	g	$A_\parallel$ (G)
CuL$^+$	2.206	185
CuL + His	2.188	183
CuL + plasma[a]	2.152, 2.181	167, 188

[a] Two distinct copper spectra are obtained using human plasma (Antholine *et al.*, 1977b).

2. MANGANESE (Mn^{2+})

Manganese ion (Mn^{2+}), isoelectronic with Fe^{3+}, has a d^5 electronic configuration. In the high-spin state ($S = 5/2$), unpaired electrons occupy the threefold degenerate t_{2g} orbital and the twofold degenerate e_g orbital. Since excited states are formed from promotion of electrons between t_{2g} and e_g orbitals, spin-orbit coupling and zero-field splitting contributions are small (Carrington and McLachlan, 1967). The energy levels of aqueous complexes of Mn^{2+} are approximately $\pm(1/2)g\beta H$, $\pm(3/2)g\beta H$, and $\pm(5/2)g\beta H$ (Figgis, 1966). This results in five fine structure transitions that are nearly degenerate. These lines are split into six hyperfine lines due to the nuclear ($I = 5/2$) spin (Fig. 8).

When Mn^{2+} is bound to a protein or nucleotide, the asymmetry in the coordination sphere results in a zero-field splitting (Carrington and McLachlan, 1967; Figgis, 1966; Abragam and Bleaney, 1970) that reduces the overlap of the five fine structure transitions. Lines may be broadened, patterns may become quite complex, and often the observable bound Mn^{2+} ESR signal intensity is more than an order of magnitude less than the free (hydrated) Mn^{2+} signal. Results depend not only on the zero-field splitting but also on the rotational correlation time of the Mn^{2+}–protein complex.

Cohn, Reed, and others have used the changes in electronic symmetry

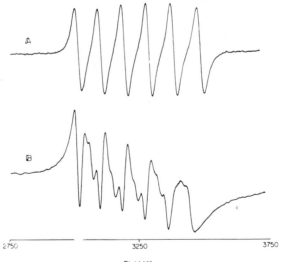

Fig. 8. X-Band spectra of aqueous complexes of Mn^{2+}. (A) 0.2 mM $MnCl_2$ at 25°C; (B) 0.2 mM $MnCl_2$ frozen in Sephadex G-25-80 to prevent aggregation ($T = 77°K$) (Reed and Cohn, 1972).

about the Mn^{2+} complex to probe the proximal biochemical environment (Buttlaire and Cohn, 1974; McLaughlin *et al.*, 1976; Reed and Cohn, 1972, 1973; Reed *et al.*, 1971; Reed and Scrutton, 1974). The work of Reed and Cohn (1972) on creatine kinase serves as a good example. Changes of the Mn^{2+} spectrum induced by the addition of substrates and anions (Figs. 9 and 10) served as the experimental observables. Changes in fine structure and line shape indicate that the binding of creatine induced a rearrangement at the active site (Fig. 10C). Nitrate and chloride anions also changed the spectrum (Fig. 10D), which showed that there was good solvent accessibility to the Mn^{2+} ion.

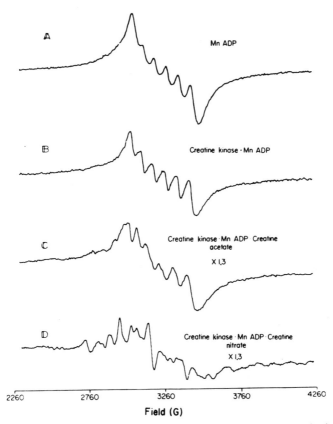

Fig. 9. X-Band spectra for Mn(II)–ADP complexes. All samples contained 0.5 mM Mn(CH$_3$CO$_2^-$)$_2$, 1.5 mM ADP, 50 mM HEPES–KOH at pH 8.0. Spectra were recorded at 1°C, and a modulation amplitude of 40 G was used. Additional components were 68 mg/ml of creatine kinase, 100 mM creatine (saturated solution), and 3 mM KNO$_3$ (Reed and Cohn, 1972).

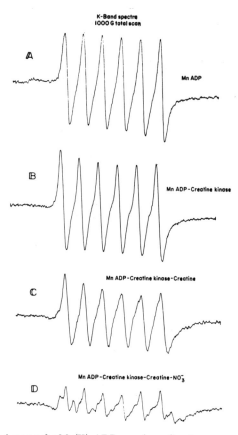

Fig. 10. K-Band spectra for Mn(II)–ADP complexes. Spectra were recorded at 26°C with a modulation amplitude of 10 G. Composition of solutions is given in the legend to Fig. 9 (Reed and Cohn, 1972).

3. Vanadyl (VO^{2+})

For VO^{2+}, the orbital angular momentum is almost completely quenched, resulting in $g_{\parallel}$ and $g_{\perp}$ values that are slightly less than the free-electron spin value. A nuclear spin $(I = 7/2)$ splits each vanadyl line into eight hyperfine lines. Typical values for $A_{\parallel}$ are 200 G and for $A_{\perp}$, 75 G. Usually the eight parallel and perpendicular hyperfine lines are well resolved (Fig. 11). In-plane chelation can be analyzed by comparison of g and hyperfine parameters with model vanadyl complexes (Chasteen et al., 1973; Fitzgerald and Chasteen, 1974; DeKoch et al., 1974).

Two disadvantages of the vanadyl ion may be the absence of ligand superhyperfine structure and the insensitivity of the ESR parameters to a sixth axial ligand.

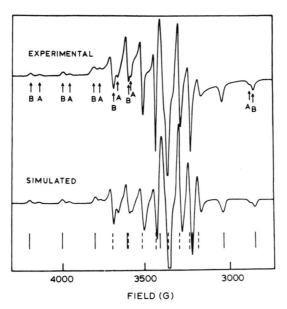

Fig. 11. Experimental and simulated first-derivative X-band ESR spectra of colorless polycrystalline vanadyl insulin at room temperature. The letters A and B designate partially resolved spectra of A and B sites. The solid and dashed vertical lines denote the parallel and perpendicular resonance fields, respectively, of the B sites. The sample was prepared by the crystal soaking method; gaussian line shape function (Chasteen *et al.*, 1973).

The vanadyl ion is particularly useful in obtaining motional information in the range of rotational correlation times of 10^{-10}–10^{-7} sec (Bruno *et al.*, 1977).

4. GADOLINIUM (Gd^{3+})

The amount of literature on bound Gd^{3+} is limited. Usually a broad Gd^{3+} signal with a linewidth of several hundred gauss is observed (Dwek *et al.*, 1971; Bielski and Gebicki, 1967). Dwek suggested that the ESR signal of the bound form may be easy to observe (relative to Mn^{2+}) due to the small zero-field splittings (Dwek, 1973, p. 184).

IV. APPLICATIONS OF THE METHOD

There is a fairly extensive and rapidly growing body of literature describing the use of this method plus two recent reviews (Eaton and Eaton, 1978; Likhtenshtein, 1976). A detailed consideration of applications is therefore not indicated here. Instead, we shall consider some general aspects and then attempt to provide a description of the specific types of applications of which we are aware.

A. Nature of the Paramagnetic Species

The actual experiments and their interpretations may be affected by the relationship of the spin probe and/or spin label to naturally occurring species. The spin label can be a naturally occurring species (e.g., the free radical in melanin), introduced at a specific site (e.g., a nitroxide spin label attached at a specific amino acid residue), or inserted less specifically (e.g., incorporation of nitroxide spin labels into membranes). Each of these possibilities introduces both experimental limitations and enhanced opportunities to obtain significant information. Similarly one can distinguish several possibilities for the spin probe: an intrinsic part of the system at a fixed position (e.g., copper in cytochrome oxidase), an analogue of a naturally occurring ion (e.g., lanthanides substituted for calcium), a paramagnetic ion introduced into specific sites (e.g., lanthanides bound to melanin), or a paramagnetic ion introduced without specific binding (e.g., in extracellular media to broaden spectra of spin labels).

B. Special Experimental Precautions

Because one often uses spin probes and/or spin labels that have active roles in the systems under investigation, we consider it worthwhile to summarize some relevant experimental precautions.

A good, generally applicable control is the use of nonparamagnetic analogues to determine whether the observed effects are due solely to magnetic interactions. If the spin probe is a lanthanide one can utilize both diamagnetic ions of the series, lanthanum and lutetium. One can do a further check on the understanding of the magnetic interaction by using lanthanides with different T_1 values. Less specifically one can often use magnesium as a nonparamagnetic substitute for manganese.

A potential source of experimental error with this method is failure to consider effects of the ions on pH. The addition of relatively large quantities of metal ions can result in the displacement of hydrogen ions with a consequent lowering of pH.

Because the usual metal ions are polyvalent, they can affect the organization of mobile anionic components of the systems, thereby perturbing the system one is attempting to probe. An example of this is the effect of polyvalent cations on the organization of some membranes, inducing clustering of membrane components.

In complex biological systems, especially those with different phases and/or barriers to free diffusion, the distribution of the paramagnetic ions can be very nonuniform. Accurate interpretations of results from such experiments require knowledge of the distribution of the paramagnetic ions. In particular, it may be an oversimplification to assume that a paramagnetic ion such as nickel, which does not readily cross cell membranes, is at negligible concentration intercellularly. Some metal ions, especially when complexed, may significantly partition into lipid phases including membranes.

In some systems in which spin-probe–spin-label experiments may be carried out, additional radicals may be inducible by metals (Sarna and Lukiewicz, 1972; Felix et al., 1978). If this occurs it must be recognized or the interpretation of the experiment will be seriously in error. The effect of the metal ion is to shift the equilibrium that exists between hydroquinone, quinone and semiquinone:

$$\text{Quinone + Hydroquinone} \rightleftharpoons \text{Semiquinone} \rightleftharpoons \text{Semiquinone-Chelate}$$

The effect is nonmagnetic and can be tested using diamagnetic cations, especially Zn^{2+}.

C. Experimental Parameters

The most frequently used parameter is the diminution of signal intensity. This has been discussed extensively in this chapter, as has the use of changes in power saturation. There are, however, other parameters that may also be of some use. The ESR spectrum of the spin probe itself may provide data that are not obtainable by other means (see Section III,D). In some experiments, critical information may be obtained by performing double integrations to distinguish between apparent and real changes of intensity. Finally, in the near future, the development and availability of time domain ESR spectrometers may enable one to measure T_{1j} directly.

D. Use of the Metal Ion as the "Label"

In the discussion to this point the free radical has been tacitly considered to be the label and the metal to be the probing species. The reverse, of course, is quite possible, and at least one such experiment has been described. A metal ion was specifically introduced into a fragment of an antibody, a spin label was placed on a hapten, and the interaction was studied (Dwek et al., 1976). This is a completely general approach applicable to any system in which the metal ion can be specifically bound. It potentially has the advantage of reducing local perturbation due to the "label" if a metal ion normally is found at the site. Experimentally one would still study effects on the free radical.

E. Types of Biochemical and Biological Applications of Spin-Probe–Spin-Label Methods

In the following paragraphs we classify actual and potential uses of the method. As far as we can determine, we include all published uses of this method as well as some that have not yet been attempted. Some of the classifications listed below overlap with each other. In most cases the applications are given in terms of metal ions probing the environment of the spin label, but in a few cases we explicitly discuss the inverse experiment, in which the metal ion is already in the system and the free radical is added.

1. DETERMINATION OF RADIAL DISTANCES

The initial formulation of the theory discussed in this chapter was developed by Leigh to determine the distance between a spin probe and a spin label. This is one of the most useful applications of this approach and perhaps the most elegant. Under the proper circumstances, one can determine the distance between the paramagnetic entities with a fairly high degree of accuracy. The experimental observable is the intensity of the free-radical signal. One usually measures the peak-to-peak height of the central line for nitroxide in the presence and absence of the spin probe. This decrease is compared with the calculated decrease expected in the presence of the particular spin probe. The calculation is based on Leigh's theoretical treatment, which was described earlier. The parameter used is C, which is defined as

$$C = \frac{g\beta\mu^2 T_{1k}}{\hbar r^6}$$

where g refers to the spin label, μ is the magnetic moment of the spin probe, r is the distance between spin probe and spin label, and T_{1k} is the relaxation time of the spin probe. If the latter is known, the only unknown in the relationship is r. Because of the sixth-power dependence of C on r, small uncertainties that occur in T_{1k} do not greatly affect the calculation of r. Leigh calculated the relationship between C and r under the conditions described in Section II,A. Dwek (1973, pp. 315–318) presented a discussion of the use of this method in enzymes, and Eaton and Eaton (1978) summarized many of the publications in which the use of this method has been described. Rao *et al.* (1979) published a more comprehensive method for the calculation of radial distances that explicitly considers rotational diffusion, which was not considered in Leigh's original treatment. This should now be used for computations of radial distances.

2. NATURE AND HETEROGENEITY OF ENVIRONMENTS OF FREE RADICALS

The nature of the location of the free radicals and whether they are in more than one type of location are two interrelated questions that can be studied by the spin-probe–spin-label method. Such knowledge is often critical, either directly or in order to interpret other experimental results (e.g., the specificity of binding of a spin label in a protein). By using different spin probes such as complexes with different sizes, charges, and solubilities, one can determine information on the nature of the environment of the spin label with regard to location behind barriers, nearby charges, and lipophilicity. Careful analysis of spectra, often in combination with different spin probes, should enable one to determine whether the spin label occurs in more than one environment.

An example of such an approach is the use of high concentrations of water-soluble metal ions such as nickel to study intracellular spin labels. The cell membrane is considered to be impermeable to the spin probe (nickel) so that, whereas the extracellular spin labels are broadened by magnetic interactions with the nickel, the intracellular spin labels are unaffected (Haak *et al.*, 1976). In a similar but somewhat more complex experiment using the naturally occurring free radical in muscle and dysprosium, we were able to determine that the free radicals are in a hydrophilic region separated from the extracellular environment by a membrane or membranes that are ordinarily impervious to dysprosium but can be disrupted by repeated freezing and thawing cycles (Antholine *et al.*, 1978).

3. PERMEABILITY OF BARRIERS SUCH AS MEMBRANES

In some ways this is a corollary of the previous use, but it is separated both for logical considerations and because of the importance of this problem. If one knows the location of the spin label and this location is separated from the environment by a barrier, it should be possible to plan a suitable experiment to test the intactness of the barrier by adding a paramagnetic metal ion. By this approach it should be possible to measure changes in membrane permeability to metal ions and/or complexing agents.

4. DETECTION AND MEASUREMENT OF PHYSICAL FLUCTUATIONS OF COMPONENTS

Many measurements provide only average positions of components in complex structures such as membranes whereas biologically important properties may depend on the point of closest approach. Because of the sixth-power dependence of the effects considered here, this approach potentially could provide this type of information. A straightforward extension of the results of Hyde and Rao (1978) indicates that the observed effect would be determined primarily by the point of closest approach of the interacting species. In the case of a membrane in which the spin label has an average position some distance from the surface, one would be able to determine its point of closest approach to the surface. Repetition of such an experiment under different conditions could provide quantitative data on how those conditions affect motion within the membrane with respect to the membrane surface. This approach could be used both for membrane components such as fatty acids and for proteins that are imbedded in the membrane.

5. Enhancement of Sensitivity in Measurement of Free Radicals

Naturally occurring free radicals and spin labels may power-saturate quite readily. This may reduce sensitivity to the point that this is the limiting factor in an experiment. This often is the case in the study of free radicals seen in tissues. Similar considerations occur in many spin-labeling experiments in which one must limit the amount of spin label to minimize perturbation of the system by the label. The addition of suitably chosen paramagnetic metal ions can significantly change T_{1j} so that much higher power levels can be used with a resulting increase in sensitivity. Success with this approach, using dysprosium, has been achieved by our group (Antholine et al., 1978). In this case the measured increase in $P_{1/2}$ was more than 20 dB. This is probably the range of increase to be expected in most biological systems, which means an increase of up to 10 in the signal-to-noise ratio.

If there are no special biological or chemical considerations, dysprosium is probably the ion of choice for this use because $\omega T_{1k} \sim 1$, which is optimal for inducing spin–lattice relaxation. This assumes that the relaxation of Dy^{3+} is unaffected by the range of experimental conditions likely to be encountered.

6. Relationship between Metal Ions and Free Radicals in Enzymes

Dipolar interactions between paramagnetic metals and free-radical intermediates of enzymes have been used productively by biochemists for many years. These useful observations may now be extended by our increasing knowledge of the theoretical basis of these interactions. The most extensive observations have been made on the flavoproteins, in which the relaxation rate of different flavin semiquinone free radicals has been used as an empirical guide as to whether a paramagnetic metal was functionally associated with the flavin; i.e., it is one of the prime criteria for differentiating between metal-containing and metal-free flavoproteins (Beinert, 1972). It may now be feasible to make detailed calculations on the radial distance between the paramagnetic components of the metal flavoproteins. But even qualitative data indicating that there are metal–free-radical interactions may be very valuable. Beinert points out that the increased spin relaxation rate of the semiquinones of these proteins is one of the most important pieces of evidence indicating that the metal and the flavin are on the same molecule.

7. Detecting "Hidden" Free-Radical–Metal-Ion Interactions

In biological systems under some normal conditions, Leigh type of effects may occur, affecting the apparent intensity of the free radical. If this is not recognized, errors in the determination of the number of free radicals may occur. One may test for the possibility of such interactions by the addition of suitable nonparamagnetic cations and observing the effects on the free radicals. [A possible source of confusion with such an approach is the failure to realize that under certain conditions diamagnetic metals can affect the shape of the free-radical spectrum and even induce free-radical formation (Felix *et al.*, 1978).]

8. MEASUREMENT OF COLLISION RATES (LOCAL VISCOSITY)

In the liquid state, the predominant mechanism of observable spin-probe–spin-label interactions is strong Heisenberg exchange (except when lanthanides are involved). This effect depends on the collision rates of the two species and can be used to study factors that bear on these rates. From a biological viewpoint, often the most interesting of these factors is the viscosity in the region of the spin probe. This property cannot be readily measured by other methods, with the possible exception of fluorescent depolarization. Collision rates in complex biological systems may also be influenced by other factors such as steric relationships. These must be ruled out before changes can be interpreted strictly in terms of local viscosity. The situation is particularly complex when the spin label is in a macromolecule or a larger organized cellular structure (e.g., membrane). Then environmental changes may indirectly change effective collision rates (e.g., by changing conformation of a protein in the region of the spin label).

A number of factors that should be considered in experiments using this approach to measure collision rates are discussed in Sections II,B and C.

9. BINDING OF METAL IONS

A rich array of information on metal binding can be obtained with this method. For example, using the spin-probe–spin-label method we were able to determine that different melanin preparations had different numbers of total binding sites and that there are several different types of binding sites in melanin. Binding was via electrostatic interactions and was reversible. We also were able to estimate the tightness of binding and the distance from the closest binding site to the free radical (Sarna *et al.*, 1976).

Much of the desired information can be obtained by titration studies in which the observable is the effect on the free radical. A graph of signal amplitude versus the concentration of metal ions is a convenient way to display the data. A number of properties of the system can be systematically studied by repeating the experiment under different conditions. Competition experiments can similarly be carried out, perhaps most conveniently by adding an excess of one metal and then stepwise adding the second metal. Another parameter of tightness of binding can be obtained by adding a complexing agent to compete for the metal ions with the system under study.

Using this approach we found that melanin competed approximately equally with EDTA for cations such as gadolinium, and we could therefore make an estimate of the binding constant.

A logical variation of this approach could be used for systems in which there is no naturally occurring free radical or a convenient way of specifically labeling some component of the system. In this situation the free radical could be added to the solution, and one would measure the release of the paramagnetic metal ions from the system by their effect on the spectrum of the free radicals. This would require a local concentration of paramagnetic metal ions of ≥ 1 mM.

10. DISTRIBUTION OF METAL IONS

In biological systems in which metals normally occur, it may not be clear whether the metal ions are randomly distributed, located only at specific sites, or clustered. The addition of free radicals to the system should provide a means to test these possibilities. Spin labels with net negative charges may be particularly useful in these experiments.

11. DETERMINATION OF T_{1k} OF PARAMAGNETIC METAL IONS

This important experimental parameter often cannot be directly measured, but it can be estimated in some spin-probe–spin-label experiments. For example, in our studies with melanin (Sarna et al., 1976) the type of effect that we observed (on signal amplitude or on $P_{1/2}$) enabled us to estimate whether ωT_{1k} was much greater or smaller than 1. For two ions (Dy^{3+} and Tm^{3+}), we concluded that in this system at $-150°C$ $\omega T_{1k} \sim 1$.

12. DIFFERENTIATION BETWEEN A CHANGE IN THE VALENCE STATE OF A PARAMAGNETIC METAL AND ITS BECOMING NONOBSERVABLE FOR OTHER REASONS

In many enzymatic systems the role of metal ions may include reversible oxidation–reduction during catalytic steps. If the ESR spectrum of a paramagnetic metal component of an enzyme disappears, it is usually assumed that the metal has added or lost an electron. There are other possibilities that could also account for its apparent disappearance (e.g., change in binding to the protein or association with another paramagnetic metal). The use of a free radical as a probe provides a potential means to distinguish between an apparent loss of paramagnetism and a real loss due to pairing of spins. Alternate means to measure this may not be feasible; e.g., the spectrum may be spread too far to permit accurate double integration, and magnetic susceptibility apparatus may not be available.

13. DETECTING THE PRESENCE OF PARAMAGNETIC IONS THAT ARE NOT DETECTABLE DUE TO SHORT T_{1k}

The presence of such metal ions may be detected by introducing or inducing free radicals into the system and looking for effects on signal amplitude or $P_{1/2}$. The latter parameter would probably be the best experimental approach because the undetectable paramagnetic ions would have T_1 values that would affect $P_{1/2}$ and this is an easy parameter to measure. A possible approach would be to expose the system under study to ionizing radiation at low temperatures. The radiation-induced free radicals would then provide the experimental observable. Such radiation-induced organic radicals ordinarily have a $P_{1/2}$ of less than 1 mW. If paramagnetic metals were present some of the radiation-induced radicals would have an unusually high $P_{1/2}$. The presence of oxygen would have to be controlled in such an experiment because it can have an effect similar to that of paramagnetic ions.

ACKNOWLEDGMENTS

Preparation of this chapter was assisted by grants 5RO1 GM22923, 1RO1 CA20737, F32 CA05530, and 5P41-RR01008 from the National Institutes of Health.

REFERENCES

Abragam, A. (1955). Overhauser effect in nonmetals. *Phys. Rev.* **98**, 1729–1735.

Abragam, A. (1961). "The Principles of Nuclear Magnetism." Oxford Univ. Press, London and New York.

Abragam, A., and B. Bleaney (1970). "Electron Paramagnetic Resonance of Transition Ions," Int. Ser. Monogr. Phys. Oxford Univ. Press (Clarendon), London and New York.

Angelici, R. J. (1973). Stability of coordination compounds. *Inorg. Biochem.* **1**, 63–101.

Antholine, W. E., K. M. Knight, and D. H. Petering (1977a). Some properties of copper and zinc complexes of 2-formylpyridine thiosemicarbazone. *Inorg. Chem.* **16**, 569–574.

Antholine, W., J. Knight, H. Whelan, and D. H. Petering (1977b). Studies of the reaction of 2-formylpyridine thiosemicarbazone and its iron and copper complexes with biological systems. *Mol. Pharmacol.* **13**, 89–98.

Antholine, W. E., J. S. Hyde, and H. M. Swartz (1978). Use of Dy^{3+} as a free-radical relaxing agent in biological tissues. *J. Magn. Reson.* **29**, 517–522.

Barry, C. D., D. R. Martin, R. J. P. Williams, and A. V. Xavier (1974). Quantitative determination of the conformation of cyclic 3′, 5′—adenosine monophosphate in solution using lanthanide ions as nuclear magnetic resonance probes. *J. Mol. Biol.* **84**, 491–502.

Beinert, H. (1972). Flavins and flavoproteins, including iron-sulfur proteins. In "Biological Applications of Electron Spin Resonance" (H. M. Swartz, J. R. Bolton, and D. C. Borg, eds.), p. 403. Wiley (Interscience), New York.

Bielski, B. H. J., and J. M. Gebicki (1967). "Atlas of Electron Spin Resonance Spectra." Academic Press, New York.

Blanstein, M. P., and D. E. Goldman (1968). The action of certain polyvalent cations on the voltage-clamped lobster axon. *J. Gen. Physiol.* **51**, 279–291.

Bloembergen, N., E. M. Purcell, and R. V. Pound (1948). Relaxation effects in nuclear magnetic resonance absorption. *Phys. Rev.* **73**, 679–712.

Bloembergen, N. (1949). On the interaction of nuclear spins in a crystalline lattice. *Physica* **15**, 386–426.

Bloembergen, N., S. Shapiro, P. S. Pershan, and J. O. Artman (1959). Cross relaxation in spin systems. *Phys. Rev.* **114**, 445–459.

Bruno, G. V., J. K. Harrington, and M. P. Eastman (1977). Electron spin resonance line shapes of vanadyl complexes in the slow tumbling region. *J. Chem. Phys.* **81**, 1111–1117.

Buckingham, D. A. (1973). Structure and stereochemistry of coordination compounds. *Inorg. Biochem.* **1**, 3–62.

Buttlaire, D. H., and M. Cohn (1974). Characterization of the active site structures of arginine kinase-substrate complexes. *J. Biol. Chem.* **249**, 5741–4748.

Campbell, I. D., C. M. Dobson, R. J. P. Williams, and A. V. Xavier (1973a). The determination of the structure of proteins in solution: Lysozyme. *Ann. N. Y. Acad. Sci.* **222**, 163–174.

Campbell, I. D., C. M. Dobson, R. J. P. Williams, and A. V. Xavier (1973b). Resolution enhancement of protein PMR spectra using the difference between a broadened and a normal spectrum. *J. Magn. Reson.* **11**, 172–181.

Carrington, A., and A. D. McLachlan (1967). "Introduction to Magnetic Resonance." Harper, New York.

Chasteen, N. D., R. J. DeKoch, B. L. Rodgers, and M. W. Hanna (1973). Use of the vanadyl (IV) ion as a new spectroscopic probe of metal binding to proteins. Vanadyl insulin. *J. Am. Chem. Soc.* **95**, 1301–1309.

Cotton, F. A., and F. R. S. Wilkinson (1972). "Advanced Inorganic Chemistry." Wiley (Interscience), New York.

Crawford, T. H., and J. O. Dalton (1969). ESR studies of copper (II) complex ions. *Arch. Biochem. Biophys.* **131**, 123–138.

DeKoch, R. J., D. J. West, J. C. Cannon, and N. D. Chasteen (1974). Kinetics and electron paramagnetic resonance spectra of vanadyl (IV) carboxypeptidase A. *Biochemistry* **13**, 4347–4354.

Dwek, R. A. (1973). "Nuclear Magnetic Resonance in Biochemistry: Applications to Enzyme Systems." Oxford Univ. Press (Clarendon), London and New York.

Dwek, R. A., R. E. Richards, K. G. Morallee, E. Nieboer, R. J. P. Williams, and A. Xavier (1971). The lanthanide cations as probes in biological systems. *Eur. J. Biochem.* **21**, 204–409.

Dwek, R. A., D. Givol, R. Jones, A. C. McLaughlin, S. Wain-Hobson, A. I. White, and C. Wright (1976). Interactions of the lanthanide- and hapten-binding sites in Fv fragment from the myeloma protein MOPC 315. *Biochem. J.* **155**, 37–53.

Eastman, M. P., R. G. Kooser, M. R. Das, and J. H. Freed (1969). Studies of Heisenberg exchange in ESR spectra. I. Linewidth and saturation effects. *J. Chem. Phys.* **51**, 2690–2709.

Eaton, S. S., and G. K. Eaton (1978). Interaction of spin labels with transition metals. *Coordination Chem. Rev.* **26**, 207–262.

Eichhorn, G. L. (1973). "Inorganic Biochemistry," Vol. 1. Elsevier, Amsterdam.

Epstein, M., A. Levitzki, and J. Reuben (1974). Binding of lanthanides and of divalent metal ions to porcine trypsin. *Biochemistry* **13**, 1777–1782.

Felix, C. C., J. S. Hyde, T. Sarna, and R. C. Sealy (1978). Interactions of melanin with metal ions. ESR evidence for chelate complexes of metal ions with free radicals. *J. Am. Chem. Soc.* **100**, 3922–3926.

Figgis, B. N. (1966). "Introduction to Ligand Fields." Wiley (Interscience), New York.

Fitzgerald, J. J., and N. D. Chasteen (1974). Electron paramagnetic resonance studies of the structure and metal ion exchange kinetics of vanadyl (IV) bovine carbonic anhydrase. *Biochemistry* **13**, 4338–4347.

Freed, J. H. (1978). Dynamic effects of pair correlation functions on spin relaxation by translational diffusion in liquids. II. Finite jumps and independent T_1 processes. *J. Chem. Phys.* **68**, 4034–4037.

Haak, R. A., F. W. Kleinhans, and S. Ochs (1976). The viscosity of mammalian nerve axoplasm measured by electron spin resonance. *J. Physiol. (London)* **263**, 115–137.

Hexem, J. G., U. Edlund, and G. C. Levy (1976). Paramagnetic relaxation reagents as a probe for translational motion of liquids. *J. Chem. Phys.* **64**, 936–941.

Horrocks, W. D., Jr. (1973). Lanthanide shift reagents and other analytical applications. *In* "NMR of Paramagnetic Molecules, Principles and Applications" (G. N. La Mar, W. D. Horrocks, Jr., and R. H. Holm, eds.), pp. 475–519. Academic Press, New York.

Hyde, J. S. (1960). Saturation of the magnetic resonance absorption in dilute inhomogeneously broadened systems. *Phys. Rev.* **119**, 1492–1495.

Hyde, J. S., and K. V. S. Rao (1978). Dipolar induced electron spin-lattice relaxation in unordered solids. *J. Magn. Reson.* **29**, 509–516.

Hyde, J. S., and T. Sarna (1978). Magnetic interactions between nitroxide free radicals and lanthanides or Cu^{2+} in liquids. *J. Chem. Phys.* **68**, 4439–4447.

Keith, A. D., W. Snipes, R. J. Mehlhorn, and T. Gunter (1977). Factors restricting diffusion of water-soluble spin labels. *Biophys. J.* **19**, 205–218.

Kivelson, D. (1960). Theory of ESR linewidths of free radicals. *J. Chem. Phys.* **33**, 1094–1106.

Kulikov, A. V., and G. I. Likhtenshtein (1977). The use of spin relaxation phenomena in the investigation of the structure of model and biological systems by the method of spin labels. *Adv. Mol. Relaxation Processes* **10**, 47–79.

Laurie, S. H., T. Lund, and J. B. Raynor (1975). Electronic absorption and electron spin resonance studies on the interaction between the biologically relevant copper (II) glycylglycine and L-histidine complexes with D-penicillamine. *J. Chem. Soc., Dalton Trans.* pp. 1389–1394.

Leigh, J. S., Jr. (1970). ESR rigid-lattice line shape in a system of two interacting spins. *J. Chem. Phys.* **52**, 2608–2612.

Lenkenski, R. E., and J. Reuben (1976). Line broadenings induced by lanthanide shift reagents: Concentration, frequency, and temperature effects. *J. Magn. Reson.* **21**, 47–56.

Likhtenshtein, G. I. (1976). "Spin Labeling Methods in Molecular Biology." Wiley (Interscience), New York.

Luk, C. K. (1971). Study of the nature of the metal-binding sites in transferrin using trivalent lanthanide ions as fluorescent probes. *Biochemistry* **10**, 2838–2843.

McConnell, H. M. (1958). Reaction rates by nuclear magnetic resonance. *J. Chem. Phys.* **28**, 430–431.

McLaughlin, A. C., J. S. Leigh, and M. Cohn (1976). Magnetic resonance study of the three-dimensional structure of creatine kinase-substrate complexes. *J. Biol. Chem.* **251**, 2777–2787.

Moeller, T. (1963). "The Chemistry of Lanthanides." Van Nostrand-Reinhold, Princeton, New Jersey.

Moeller, T. (1972). Complexes of the lanthanides. *Inorg. Chem., Ser. One* **7**, 275–598.

Nonaka, Y., T. Tokii, and S. Kida (1974). Factors affecting the line-width of nitrogen superhyperfine structure in the ESR spectra of copper (II) complexes. *Bull. Chem. Soc. Jpn.* **47**, 312–315.

Pake, G. E., and T. R. Tuttle, Jr. (1959). Anomalous loss of resolution of paramagnetic resonance hyperfine structure in liquids. *Phys. Rev.* **3**, 423–425.

Peisach, J., and W. E. Blumberg (1969). A mechanism for the action of penicillamine in the treatment of Wilson's disease. *Mol. Pharmacol.* **5**, 200–209.

Peisach, J., and W. E. Blumberg (1974). Structural implications derived from the analysis of electron paramagnetic resonance spectra of natural and artificial copper proteins. *Arch. Biochem. Biophys.* **165**, 691–708.

Peisach, J., W. G. Levine, and W. E. Blumberg (1967). Structural properties of stellacyanin, a copper mucoprotein from rhus vernicifera, the Japanese lac tree. *J. Biol. Chem.* **242**, 2847–2858.

Portis, A. M. (1953). Electronic structure of F centers: Saturation of the electron spin resonance. *Phys. Rev.* **91**, 1071–1078.

Rao, K. V. S., J. S. Hyde, and J. H. Freed (1979). EPR line shape in a system containing two interacting spins and undergoing slow rotational diffusion. (Submitted for publication.)

Redfield, A. G. (1955). Nuclear magnetic resonance saturation and rotary saturation in solids. *Phys. Rev.* **98**, 1787–1809.

Redfield, A. G. (1957). On the theory of relaxation processes. *IBM J. Res. Develop.* **1**, 19–31. See also A. G. Redfield (1965). The theory of relaxation processes. *In* "Advances in Magnetic Resonance Volume 1" (J. S. Waugh, ed.), pp. 1–32. Academic Press, New York.

Reed, G. H., and M. Cohn (1972). Structural changes induced by substrates and anions at the active site of creatine kinase. *J. Biol. Chem.* **247**, 3073–3081.

Reed, G. H., and M. Cohn (1973). Electron paramagnetic resonance studies of manganese (II)-pyruvate kinase-substrate complexes. *J. Biol. Chem.* **248**, 6436–6442.

Reed, G. H., and M. C. Scrutton (1974). Pyruvate carboxylase from chicken liver. *J. Biol. Chem.* **249**, 6156–6162.

Reed, G. H., J. S. Leigh, Jr., and J. E. Pearson (1971). Electron paramagnetic relaxation and EPR line shapes of manganous ion complexes in aqueous solutions. Frequency and ligand dependence. *J. Chem. Phys.* **55**, 3311–3316.

Reuben, J. (1971). Gadolinium (III) as a paramagnetic probe for proton relaxation studies of biological macromolecules. Binding to bovine serum albumin. *Biochemistry* **10**, 2834–2838.

Reuben, J., and D. Fiat (1969). Nuclear magnetic resonance studies of solutions of the rare-earth ions and their complexes. IV. Concentration and temperature dependence of the oxygen-17 transverse relaxation in aqueous solutions. *J. Chem. Phys.* **51**, 4918–4927.

Rist, G. H., and J. S. Hyde (1969). Ligand ENDOR of Cu-8-hydroxyquinolinate substituted into organic single crystals. *J. Chem. Phys.* **50**, 4532–4542.

Salikhov, K. M., A. B. Doctorov, Yu. N. Molin, and K. I. Zamaraev (1971). Exchange broadening of ESR lines for solutions of free radicals and transition metal complexes. *J. Magn. Reson.* **5**, 189–205.

Sarna, T., and S. Lukiewicz (1972). Electron spin resonance studies on living cells. IV. Pathological changes in amphibian eggs and embryos. *Folia Histochem. Cytochem.* **10**, 265–278.

Sarna, T., J. S. Hyde, and H. M. Swartz (1976). Ion-exchange in melanin: An electron spin resonance study with lanthanide probes. *Science* **192**, 1132–1134.

Skubnevskaya, G. I., and Yu. N. Molin (1967). Exchange interactions of aquo-complexes of ions of the iron group with paramagnetic particles in solutions. *Kinet. Katal.* **8**, 1192–1197.

Skubnevskaya, G. I., and Yu. N. Molin (1972). Temperature dependence of constants for strong and weak spin exchange between free radicals and paramagnetic complexes. *Kinet. Katal.* **13**, 1383–1388.

Slichter, C. P. (1963). "Principles of Magnetic Resonance," p. 148. Harper, New York.

Taketa, M., W. F. Pickard, J. Y. Lettvin, and J. W. Moore (1966). Ionic conductance changes in lobster axon membrane when lanthanum is substituted for calcium. *J. Gen. Physiol.* **50**, 461–471.

Taylor, J. S., J. S. Leigh, Jr., and M. Cohn (1969). Studies of spin-labeled creatine kinase system and interaction of two paramagnetic probes. *Proc. Natl. Acad. Sci. U.S.A.* **64**, 219–226.

Vänngård, T. (1972). Copper proteins. *In* "Biological Applications of Electron Spin Resonance" (H. M. Swartz, J. R. Bolton, and D. C. Borg, eds.), pp. 411–447. Wiley (Interscience), New York.

Van Vleck, J. H. (1932). "The Theory of Electric and Magnetic Susceptibilities." Oxford Univ. Press, London and New York.

Van Vleck, J. H. (1948). The dipolar broadening of magnetic resonance lines in crystals. *Phys. Rev.* **74**, 1168–1183.

Walker, F. A., H. Siegel, and D. B. McCormick (1972). Spectral properties of mixed ligand copper (II) complexes and their corresponding binary parent complexes. *Inorg. Chem.* **11**, 2756–2763.

Wayland, B. B., and V. K. Kapur (1974). Electron paramagnetic resonance and electronic spectral evidence for isomers resulting from basal and axial ligation of bis (hexafluoroacetylacetonato) copper (II) by triphenylphosphine. *Inorg. Chem.* **13**, 2517–2520.

Williams, R. J. P. (1970). The biochemistry of sodium, potassium, magnesium, and calcium. *Q. Rev., Chem. Soc.* **24**, 331–365.

Wong, S. K., and J. K. S. Wan (1971). EPR of trapped H and D atoms in γ-irradiated acidic glasses. Anomalous differential saturation of hyperfine components via cross relaxation induced by paramagnetic metal ions. *J. Chem. Phys.* **55**, 4940–4947.

3

New Aspects of Nitroxide Chemistry

JOHN F. W. KEANA

DEPARTMENT OF CHEMISTRY
UNIVERSITY OF OREGON
EUGENE, OREGON

I.	Introduction	115
II.	Structural Features That Make a Nitroxide Stable	116
III.	Characterization of New Nitroxide Spin Labels and Nitroxide Spectra	126
IV.	Synthesis of New Stable Nitroxide Spin Labels	129
	A. Doxyl Nitroxides	130
	B. Proxyl Nitroxides	132
	C. Azethoxyl Nitroxides (Minimal Perturbation Spin Labels)	134
	D. Imidazolidine-Derived Nitroxides	137
	E. Δ^3-Imidazoline-Derived Nitroxides	139
	F. Δ^1-Imidazoline-Derived Nitroxides (Nitronyl Nitroxides)	140
	G. Tetrahydrooxazine-Derived Nitroxides	141
	H. Other Nitroxides in Which the Nitroxyl Nitrogen Is Part of the Skeleton	142
	I. New Nitroxide Spin Labels **I** and **II** Derived from 2,2,6,6-Tetramethyl-piperidine-*N*-oxyl and 2,2,5,5-Tetramethylpyrrolidine-*N*-oxyl	144
V.	Chemical Reactions Involving the Nitroxide Grouping	147
	A. Nitroxide Overoxidation	147
	B. Nitroxide Reduction	148
	C. Nitroxide Photolysis and Thermolysis	151
VI.	Chemical Reactions Not Involving the Nitroxide Grouping	152
VII.	Dinitroxide Spin Labels	152
VIII.	Experimental Procedures for the Synthesis of Selected Proxyl and Imidazolidine-Derived Nitroxide Spin Labels	154
	References	166

I. INTRODUCTION

The continuing successful development of the nitroxide spin-labeling technique for studying biological systems depends in a large measure on the ability of the experimenter to design and carry out the synthesis of new

115

spin-labeled molecules that incorporate the desired structural features and biochemical properties. For many biochemical–biophysical laboratories wishing to do spin-labeling studies, a major hurdle has been the chemical synthesis of the requisite labels. Fortunately, many of the heavily exploited nitroxide spin labels now are available commercially. Others must be synthesized, often by multistep routes. One purpose of this chapter is to apprise the scientific community performing spin-labeling studies of the array of available stable nitroxide free radicals. Another aim is to provide guidance for the synthesis of new nitroxide spin labels, especially for those laboratories not accustomed to doing moderately involved organic synthesis. Occasionally, possibilities for new nitroxide spin labels are described. Considerable emphasis therefore is placed, when appropriate, on the practical side of nitroxide spin-label synthesis. Selected recent advances in the chemistry of stable nitroxide free radicals pertinent to the synthesis and covalent attachment of new or existing spin labels serve as the framework for discussion.

Aspects of the chemistry of nitroxide spin labels have been conveniently reviewed by Gaffney (1976), who also describes procedures for the synthesis and covalent attachment of several common spin labels. A comprehensive review of new aspects of the chemistry of nitroxide spin labels recently has been completed (Keana, 1978). The reader is referred to these two reviews and the references cited therein for an overall picture of nitroxide spin-label chemistry.

II. STRUCTURAL FEATURES THAT MAKE A NITROXIDE STABLE

The design of a new stable nitroxide spin label normally takes into account the following considerations. First, the new nitroxide should be sufficiently stable to enable purification and subsequent chemical and spectral characterization and sufficiently stable so that appreciable signal loss does not occur during the spin-labeling experiment, unless, of course, that is the intent of the experiment. Second, the chemical structure of the new nitroxide should be such that the physical and chemical behavior of the spin-label system is minimally perturbed by the presence of the label. Third, one would like to have a choice as to the molecular shape and chemical reactivity of the nitroxide moiety in order, for example, to provide an experimental check on the extent to which the system is perturbed by the presence of a given label or to vary the susceptibility of the nitroxide moiety toward, say, chemical reduction.

The great majority of nitroxide spin labels in use today are members of one of the following three classes of nitroxides: the 2,2,6,6-tetramethylpiperidine-N-oxyl nitroxides(**I**), in which the substituents are typically at the 4 position; the 2,2,5,5-tetramethylpyrrolidine-N-oxyl nitroxides(**II**), in which the substituents are typically at the 3 position; and the 4,4-dimethyl-oxazolidine-N-oxyl(doxyl) nitroxides(**III**), in which the moiety is attached to the parent compound at the 2 position.

Recently, two other versatile classes (see below) of nitroxide spin labels have been introduced. These are the side-chain-substituted 2,2,5,5-tetra-methylpyrrolidine-N-oxyl(proxyl) nitroxides(**IV**) (Keana *et al.*, 1976) and the *cis*- and *trans*-azethoxyl nitroxides(**V**) (Lee *et al.*, 1977). Nitroxides **IV** and **V**, like **II**, also are pyrrolidine-N-oxyl nitroxides. However, in **IV** both methyl groups attached to the 2 position are substituted, whereas in **V** one methyl group on either side of the nitroxyl nitrogen bears a substituent. It is apparent with **V** that *cis,trans* isomers are possible, offering an added versatility to these labels (see Section IV,C).

I

(a piperidine nitroxide)

II

(a pyrrolidine nitroxide)

III

(a doxyl nitroxide)

IV

(a proxyl nitroxide)

Va

(a *cis*-azethoxyl nitroxide)

Vb

(a *trans*-azethoxyl nitroxide)

It is immediately apparent that all these stable nitroxides are secondary amine N-oxides that bear no hydrogen atoms on the carbons attached to the nitroxyl nitrogen, i.e., the α-carbon atoms. The term "stable nitroxide" refers to a nitroxide that can be obtained in pure form, stored, and handled in the laboratory with no more precaution than that normally observed when one is working with most organic substances. Some "stable" nitrox-ides are, of course, more "stable" than other stable nitroxides. This definition parallels that of Griller and Ingold (1976) for carbon-centered radicals.

When one or more hydrogen atoms are present on the α-carbon atoms of the nitroxide, the nitroxide is normally not stable, although it may "persist" (Griller and Ingold, 1976) for hours, days, or longer in dilute solution. Such nitroxides typically undergo a disproportionation reaction (Bowman *et al.*, 1971; Briere and Rassat, 1976), producing the corresponding nitrone and

N-hydroxyamine (**VI → VII + VIII**), either or both of which may undergo further chemical reaction.

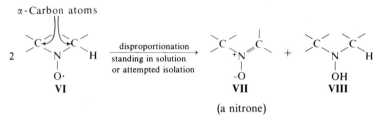

(a nitrone)

It is instructive to consider in some detail what happens to a representative nitroxide that bears an α-hydrogen, namely, *tert*-butylisopropyl nitroxide(**IX**). Briere and Rassat (1976) studied the decomposition of **IX** as a concentrated solution in cyclohexane. The reaction shows a large primary isotope effect ($k_H/k_D \cong 9$), consistent with breakage of the α C—H bond in the slow step of the decomposition. Reaction products were isolated and identified as **X**, **XI**, **XII**, **XIII**, **XIV**, and **XV**. Scheme I was proposed to account for these observations.

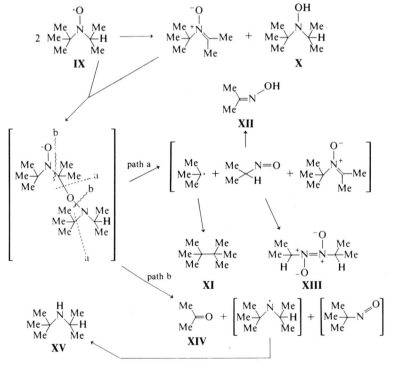

Scheme I (Briere and Rassat, 1976)

The kinetics of the self-reactions of several other dialkyl nitroxides were examined by Bowman et al. (1971). The rate of the disproportionation reaction strongly depends on the degree of substitution on the α-carbons and on the solvent.

There are occasional surprises. For example, Lin et al. (1974) synthesized proline nitroxide **XVI** by oxidation of N-hydroxyproline with tert-butyl hydroperoxide in ethanol. Despite the fact that **XVI** has three α-hydrogens, it is a light-yellow, crystalline solid that is stable indefinitely in the solid state and shows a half-life of about 16 hr in pH 7 phosphate buffer at 24°C. Undoubtedly, one factor in the unusual stability of **XVI** is the presence of the carboxylate anion at pH 7. This would slow down the disproportionation since two anions would not approach one another.

XVI

A potentially general method for attaching an oxazolidine nitroxide grouping to a molecule at the site of a carbon–carbon double bond is that of Williams et al. (1971). The synthetic method is shown in reaction (1) and results in nitroxides(**XVII**) that bear one hydrogen on the α-carbon atom and that undergo motion about the X principal axis of the nitroxide. The nitroxides showed a half-life of about 2 hr in ethyl oleate solution at 40°C. It is highly likely that a mixture of positional and stereoisomeric nitroxides is produced in the reaction. Another difficulty is that experimental details for the last step, i.e., the oxidation to the nitroxide(s), were not provided, and the intermediates and products were characterized incompletely. The method, therefore, should be confirmed.

$$(1)$$

XVII

There are several interesting stable nitroxides besides proline nitroxide **XVI** that bear hydrogen atoms on the α-carbon atoms. These are the bicyclic nitroxides, the first example of which (**XVIII**) was synthesized by Dupeyre and Rassat (1966). The stability of the bicyclic nitroxides toward disproportionation is attributed to the fact that the nitrone(**XXII**) that would be formed would have a double bond involving a bridgehead carbon atom, in violation of the well-known Bredt's rule.

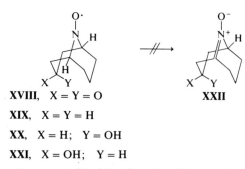

XVIII, X = Y = O

XIX, X = Y = H

XX, X = H; Y = OH

XXI, X = OH; Y = H

Several other known stable bicyclic nitroxides are the following: **XIX**, **XX**, **XXI**, and **XXIII** (Mendenhall and Ingold, 1973a); **XXIV** and **XXV** (Rassat and Ronzaud, 1976); **XXVI**, **XXVII**, and **XXVIII** (Dupeyre and Rassat, 1973); and **XXIX** and **XXX** (Rozantsev and Sholle, 1969, 1971).

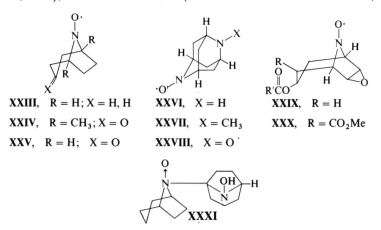

XXIII, R = H; X = H, H   XXVI, X = H   XXIX, R = H

XXIV, R = CH$_3$; X = O   XXVII, X = CH$_3$ XXX, R = CO$_2$Me

XXV, R = H; X = O   XXVIII, X = O

XXXI

In general, these bicyclic nitroxides have not enjoyed wide use as spin labels. Reasons for this may include the fact that the electron spin resonance (ESR) spectra are complicated by additional splitting due to the α-hydrogens, the fact that some are not as easily prepared as nitroxides **I–VII**, and the fact that, as a group, the bicyclic nitroxides tend not to be quite as stable as nitroxides **I–VII**. For example, solutions of **XXIII** slowly and irreversibly form a dimeric structure upon standing, which was shown to be **XXXI** by X-ray analysis (Mendenhall and Ingold, 1973b).

Nevertheless, it is worth pointing out that there are bicyclic nitroxides such as **XXIV** and others (see below) that do not have α-hydrogens. It also should be recognized that the overall molecular shape of these nitroxides differs significantly from that of conventional nitroxides. Thus, these nitroxides could become important spin labels as part of a comparative study probing, for example, the dependence of a spin-labeling experimental result on the structure of the nitroxide spin label employed.

Studies particularly with the bicyclic nitroxides reveal that in solution even relatively unhindered nitroxides show little tendency toward either reversible or irreversible dimerization. Thus, nitroxide **XVIII** was shown by its infrared (IR), ultraviolet (UV), and ESR spectra to be monomeric in solution (Dupeyre and Rassat, 1966). Rare exceptions do occur, however, as illustrated by the observation that dinitroxide **XXVIII** exists as a dimer or higher aggregate in dimethylformamide solution (0.02 M) at room temperature (Dupeyre et al., 1974).

The situation is different in the solid state. Single crystals of **XXIII** are diamagnetic, whereas those of **XIX** are strongly paramagnetic (Mendenhall and Ingold, 1973b).

Two thionitroxides(**XXXIV** and **XXXV**) were reversibly generated by heating the corresponding disulfides(**XXXII** and **XXXIII**) in iodobenzene solution at 90°–200°C (Bennett et al., 1967). The thionitroxides exist in the dimeric form at room temperature despite the presence of the four bulky methyl groups. Apparently, both the larger size of sulfur as compared to oxygen and the greater strength of the S—S bond as compared to an O—O bond are significant factors in determining the position of the dimer–monomer equilibrium.

It should be possible to capitalize on this reversible dimerization in the design of new thionitroxide or thionitroxide–nitroxide spin labels. For example, in view of the fact that **XXXII** begins to dissociate at 90°C whereas **XXXIII** begins at 140°C, one could envisage other thionitroxide dimers or nitroxide–thionitroxide complexes that might dissociate reversibly at slightly above ambient temperature. These interesting possibilities are open for exploration.

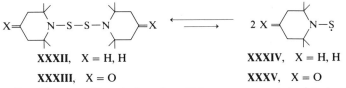

XXXII, X = H, H   XXXIV, X = H, H

XXXIII, X = O   XXXV, X = O

The nitroxides considered thus far all have saturated (sp³-hybridized) carbons attached to the nitroxyl nitrogen. In a large number of stable nitroxides, the nitroxide group is stabilized by conjugation to an aromatic ring or to other unsaturated linkages (Keana, 1978; Rozantsev, 1970). As with the bicyclic nitroxides, the conjugated stable nitroxides have two unattractive features as spin labels, namely, more complicated ESR spectra in general and a somewhat reduced stability in certain cases (see below). In view of the current interest in polynuclear hydrocarbon–cellular component interactions, it is somewhat surprising that more has not been done to date with aryl-substituted stable nitroxides as spin labels.

Several of the more recently prepared stable conjugated nitroxides are the following (Fig. 1): **XXXVI** (Forrester et al., 1974); **XXXVII** (Baldry et

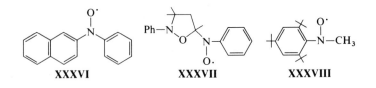

XXXVI　　　　　XXXVII　　　　　XXXVIII

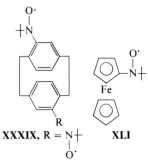

XXXIX, R = N+ O·

XL, R = H

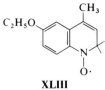

XLI

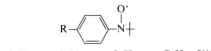

XLII, R = C$_2$H$_5$, i = C$_3$H$_7$, n = C$_4$H$_9$, t − C$_4$H$_9$, Si(CH$_3$)

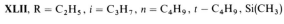

XLIII　　　　　XLIV, R = CH$_3$, C$_2$H$_5$, Ph, PhCH$_2$　　　　　XLV

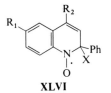

XLVI

R$_1$ = R$_2$ = H

R$_1$ = H; R$_2$ = OEt

$\qquad\qquad\quad$ OH
$\qquad\qquad\quad$ |
R$_1$ = H; R$_2$ = C(Ph)$_2$

R$_1$ = OMe; R$_2$ = H

R$_1$ = OMe; R$_2$ = Cl

X = CH$_3$, C$_2$H$_5$, Ph

Fig. 1. Several stable conjugated nitroxides.

XLVIIa, R = Ph

XLVIIb, R = H, CH$_3$, C$_2$H$_5$, *n*-C$_{13}$H$_{27}$, CH$_2$OH, CH(OC$_2$H$_5$)$_2$

XLVIIc, R = I, Br, CN, C—OCH$_3$
                              ‖
                              NH

XLVIId, R = *p*-C$_6$H$_4$NO$_2$, *p*-C$_6$H$_4$OH, *p*-C$_6$H$_4$N(CH$_3$)$_2$, CH(CH$_3$)$_2$, CH$_2$CH(CH$_3$)$_2$,

CH$_2$Ph, CH=CH$_2$, CH=CH—CH$_3$, CH=CHPh, C≡CPh,  CH$_2$NHTs,

CH$_2$Cl, CH$_2$Br, CH$_2$I, CF$_3$,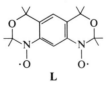

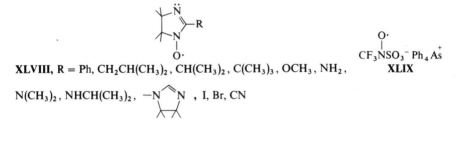

XLVIII, R = Ph, CH$_2$CH(CH$_3$)$_2$, CH(CH$_3$)$_2$, C(CH$_3$)$_3$, OCH$_3$, NH$_2$,

N(CH$_3$)$_2$, NHCH(CH$_3$)$_2$, [imidazoline ring], I, Br, CN

XLIX

L

Fig. 1. (Continued.)

al., 1976); **XXXVIII** (Martinie-Hombrouck and Rassat, 1974); **XXXIX** (Forrester and Ramasseul, 1975); **XL** (Forrester and Ramasseul, 1971); **XLI** (Forrester and Hepburn, 1969); **XLII** (Torsell *et al.*, 1973); **XLIII** (Lin and Olcott, 1975); **XLIV** (Berti *et al.*, 1975); **XLV** (Ramasseul and Rassat, 1970); **XLVI** (Berti *et al.*, 1976); those nitronyl nitroxides shown as **XLVIIa** (Osiecki and Ullman, 1968), **XLVIIb** (Boocock *et al.*, 1968), **XLVIIc** (Boocock and Ullman, 1968), and **XLVIId** (Ullman *et al.*, 1972); those conjugated imino nitroxides shown under structure **XLVIII** (Ullman *et al.*, 1970); the fluorocarbon analogue of Fremy's salt **XLIX** (Banks *et al.*, 1974); and dinitroxide **L** (Rassat and Sieveking, 1972).

Despite the fact that all the nitroxides shown in Fig. 1 can be isolated and characterized, several undergo slow decomposition upon standing. As might be expected, for example, *N*-methyl-*N*-tri-*tert*-butylphenyl nitroxide(**XXXVIII**) undergoes a disproportionation reaction upon standing overnight, affording the corresponding nitrone and *N*-hydroxyamine (Martinie-Hombrouck and Rassat, 1974).

Calder and Forrester (1969) showed that aryl-*tert*-butyl nitroxides tend to dimerize by coupling of the nitroxide oxygen atom to the para position of another nitroxide molecule, if that para position is not too hindered. The isolated products are *p*-benzoquinone-*tert*-butylimine *N*-oxides and *tert*-butylanilines [reaction (2)]. The initial dimerization is inhibited either by the presence of bulky ortho substituents, which cause the aryl ring to be twisted out of conjugation with the nitroxide (Forrester and Hepburn, 1970), or by the presence of certain substituents such as methoxy or *tert*-butyl at the para position (Forrester and Hepburn, 1974). The decomposition of 1- and 2-naphthyl-*tert*-butyl nitroxides (Calder *et al.*, 1974) and several halogenophenyl-*tert*-butyl nitroxides (Calder *et al.*, 1973) also has been studied.

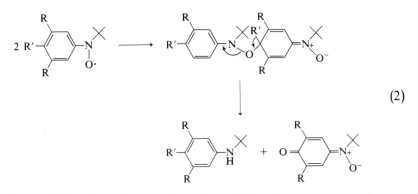

$$(2)$$

To conclude this section on nitroxide stability, it is appropriate for the purpose of comparison, to show representative conjugated nitroxides that are not stable. Much more is discussed below concerning the stability of

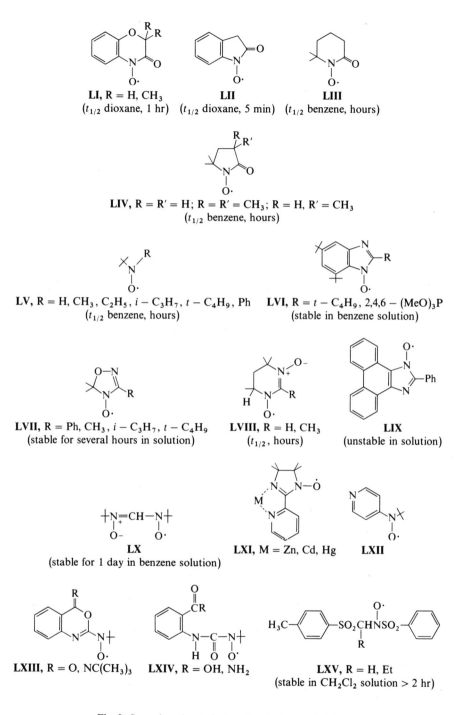

Fig. 2. Several conjugated nitroxides that are not stable.

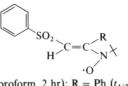

LXVI, R = H ($t_{1/2}$ chloroform, 2 hr); **R = Ph** ($t_{1/2}$ chloroform, 10 min)

Fig. 2. (Continued.)

nitroxides under a variety of circumstances, particularly regarding synthetic transformations on remote portions of nitroxide-bearing molecules and new methods of covalent attachment of nitroxides to other molecules. Figure 2 shows the following unstable nitroxides and gives in many instances some indication of their persistence: **LI** and **LII** (Balaban and Pascaru, 1972); **LIII, LIV,** and **LV** (Aurich and Trösken, 1973); **LVI** and **LVII** (Aurich and Stork, 1975); **LVIII** (Ghriofa *et al.*, 1977); **LIX** (Volkamer and Zimmerman, 1970); **LX** (Aurich and Höhlein, 1974); **LXI** (Wagner *et al.*, 1976); **LXII** (Cazianis and Eaton, 1974); **LXIII** and **LXIV** (Aurich and Weiss, 1976); **LXV** (Rawson and Engberts, 1970); and **LXVI** (Aurich *et al.*, 1975).

III. CHARACTERIZATION OF NEW NITROXIDE SPIN LABELS AND NITROXIDE SPECTRA

Clearly, proper characterization of a new nitroxide spin label or spin-labeled molecule is of utmost importance in order to eliminate any uncertainty regarding the nature of the spin label employed in a given study. Yet one finds in the literature multistage syntheses in which intermediates are not characterized and, even more unsettling, the "structural" evidence for the newly synthesized nitroxide may be limited to observing one spot on a thin-layer chromatography plate and a three-line ESR spectrum or other combinations of flimsy evidence. Often, the chemical synthesis (without experimental detail) and characterization of the spin label are relegated to a very minor, almost incidental, segment of such papers. One should note, incidentally, that both the common buffer Tris and EDTA give a well-defined nitroxide-like ESR signal upon oxidation with H_2O_2–sodium tungstate (Murphy *et al.*, 1974).

The following suggestions are offered. Evidence of the purity of each intermediate that can be isolated and of the final product should be given. An interpreted nuclear magnetic resonance (NMR) spectrum or a listing of characteristic IR peaks is often useful. An elemental analysis or a high-resolution mass spectrum is essential for any final product showing the exact mass of the molecular ion. Thin-layer chromatographic behavior of a substance can also be helpful to other investigators.

Conventional NMR spectra of molecules containing nitroxide groups tend to be quite broad (La Mar et al., 1973). This is caused by the intramolecular and intermolecular interaction of the unpaired electron with nuclear spins, leading to considerably shortened lifetimes and therefore broad spectral lines. Nevertheless, in the author's laboratory 100-MHz NMR spectra are recorded routinely on dilute solutions (5–10 mg in 0.3 ml) of nitroxides in $CDCl_3$ as a valuable aid in monitoring structural changes during a chemical synthesis on nitroxide-containing molecules. For example, 12-proxylpalmityl alcohol(**LXVII**) shows a broad peak (peak width at half-height = 14 Hz) at 3.76 ppm integrating very approximately for two protons and corresponding to the protons on the carbon bearing the hydroxy group. The methanesulfonate derivative **LXVIII**, on the other hand, shows a somewhat broadened three-proton singlet at 3.00 ppm corresponding to the methanesulfonate group and a broad peak at 4.24 ppm due to the protons on the carbon bearing the methanesulfonyloxy group. In both substances, the protons of the terminal methyl group are clearly seen as a somewhat broadened hump on the high-field side of the methylene envelope at about 0.9 ppm, the expected place.

$$CH_3CH_2CH_2CH_2CCH_2CH_2CH_2CH_2CH_2CH_2CH_2CH_2CH_2CH_2CH_2OR$$

LXVII,   R = H

LXVIII,   R = SO_2CH_3

Protons very close to a nitroxide group experience a large chemical shift (relative to a diamagnetic molecule of similar structure) in a direction depending on the sign of the electron–nuclear hyperfine splitting constant. The effect is illustrated by the NMR spectrum (Briere et al., 1970) of nitroxide **LXIX**.

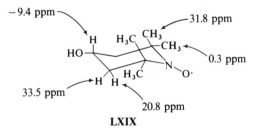

LXIX

Typically, when the shifts (and attendant broadening) are large, the NMR spectra are run at concentrations as high as 3 M so that the electron spin-relaxation time becomes shorter than the inverse of the proton hyperfine coupling constant (Carrington and McLachlan, 1967). An alternative, although less convenient, procedure is to employ a paramagnetic solvent such as di-*tert*-butyl nitroxide (**LXX**), 2-doxylpropane (**LXXI**), or the completely deuterated analogues (Chiarelli and Rassat, 1973) as spin relaxer (Kreilich, 1968) solvents so that solute spectral lines become sharp enough to detect.

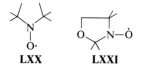

         LXX   **LXXI**

Probably the most convenient way to acquire structural information on a nitroxide by NMR spectroscopy is to reduce the $CDCl_3$ solution of the nitroxide *in situ* in the NMR tube to the corresponding diamagnetic *N*-hydroxy derivative with 1.5 equivalents of phenylhydrazine (for example, **LXXII → LXXIII**) (Lee and Keana, 1975). After observation of the spectrum in the normal way, the nitroxide can be recovered by air oxidation in methanol containing a trace of Cu^{2+} ion. Rassat and Rey (1974a) independently used the method in order to acquire structure and stereochemical information on several bicyclic and polycyclic tetrahydrooxazine-derived nitroxides.

      LXXII   **LXXIII**

The rate of the phenylhydrazine reduction depends on the structure. Thus, doxyl nitroxides are reduced in seconds, proxyl nitroxides require minutes, and azethoxyl nitroxides are reduced too slowly by phenylhydrazine under the above conditions for the method to be useful (Lee *et al.*, 1977).

One additional technique applicable to azethoxyl nitroxides as well as to all the others is the conversion of the nitroxide into the corresponding *N*-acetoxy derivative. This is easily accomplished by catalytic reduction of the nitroxide in tetrahydrofuran (THF) solution over 10% Pd/C followed by an immediate quench with acetyl chloride–triethylamine. Thus, Lee (1977) prepared a series of *cis*-azethoxyl nitroxides(**Va**) and *trans*-azethoxyl nitroxides(**Vb**) and converted them to the corresponding *N*-acetoxy derivatives(**LXXIV**). Not only were the NMR spectra easily recorded, but also the cis and trans isomers could be separated easily in the form of the *N*-acetoxy derivatives. The nitroxides could be regenerated in high yield by hydrolysis of the *N*-acetoxy derivatives followed by air oxidation.

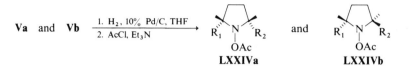

The ^{13}C NMR spectra of several nitroxides have been measured and interpreted (Hatch and Kreilich, 1971; Neely *et al.*, 1974). In the IR region, most nitroxides show a band in the range 1310–1370 cm^{-1}. Morat and Rassat (1972) established the frequency of the N—O stretch in a series of piperidine nitroxides to be 1373 ± 7 cm^{-1}. The corresponding ^{15}N radicals were used to aid in the assignment.

Although the mass spectra of numerous nitroxides have been routinely determined, relatively few detailed studies of the fragmentation patterns have been reported. Pyrrolidine nitroxides(**II**) substituted with CH$_2$OH, OH, and NH$_2$ show [M + 1]‡ ions together with ions resulting from sequential loss of methyl, isobutylene, and nitric oxide (Davies *et al.*, 1974). Piperidine nitroxides, however, behave somewhat differently, although they, too, show a molecular ion, [M + 1]$^{+}$, and loss of a methyl group from the [M + 1] moiety (Morrison and Davies, 1970).

Simple doxyl nitroxides typically show an *m/e* peak corresponding to the protonated ketone as the most intense peak (i.e., the base peak) in the mass spectrum [reaction (3)] (Chou *et al.*, 1974). Doxylstearic acids and methyl esters, on the other hand, both show *m/e* 281 as the most intense peak in the mass spectrum. This observation was made in connection with the development of a gas–liquid chromatography–mass spectrometry method of identifying a series of doxylstearic acids. The method (Marai *et al.*, 1976) may be useful in determining the fate of biologically incorporated doxyl fatty acid spin labels.

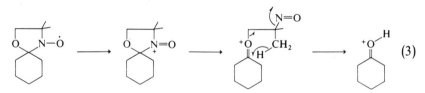

(3)

IV. SYNTHESIS OF NEW STABLE NITROXIDE SPIN LABELS

This section focuses on various new ways of constructing a stable nitroxide free-radical grouping from nonparamagnetic precursors. The resulting nitroxide itself may be a spin label or spin probe. More commonly, it will bear a functional group capable of forming a covalent bond with another molecule, thus leading to a spin label or spin probe. The reader is referred to two recent reviews (Keana, 1978; Morrisett, 1975) that contain a listing of most of the functionalized nitroxides known to date. They are grouped as nitroxide alkylating agents, acylating agents, sulfonating agents, phosphorylating agents, and miscellaneous reagents.

It is frequently desirable to perform further chemical transformations on the new nitroxide in order to manipulate the molecule into a more useful derivative. This topic constitutes Section V.

A. Doxyl Nitroxides

When a ketone is condensed with 2-amino-2-methylpropanol(**LXXV**), an oxazolidine(**LXXVI**) results which can be oxidized with *m*-chloroperbenzoic acid (MCPA) to a stable doxyl nitroxide(**LXXVII**) [reaction (4)] (Keana *et al.*, 1967).

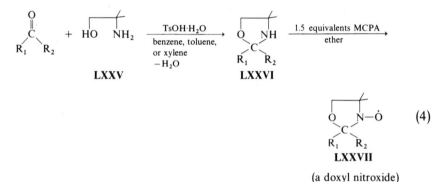

(a doxyl nitroxide)

Two important features of doxyl nitroxides are the ease with which the doxyl group can be introduced into a molecule at the site of a ketone group and the rigid nature of the attachment of the nitroxide to the molecule via a spiral linkage. It should be recognized, however, that the formerly achiral carbon atom of the ketone is converted to a chiral atom when it becomes part of the doxyl ring. Thus, when a doxyl group is introduced into an already chiral molecule such as a steroid or phospholipid, two diastereomers are usually formed and in unequal amounts (see, for example, P. Michon and A. Rassat, 1974; Moutin and Rassat, 1976). Normally, with nicely crystalline molecules such as steroids the two diastereomers can be separated by crystallization at the oxazolidine stage or by crystallization and/or chromatography at the nitroxide stage.

When the doxyl group is introduced into a nonchiral molecule such as 12-ketostearic acid, a racemic mixture of doxyl nitroxides is produced that normally is not resolved. This racemic mixture affords a virtually inseparable mixture of diastereomers when it is employed, for example, for the acylation of naturally occurring chiral molecules such as lysophospholipids.

There is good reason to believe that the two phospholipid diastereomers will behave essentially identically during a spin-labeling experiment. When spin labels are used as substrates of enzymes or other chiral assemblages, one must pay closer attention to stereoisomers. Thus, Flohr *et al.* (1975) observed enantiomeric (mirror-image) specificity in the hydrolysis of the chiral ester **LXXVIII** catalyzed by cyclohexaamylose. The absolute configuration of the acid **LXXIX** is known (Wetherington *et al.*, 1974).

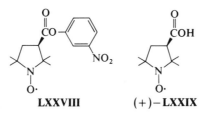

LXXVIII   (+)−**LXXIX**

Many hindered ketones form the oxazolidine derivative with great difficulty. One technique for improving the rate of oxazolidine formation is to use the amino alcohol **LXXV** as the solvent for the reaction (Nelson *et al.*, 1975). Another is to employ xylene as the solvent and a Dean–Stark water separator that contains excess amino alcohol **LXXV**, xylene, and anhydrous K_2CO_3 (J. F. W. Keana, unpublished results). Unlike other drying agents, K_2CO_3 does not absorb the amino alcohol, which slowly codistills into the trap with the xylene.

Recently, a second method of synthesizing doxyl nitroxides has been reported (Keana and Lee, 1975; Lee and Keana, 1976). This procedure takes advantage of the ready availability of a wide variety of oxazolines stemming from the work of Meyers *et al.* (1974). Oxidation of the representative oxazoline **LXXX** by MCPA gave the oxaziridine **LXXXI**, which underwent smooth isomerization to nitrone **LXXXII** upon silica gel chromatography. Addition of an organometallic reagent to **LXXXII** followed by air oxidation afforded the doxyl nitroxide **LXXXIII**. Because the MCPA oxidation step

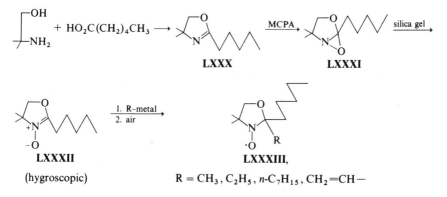

LXXX   **LXXXI**

LXXXII   **LXXXIII**,

(hygroscopic) $R = CH_3, C_2H_5, n\text{-}C_7H_{15}, CH_2\!=\!CH-$

takes place before addition of the second grouping, this last grouping conveniently may contain carbon–carbon double bonds or other MCPA-reactive functional groups. The synthesis of **CXLII** described below, however, illustrates that in certain instances, at least, one is able to oxidize a secondary amine to a nitroxide in the presence of carbon–carbon double bonds.

B. Proxyl Nitroxides

Keana *et al.* (1976) have introduced a new series of nitroxide lipid spin labels(**IV**) based on the pyrrolidine-*N*-oxyl structure. These nitroxides are termed proxyl nitroxides. They retain the rigid-attachment advantages of the doxyl nitroxides and, additionally, proxyl nitroxides are chemically more stable and are of significantly lower polarity than the corresponding doxyl nitroxides. Most of the new nitroxide spin labels currently under investigation in the author's laboratory are either proxyl or azethoxyl (see Section IV,C) nitroxides.

Proxyl nitroxides are readily assembled starting with the commercially available (Aldrich Co.) nitrone **LXXXIV**. Reaction with a Grignard reagent affords the *N*-hydroxyamine **LXXXV** (not normally isolated). Next, advantage is taken of the tendency of nitroxides with α-hydrogen atoms (e.g., **LXXXVI**), to undergo a disproportionation reaction (see Section II). Thus, *in situ* Cu^{2+}-catalyzed air oxidation of **LXXXV** in MeOH quickly affords a new nitrone(**LXXXVII**) in high yield, probably via the intermediate nitroxide **LXXXVI**. Reaction of **LXXXVII** with a second organometallic reagent followed by Cu^{2+}–air oxidation gives the desired proxyl nitroxide.

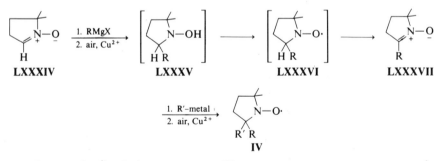

One or both of the organometallic reagents may possess a second remote, protected functional group. Then, after the synthesis of the proxyl group is completed, the protecting group may be cleaved, with the new functional group now available for further elaboration. Several examples of proxyl nitroxides that have been synthesized are the following (Fig. 3): **LXXXVIII** and **XC** (Keana *et al.*, 1976); **LXXXIX** and **XCIII** (Birrell *et al.*, 1978; J. F. W. Keana, R. Roman, and E. M. Bernard, unpublished results); and **XCI** and **XCII** (Keana and LaFleur, 1978).

LXXXVIII, X = CH_2OH^*, CHO^*, CO_2H^*, CO_2Me, $CH_2OSO_2CH_3^*$, $CH_2\overset{+}{N}(Me)_3$ $\overset{-}{O}SO_2CH_3^*$

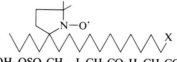

LXXXIX, X = OH, OSO_2CH_3, I, CH_2CO_2H, CH_2COCl, CH_2CH_2OH

XC, X = CN, CO_2H, CO_2Me

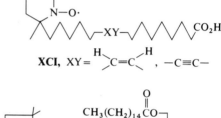

XCI, XY = $\overset{H}{}\underset{}{C}=\overset{H}{}\underset{}{C}$, $-C\equiv C-$

XCII, X = OH, OSO_2CH_3, I, CH_2CO_2H

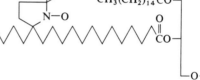

XCIII, X = OH, choline derivative, glycerol derivative, serine derivative, ethanolamine derivative

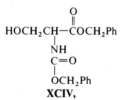

$$HOCH_2CH-\overset{\overset{O}{\parallel}}{C}OCH_2Ph$$

XCIV,

Fig. 3. Examples of proxyl nitroxides that have been synthesized. Synthetic Procedures are provided in Section VIII for those structures marked with an asterisk.

The inertness of the proxyl group during further chemical transformations on remote portions of the molecule is illustrated by the synthesis of the numerous derivatives shown in Fig. 3. Proxyl nitroxide acid chlorides can be prepared using dry chloroform–pyridine–oxalyl chloride (for example, **LXXXIX**, X = CH_2COCl). Protecting groups such as benzyloxy that can be cleaved by catalytic hydrogenation can be cleaved in the presence of proxyl nitroxides, whereas with doxyl nitroxides catalytic reduction leads to the hydrolytically unstable *N*-hydroxyoxazolidine derivative. Thus, the first preparation in the author's laboratory of spin-labeled phosphatidylserine utilized the benzyloxy-protected serine derivative **XCIV**. Now, however, spin-labeled phosphatidylserine is best prepared from spin-labeled phosphatidylcholine by reaction with excess serine in the presence of phospholipase D by a modification of the method of Comfurius and Zwaal (1977).

In order to determine the relative polarity of proxyl versus doxyl nitroxides, nitroxides **XCV** and **XCVI** were synthesized and allowed to partition

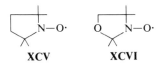

between dodecane and water. Partition coefficients indicated that the proxyl group was significantly less polar (preferred the dodecane phase to a greater extent) than the doxyl nitroxide. As an added dividend, the ESR spectrum of the proxyl nitroxide was more sensitive to changes in polarity of the medium than that of the doxyl nitroxide (Keana *et al.*, 1976).

C. Azethoxyl Nitroxides (Minimal Perturbation Spin Labels)

One criticism of the spin-labeling method is that the steric bulk of the nitroxide may significantly perturb the system under investigation. With the aim of at least partially meeting this criticism, a new series of spin labels called azethoxyl nitroxides (**Va** and **Vb**) was developed by Lee *et al.* (1977). The synthesis (Lee and Keana, 1978) of azethoxyl nitroxides parallels that of the proxyl nitroxides except that a different nitrone is used as the starting point. Reaction of nitrone **XCVII** with an organometallic reagent followed by air oxidation gives the new nitrone **XCVIII**, in which the carbon–nitrogen double bond now involves the other α-carbon atom. Addition of a second organometallic reagent followed by oxidation affords a mixture of *cis*- and *trans*-azethoxyl nitroxides (**Va** and **Vb**).

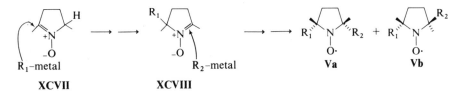

The trans isomer **Vb** normally predominates in this sequence, probably because the preferred pathway (for steric reasons) for the second addition of the organometallic reagent is from the side opposite the bulky R_1 group. By starting with yet another nitrone(**XCIX**), the cis isomer **Va** can be made to predominate [reaction (5)]. The *cis*- and *trans*-azethoxyl nitroxides are separated from each other most easily by (a) conversion to the *N*-acetoxy derivatives by hydrogenation in THF over 10% Pd/C followed by an acetyl chloride–triethylamine quench; (b) silica gel chromatography; and then (c) regeneration of the nitroxides by base hydrolysis in the presence of air.

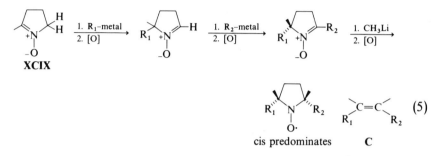

$$ \text{cis predominates} \qquad \mathbf{C} \tag{5} $$

    An inspection of molecular models indicates that the orientation of the two R groups of the cis isomer in the vicinity of the pyrrolidine ring resembles a cis carbon–carbon double-bond **C**. Thus, these molecules may be viewed as an analogue of a cis monounsaturated hydrocarbon chain with a nitroxide group at the position of the double bond. Interesting information may be forthcoming from these labels concerning the motion probably experienced by a hydrocarbon chain in the region of a cis double bond.
    The models indicate that the trans isomer, on the other hand, resembles a normal hydrocarbon chain quite closely. The nitroxyl nitrogen atom is, in fact, part of the chain. Direct evidence showing the smaller steric size of the azethoxyl labels comes from the observation (Lee, 1977; Lee *et al.*, 1978) that several *cis*- and *trans*-azethoxyl nitroxides can be included in the tubular cavities of thiourea crystals whereas doxyl and proxyl nitroxides do not fit.
    The chemical stability of azethoxyl nitroxides is comparable to that of the proxyl nitroxides except that azethoxyl nitroxides are much more resistant to reduction with phenylhydrazine and other reagents (see Section III). Quite possibly, this increased resistance to reduction will enable spin-labeling experiments to be performed under conditions in which doxyl and the more reduction-resistant proxyl (Keana *et al.*, 1976) nitroxides fail to give strong ESR signals because of reduction.
    Figure 4 shows representative azethoxyl nitroxides that have been prepared (Lee *et al.*, 1977; Lee and Keana, 1978).
    The synthetic approach to the long-chain azethoxy acids is illustrated by the route to **CIV** (R = CO$_2$H) [reaction (6)]. The key step is the displacement of iodide **CII** (R = I) with the Meyers carbanion **CVI** (Meyers *et al.*, 1974) to give oxazoline (**CVII**). Methyl iodide converts **CVII** to the methio-

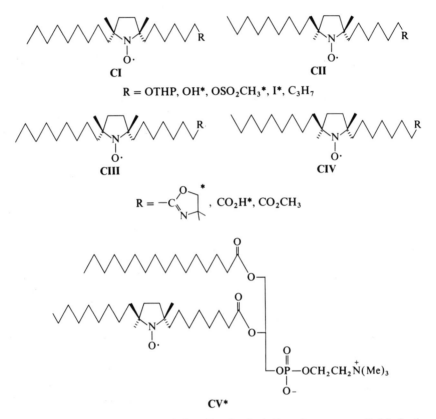

$$R = OTHP, OH^*, OSO_2CH_3^*, I^*, C_3H_7$$

$$R = -C\overset{O}{\underset{N}{\diagdown}}\overset{*}{\diagup}\!, \ CO_2H^*, CO_2CH_3$$

Fig. 4. Representative azethoxyl nitroxides. Synthetic Procedures are provided in Section VIII for those structures marked with an asterisk.

dide salt **CVIII**, which is easily hydrolyzed by sodium hydroxide to the carboxylic acid.

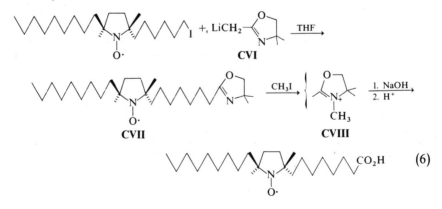

(6)

Collective experience with numerous additions of organometallic reagents to nitrones leads to two useful generalizations and one cautionary note:

(1) When proxyl or azethoxyl nitroxides are assembled by sequential addition of two different alkyl groups, higher overall yields usually result when the larger of the two alkyl groups is added first.

(2) When an organometallic reagent is added to nitrones containing a methyl group at the 2 position, Grignard reagents seem to afford better yields than organolithium reagents. When an alkyl group other than methyl is at the 2 position, organolithium reagents are preferable.

Cautionary note: 2-Methylnitrones also may undergo self-condensation followed by addition of the organometallic reagent, leading to nitroxides that can be mistaken for the desired one.

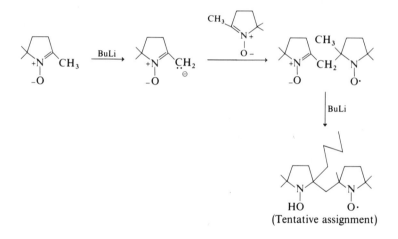

(Tentative assignment)

D. Imidazolidine-Derived Nitroxides

In addition to a doxyl nitroxide synthesis, there are three other general methods for rigidly attaching a stable nitroxide free radical to a molecule at the site of a ketone group via a spiral linkage. These are the synthesis of imidazolidine-derived nitroxides discussed in this section and the synthesis of Δ^3-imidazoline-derived nitroxides and tetrahydrooxazine nitroxides discussed in succeeding sections.

Imidazolidine nitroxides can be readily prepared by condensation of a ketone with 2,3-dimethyl-2,3-diaminobutane (**CIX**) to give the imidazolidine **CX**, which can be oxidized with MCPA to the mononitroxide **CXI** (Keana *et* *al.*, 1978b) [reaction (7)]. Overoxidation, leading to a dinitroxide (see Section VII), was not a problem.

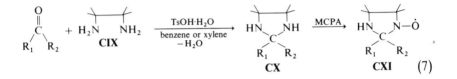

$$\qquad\qquad (7)$$

All the nitroxides **CXI** prepared so far are nicely crystalline substances. These together with their precursor imidazolidines are shown in Fig. 5.

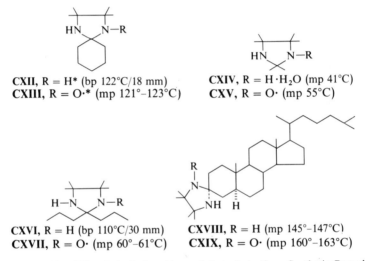

Fig. 5. Imidazolidine-derived nitroxides and their derivatives. Synthetic Procedures are provided in Section VIII for those structures marked with an asterisk.

Note that these nitroxides contain an amino group, which might be used to attach these nitroxides to another molecule. Preliminary experiments, however, indicated that the remaining amino function is quite unreactive toward alkylating or acylating agents, undoubtedly due both to steric hindrance and to the proximal nature of the transannular nitroxide group, rendering the amine less nucleophilic (does not react with excess MCPA, for example).

Another imidazolidine nitroxide(**CXX**) has been prepared [reaction (8)] (J. F. W. Keana and S. Magnuson, unpublished results) which should more readily allow covalent bond formation at the free amino group. These possibilities are under active investigation in the author's laboratory.

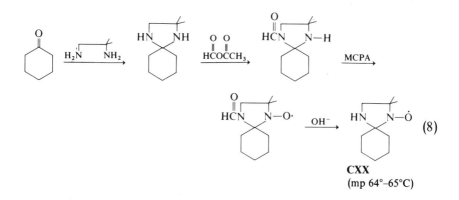

(8)

CXX
(mp 64°–65°C)

E. Δ³-Imidazoline-Derived Nitroxides

This series of stable nitroxides has been developed over the past several years in the USSR in L. B. Volodarskii's laboratory. Condensation of acetone, for example, with 1 equivalent of 2-hydroxyamino-2-methylbutanone (**CXXI**) in the presence of ammonia or ammonium acetate leads to the *N*-hydroxy-Δ³-imidazoline **CXXII**, which undergoes smooth oxidation with PbO_2 to the corresponding nitroxide **CXXIII** (Sevastyanova and Volodarskii, 1972). Other nitroxides prepared in this way include **CXXIV, CXXV,** and **CXXVI**.

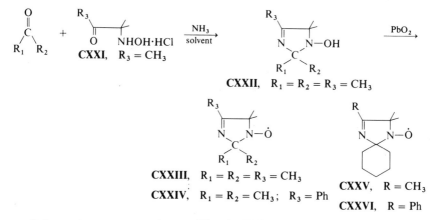

CXXII, $R_1 = R_2 = R_3 = CH_3$

CXXIII, $R_1 = R_2 = R_3 = CH_3$
CXXIV, $R_1 = R_2 = CH_3$; $R_3 = Ph$

CXXV, $R = CH_3$
CXXVI, $R = Ph$

It is pertinent to note that, unlike the $TsOH \cdot H_2O$-catalyzed condensations described above, the formation of **CXXII** takes place under mildly alkaline conditions. Thus, the method may prove especially valuable in converting ketone groups to stable nitroxide free radicals in the presence of acid-sensitive groups. Presumably, oxidizing agents milder than PbO_2 such as air also would oxidize **CXXII** to the nitroxide.

A number of functionalized nitrones based on the Δ^3-imidazoline nitroxide structure have also been reported. These include **CXXVII** and **CXXVIII** (Volodarskii *et al.*, 1971); **CXXIX** (Grigorev and Volodarskii *et al.*, 1973); **CXXX** and **CXXXI** (Volodarskii *et al.*, 1974); and **CXXXII** (Grigorev *et al.*, 1976).

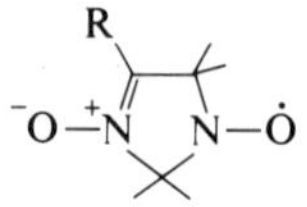

CXXVII, $R = CONH_2$

CXXVIII, $R = CO_2H$

CXXIX, $R = CH_3CO$

CXXX, $R = CH_2Br$

CXXXI, $R = PhCH{=}CH-$

CXXXII, $R = CH{=}NOH$

It should be noted that the above abbreviated listing contains nitroxides that are alkylating agents and some that are acylating agents. The high polarity of the nitrone grouping and its proximity to the nitroxide may lead to new spin labels in which the nitroxide moiety should clearly prefer an aqueous phase over a hydrocarbon-like phase. Moreover, unlike those of the nitronyl nitroxides (see Section IV,F), the ESR spectra of the Volodarskii nitrone nitroxides should consist essentially of three lines.

F. Δ^1-Imidazoline-Derived Nitroxides (Nitronyl Nitroxides)

A general method for converting aldehydes to stable nitronyl nitroxide free radicals has been developed and the chemistry of these nitroxides extensively exploited in the laboratory of E. F. Ullman (for an excellent summary paper, see Ullman *et al.*, 1972). The method consists of a condensation of the aldehyde with 2,3-hydroxyamino-2,3-dimethylbutane followed by oxidation of the resulting di-*N*-hydroxy intermediate with PbO_2 or $NaIO_4$ [reaction (9)]. Several of the nitronyl nitroxides prepared to date are listed under structure **XLVII** above.

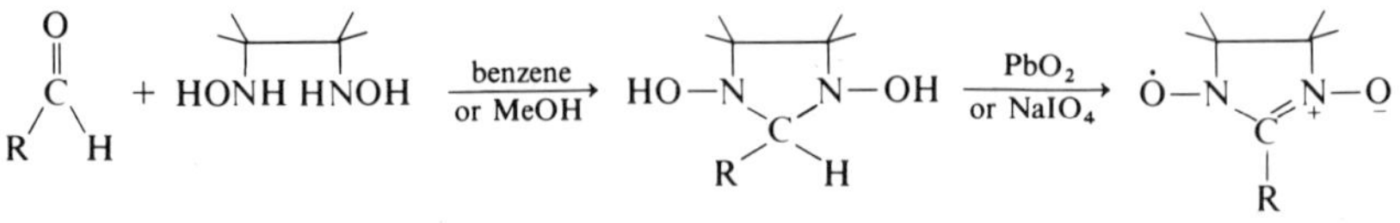

Treating nitronyl nitroxides with triphenylphosphine in benzene or with nitrous acid leads to 4,4,5,5-tetramethylimidazoline-1-oxyls in high yield [reaction (10)] (Ullman *et al.*, 1970). Several such nitroxides prepared by the Ullman group are shown under structures **XLVIII** above.

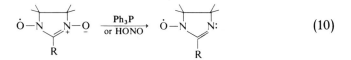

$$(10)$$

One limitation of the nitronyl nitroxide series as spin labels is the rather complex nature of the ESR spectra of the radicals. Typically, one observes a five-line spectrum caused by splitting of two equivalent nitrogen atoms. Each line may be further split by coupling to protons of the substituent attached to the 2 position.

Weinkam and Jorgensen (1971a) utilized the Ullman procedure to synthesize two free-radical analogues (**CXXXIII** and **CXXXIV**)of histidine [reaction (11)]. These radicals were then used to study the conformation of the carboxyl terminal region of angiotensin II polypeptide analogues (Weinkam and Jorgensen, 1971b).

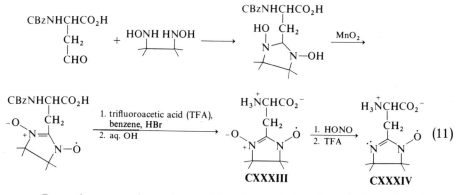

$$(11)$$

Recently, a new nitronyl nitroxide spin-labeled analogue of nicotinamide adenine dinucleotide was described by Abdallah *et al.*, (1977).

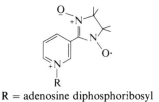

R = adenosine diphosphoribosyl

G. Tetrahydrooxazine-Derived Nitroxides

A variation on the doxyl nitroxide synthesis was reported by Rassat and Rey (1974a). Instead of using the amino alcohol **LXXV**, they prepared 2-amino-2-methylpentan-4-ol(**CXXXV**) and allowed it to react with cyclohexanone, producing the tetrahydrooxazine **CXXXVI**. Oxidation of the latter with MCPA gave the nitroxide **CXXXVII**. An alternative route to tetrahydrooxazine nitroxides that parallels the conversion **LXXXII → LXXXIII** was described by Lee and Keana (1976).

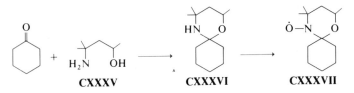

| | **CXXXV** | **CXXXVI** | **CXXXVII** |

  (+)-Pulegone was converted to the four amino alcohols **CXXXVIII–CXLI** (actually seen to be derivatives of **CXXXV**), which in turn were allowed to react with several ketones, producing a new series of bicyclic and polycyclic nitroxides, two of which are shown below (Rassat and Rey, 1974a). One notes that the nitroxide group is an integral part of the bicyclic system, and therefore the motion of the ring system would be expected to be reflected quite accurately by the motion of the nitroxide moiety, especially when the rings are trans fused.

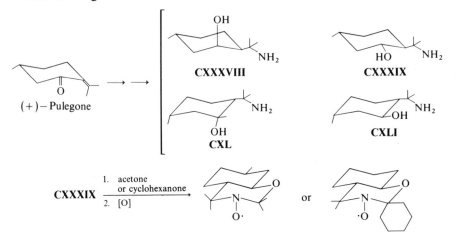

H. Other Nitroxides in Which the Nitroxyl Nitrogen Is Part of the Skeleton

  A potentially general method for converting an appropriately substituted six-membered ring lactam to a stable nitroxide was described by Ramasseul and Rassat (1971) with the synthesis of steroid nitroxide **CXLII**. Their route is shown in reaction (12). Note that the two double bonds of the allyl groups were not epoxidized during the MCPA oxidation step. An ESR study of the interaction of steroid **CXLII** with bovine serum albumin has been described (Chambaz *et al.*, 1971).

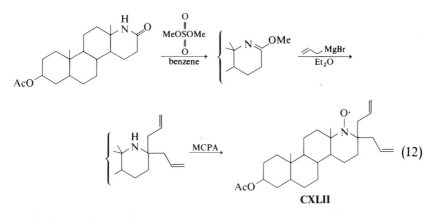

CXLII

An interesting series of isoquinuclidine nitroxides was synthesized by Rassat and Rey (1972). They allowed the readily available (Beereboom, 1966) piperitenone **CXLIII** to react with ammonia, producing ketone **CXLIV** in 89% yield. This substance could be reduced to the alcohols **CXLVI** and **CXLVIII** using lithium in liquid ammonia. All three substances readily afforded the corresponding stable nitroxides **CXLV**, **CXLVII**, and **CXLIX** upon oxidation with H_2O_2–phosphotungstic acid.

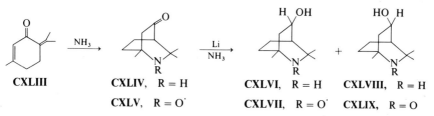

CXLIII	**CXLIV,** R = H	**CXLVI,** R = H **CXLVIII,** R = H
	CXLV, R = O·	**CXLVII,** R = O· **CXLIX,** R = O

Even though the nitroxides are fully substituted on the α-carbon, their ESR spectra nevertheless show additional splitting due to long-range coupling to distant protons. For example, the ESR spectrum of **CXLV** consists of a triplet ($a_N \cong 16$ Oe) of triplets ($a_H \cong 3.5$ Oe) (Rassat and Rey, 1972).

Rassat and Rey (1971) prepared the parent isoquinuclidene nitroxide **CLI** by a photochemical route [reaction (13)].

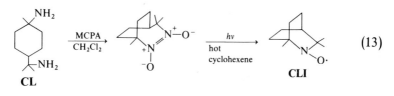

(13)

A more highly functionalized bicyclic stable nitroxide(**CLII**) was prepared by Quon *et al.* (1974) by a photochemical addition of *N*-nitroso-piperidine to α-pinene [reaction (14)].

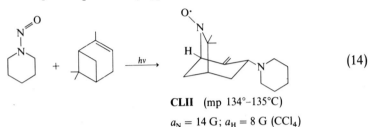

$$(14)$$

CLII (mp 134°–135°C)

$a_N = 14 \text{ G}; a_H = 8 \text{ G (CCl}_4)$

I. New Nitroxide Spin Labels I and II Derived from 2,2,6,6-Tetramethylpiperidine-*N*-oxyl and 2,2,5,5-Tetramethylpyrrolidine-*N*-oxyl

Nitroxides of general structure **I** and **II** continue to constitute the majority of nitroxides commonly used in spin-labeling studies. New derivatives of **I** and **II** continue to appear, many of which offer added versatility to the spin labeler's armament. Several recent developments are discussed in this section.

Despite the commercial availability of many of these substances, an improved method for the synthesis of the key intermediate triacetone amine **CLIII** has been published (Sosnovsky and Konieczny, 1977c). A high-yield method for the oxidation of amine **CLV** to the nitroxide **CLVI** also has been described (Sosnovsky and Konieczny, 1976).

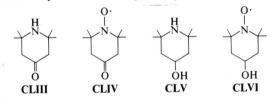

| CLIII | CLIV | CLV | CLVI |

Rauchman *et al.* (1975) provide a brief summary of the methods of oxidizing secondary amines to nitroxides. They favor an acetonitrile–methanol–hydrogen peroxide–sodium tungstate–sodium hydrogen carbonate combination for the oxidation of **CLIII** → **CLIV** and **CLVII** → **CLVIII** in > 85% yield, for example.

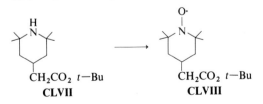

$CH_2CO_2 \ t\text{—Bu}$ $CH_2CO_2 \ t\text{—Bu}$

CLVII   **CLVIII**

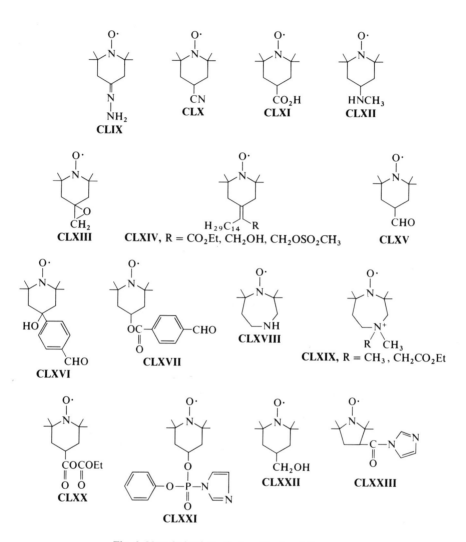

Fig. 6. New derivatives of nitroxides **I** and **II**.

Figure 6 shows several derivatives of nitroxides **I** and **II**, which have been either newly prepared or prepared by an improved route. The reader is referred to a recent review (Keana, 1978) for a more detailed discussion of the synthetic routes to these molecules and some of their applications.

The salient points regarding the spin labels shown in Fig. 6 are the following. Hydrazone **CLIX** is a new reagent designed to convert aldehydes and ketones to spin-labeled derivatives via formation of the corresponding azine derivative (Schlude, 1976). Nitroxide acid **CLXI**, mp 171°–172°C, was obtained by the reaction of tosylmethyl isocyanide with ketone **CLIV** to give nitrile **CLX** followed by hydrolysis of the latter (Rauchman *et al.*, 1976a). The synthesis of **CLXI** originally was claimed by Wong *et al.* (1974); however, in view of the work of Rauchman *et al.* (1976a) it now appears that Wong *et al.* in fact prepared the α,β-unsaturated derivative, mp 195°–196°C, of **CLXI**.

Nitroxide amine **CLXII** (Rosen, 1974) and a host of other amine derivatives (Sinha and Chignell, 1975; Rosen and Abou-Donia, 1975) of nitroxide ketone **CLIV** have been prepared by a reductive amination reaction using the appropriate amine and sodium cyanoborohydride (NaBH₃CN) (Borch *et al.*, 1971). In contrast to the familiar sodium borohydride, NaBH₃CN can be used in aqueous solutions of pH as low as 5 as well as pH 7, and the nitroxide function is not reduced. Rauchman *et al.* (1976b), however, noted that reductive aminations involving NaBH₃CN and high molecular weight amines give unsatisfactory yields in some instances, and they recommended alternative routes involving nucleophilic displacements with nitroxide amine **CLXXIV** [see reaction (15), for example].

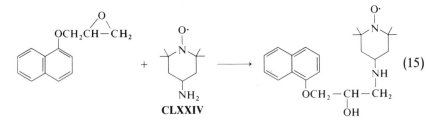

$$(15)$$

Epoxide **CLXIII** is a useful alkylating agent prepared from nitroxide ketone **CLIV** by reaction with the ylid derived from trimethylsulfonium iodide (Rauchman *et al.*, 1976c). The epoxide was prepared earlier by Schlude (1973) by the reaction of **CLIV** with diazomethane. The several nitroxides listed under structure **CLXIV** were prepared by Dvolaitzky *et al.* (1975). The first member (R = CO₂Et) was obtained by the reaction of the modified Wittig reagent $(C_2H_5O)_2POC(CO_2C_2H_5)(C_{14}H_{29})$ with **CLIV**. Subsequent members were obtained by conventional chemical methods not involving the nitroxide group.

The nitroxide aldehydes **CLXV**, **CLXVI**, and **CLXVII** were reported by Schlude (1973). The aldehydes were obtained by acid-catalyzed hydrolysis of the precursor acetals. Schlude recommended reduction of the nitroxide group to the N—OH stage before the acid hydrolysis step in order to prevent disproportionation of the nitroxide group, and then regeneration of the nitroxide with Fremy's salt. This all seems quite unnecessary in view of the several successful acid-catalyzed hydrolyses of nitroxide-bearing molecules in our own laboratory. Schlude (1975) also reported the synthesis of a number of nitroxide enol ethers by the reaction of **CLIV** with oxygenated Wittig reagents.

An alternative although somewhat lengthy preparation of amino nitroxide **CLXVIII** was described by Ramasseul *et al.* (1976). The use of the well-known β,β,β-trichloroethoxycarbonyl protecting group was illustrated in this synthesis. That amine **CLXVIII** is readily quaternized was shown by the synthesis of nitroxides **CLXIX** (Ramasseul *et al.*, 1975).

Mixed anhydride **CLXX** (Rauchman and Rosen, 1976) is another example of the class of nitroxide mixed-anhydride acylating agents introduced by Griffith *et al.* (1967) some time ago. The phosphorylating agent **CLXXI** was recommended by Sosnovsky and Konieczny (1977a,b) as the reagent of choice for the synthesis of spin-labeled phosphates and phosphoramidates. Other nitroxide spin-labeled phosphorylating and phosphonylating agents have been reviewed by Morrisett (1975) and Keana (1978).

Nitroxide alcohol **CLXXII** was synthesized by Rauchman and Rosen (1976) by reduction of authentic nitroxide acid **CLXI** with lithium aluminum hydride (LAH) in THF. The point was made that the nitroxide grouping was not reduced by the LAH.

Nitroxide acylimidazole **CLXXIII** is a new, stable, crystalline (mp 128°–129°C), racemic acylating agent that can be stored for months at 4°C without decomposition (Adackaparayil and Smith, 1977). These properties are in contrast to the unstable nature of the corresponding α,β-unsaturated derivative. Unfortunately, the properties of the acylimidazole derived from the optically active acid **LXXIX** were not described because when racemic **CLXXIII** is used with a chiral alcohol such as testosterone a diastereomeric mixture of nitroxide esters results.

V. CHEMICAL REACTIONS INVOLVING THE NITROXIDE GROUPING

A. Nitroxide Overoxidation

It is well known (or in some instances strongly suspected) that oxidizing agents such as Cl_2, Br_2 (Rozantsev, 1970), moist Ag_2O (Rassat and Rey, 1974b), or CrO_3 (J. F. W. Keana and E. M. Bernard, unpublished results) and electrochemical methods (Sümmerman and Deffner, 1975) convert stable nitroxides to the corresponding oxoammonium salts, which may then decompose via several pathways [reaction (16)].

$$\hspace{6cm} \text{etc.} \hspace{2cm} (16)$$

An oxoammonium salt

It has been observed that oxidation of amino alcohol **CLV** with MCPA leads not only to the expected nitroxide **CLVI** but also to ketone nitroxide **CLIV** (Cella *et al.*, 1974, 1975a; Ganem, 1975). The reaction is thought to proceed by reaction of the oxoammonium salt with the alcohol [reaction (17)].

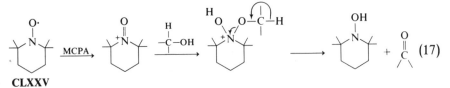

CLXXV

The relevance of the reaction to spin-label synthesis is the realization that a given MCPA oxidation may not be straightforward. As suggested by Cella *et al.* (1975b), overoxidation of the nitroxide by MCPA during a doxyl nitroxide synthesis (Keana *et al.*, 1967) is the likely reason that yields are typically only about 30% in this reaction.

Interestingly, pyrrolidine nitroxides (presumably including proxyl and azethoxyl nitroxides) apparently do not undergo overoxidation with MCPA (Cella *et al.*, 1975a). Confirmation of this statement is seen in the work of Kavanagh-Caron *et al.* (1976), in which an epoxidation is performed with MCPA on an unsaturated pyrrolidine intermediate.

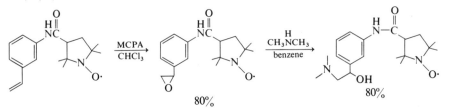

An efficient oxidative cleavage of a doxyl nitroxide back to the parent ketone was achieved by treatment of the doxyl nitroxide with commercial NO (containing NO_2) in cyclohexane or ethanol solvent (Nelson *et al.*, 1971; Chou *et al.*, 1974). The active oxidant is the NO_2. The reaction is thought to proceed by oxidation of the nitroxide group to the oxoammonium intermediate, followed by a fragmentation reaction.

B. Nitroxide Reduction

Stable nitroxide free radicals are able to oxidize strongly basic carbanions to the corresponding free radicals, which can then undergo typical free-radical reactions. In doing so, the nitroxide of course suffers reduction, i.e., a gain of one electron. Thus, the reaction of Grignard reagents with nitroxide **CLXXV** leads to the products shown in reaction (18) (Sholle *et al.*, 1971).

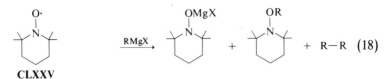

CLXXV

$R = Ph, n\text{-}C_6H_{13}, C_2H_5, CH_3$

Harcus et al. (1976) reported that lithium derivative **CLXXVI** reduced the nitroxide group of **CLIV** rather than add to the carbonyl group. In our laboratory (J. F. W. Keana and L. E. LaFleur, unpublished results), we have observed similar complications in reactions involving acetylene anions and nitroxides.

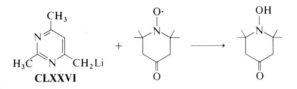

CLXXVI

The rapid and essentially quantitative reaction of nitroxide **CLXXV** with butyl lithium in hexane at $-70°C$ was examined carefully by Whitesides and Newirth (1975). The products after a quench with dimethyl sulfate are as shown below.

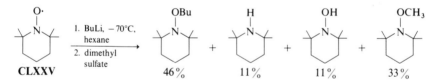

CLXXV

1. BuLi, $-70°C$, hexane
2. dimethyl sulfate

46%   11%   11%   33%

It is clear from the above discussion that one should not expect to observe the usual clean organometallic addition reactions to carbonyl functional groups of molecules that also bear a nitroxide group. As a way of circumventing the problem, it should be feasible, however, to reduce the nitroxides to the N-hydroxy group with 10% Pd/C in THF, for example (see **LXXIVa** and **LXXIVb** above), and then proceed with the addition of excess organometallic reagent, 1 equivalent forming the salt at the N—OH grouping and the remainder free to react in the usual way.

One also could consider prior reduction of the nitroxide to the amine stage using sodium sulfide (Kornblum and Pinnick, 1972) before proceeding with the organometallic addition. In this case, reoxidation to the nitroxide requires a bit more than just air (Rauchman et al., 1975).

For the reductive destruction of a nitroxide during spin-labeling studies ascorbic acid has been used extensively (Hubbell and McConnell, 1969;

Kornberg and McConnell, 1971; Yu *et al.*, 1974; Setaka and Lagercrantz, 1975; Busby *et al.*, 1976). Paleos and Dais (1977) developed the ascorbic acid nitroxide reduction reaction into an analytical method for nitroxides. Daniel and Cohn (1976), on the other hand, used phenylhydrazine instead of ascorbic acid to destroy the spin label since ascorbic acid obscures the methyl region in the NMR spectrum of tRNA, the system under study.

Sulfhydryl groups have been implicated in several spin-labeling studies in which ESR signal loss is observed as a result of reductase activity present in the preparation (Giotta and Wang, 1972; Baldassare *et al.*, 1974). Other reductase activity also may lead to ESR signal loss (Maruyama and Ohnishi, 1974; Butterfield *et al.*, 1975).

In one study, reduction of the nitroxide spin label occurred by passage of a solution through a stainless steel syringe needle. The use of platinum needles and Teflon-tipped plungers was recommended (Wenzel *et al.*, 1977). In our laboratory, however, we routinely have used conventional syringe equipment without untoward effects on the nitroxide spin labels.

The relative rates of reduction of 12-doxylstearic acid, 14-proxylstearic acid, and *trans*-azethoxylstearic acid (**CIV**, $R = CO_2H$) toward sodium ascorbate and dithiothreitol were determined (Lee *et al.*, 1977, 1978). The nitroxides ($1.1 \times 10^{-4}\ M$) were dissolved in 0.1 M phosphate buffer, pH 7.5, containing sucrose (0.25 M), EDTA ($10^{-3}\ M$) and either sodium ascorbate (0.011 M) or dithiothreitol (0.11 M). After the nitroxides had been dissolved for 20 min in the ascorbate solution, only 5% of the original ESR signal intensity remained for the doxyl nitroxide, whereas 88% remained for the proxyl nitroxide and 93% remained for the azethoxyl nitroxide. After the nitroxides had been dissolved the same period in the dithiothreitol solution, 76% of the original signal intensity remained for the doxyl nitroxide, whereas 93% remained for the other two. Thus, it is clear that the doxyl nitroxide is much more susceptible to reduction than proxyl or azethoxyl nitroxides, and the latter two behave quite similarly.

It may be worth emphasizing here that these one-electron reductions, whether chemically or biochemically affected, of proxyl and azethoxyl nitroxides to the corresponding N-hydroxy compounds normally can be easily reversed by the addition of fresh oxidizing agent (for example, air or Cu^{2+}). This has the effect of regenerating the original ESR signal.

With doxyl nitroxides, however, the analogous one-electron reduction leads to an N-hydroxyoxazolidine. It has been observed, on a limited basis, in the author's laboratory that the latter compounds tend to undergo an essentially irreversible hydrolysis back to the ketone from which the original doxyl nitroxide was derived. Like ketals, the N-hydroxyoxazolidines are susceptible to hydrolysis because of the presence of two heteroatoms (O and N) with unshared electrons bonded to the same carbon atom.

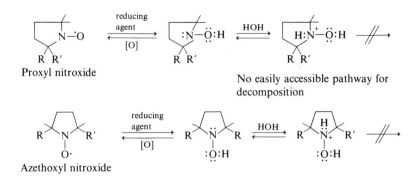

Proxyl nitroxide

No easily accessible pathway for decomposition

Azethoxyl nitroxide

One possible abbreviated hydrolysis mechanism is shown below. The hydrolysis can be safely predicted to occur more readily at acidic or neutral pH than at alkaline pH. It might be added that preliminary attempts at synthesizing N-hydroxyoxazolidines by reaction of ketones with 2-hydroxy-amino-2-methylpropanol under several sets of conditions were not successful (J. F. W. Keana, unpublished results). Nevertheless, it has been possible to generate N-hydroxyoxazolidines by other routes and to study them spectroscopically (Lee and Keana, 1975, 1976).

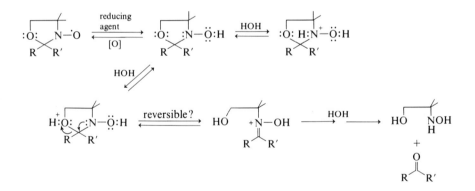

C. Nitroxide Photolysis and Thermolysis

Selective photolysis of a nitroxide spin label incorporated into a biological sample with long-wavelength ultraviolet–visible light would be expected to lead to destruction of the spin label. The rate of the reaction would depend on the structure of the nitroxide (Keana *et al.*, 1971; Keana and Baitis, 1968; Call and Ullman, 1973). The reader is referred to a recent review (Keana, 1978) for more details regarding the photolysis and thermolysis of nitroxides.

VI. CHEMICAL REACTIONS NOT INVOLVING THE NITROXIDE GROUPING

At the heart of this section is the wealth of information regarding the covalent attachment of small nitroxide molecules to other, usually much larger molecules. The many spin-labeling reviews cited by Gaffney (1976) and Keana (1978) cover the topics well. Only a few additional important points should be made here.

As noted above, many conventional oxidizing agents attack the nitroxide grouping. The following examples, however, illustrate the use of several oxidizing agents that apparently do not affect the nitroxide grouping. Dupeyre and Rassat (1975) reported that alkaline potassium permanganate converts several tertiary and secondary amines to the stable nitroxide free radical. Since MnO_2 is a by-product of the reaction, nitroxides are also stable to this reagent (oxidizes allylic alcohols selectively).

Keana et al. (1976) oxidized proxyl alcohol **CLXXVI** to aldehyde **CLXXVII** using the N-chlorosuccinimide dimethyl sulfide reagent of Corey and Kim (1972). The aldehyde then was oxidized to 7-proxylstearic acid(**CLXXVIII**) with Ag^+ ion (Tollen's reagent).

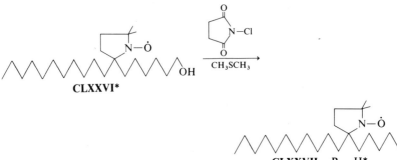

CLXXVI*

CLXXVII, R = H*

CLXXVIII, R = OH*

A reagent that we have found useful in our laboratory (Keana et al., 1978a) for the selective reduction of nitroxide carboxylic acids to nitroxide alcohols is borane–methyl sulfide (BMS) (Aldrich Co.). Unlike $LiAlH_4$, which reduces most other functional groups more readily than a carboxyl group, BMS reduces free carboxyl groups most rapidly (Lane, 1975).

VII. DINITROXIDE SPIN LABELS

The use of biradicals as spin probes was reviewed by Luckhurst (1975). Another recent review (Keana, 1978) provides references for several molecules that contain two (or more) nitroxide groups. Most of these are prepared by attaching a nitroxide-containing residue such as **I** or **II** to a polyfunctional derivative. Others are prepared by attaching doxyl groups to a molecule containing two or more ketone groups (see, for example, Metzner *et al.*, 1977). Still others, such as **XXVIII** and **L**, are prepared by multistage sequences.

Two general methods are available for the relatively rigid attachment of a dinitroxide spin label to a molecule at the site of a ketone group. The first involves condensation of diamine alcohol **CLXXIX** with the ketone in xylene followed by oxidation of the resulting diamine **CLXXX** with MCPA to give the dinitroxide **CLXXXI** (Keana and Dinerstein, 1971). J. Michon and A. Rassat (1974) prepared other derivatives bearing this label and used one in a spin-label study of cyclodextrin inclusion compounds (Martinie *et al.*, 1975).

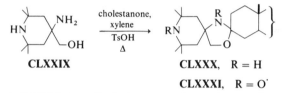

CLXXIX   CLXXX, R = H

CLXXXI, R = O·

Keana *et al.* (1978b) described the route shown in reaction (19) to two dinitroxide ketone spin labels (**CLXXXII** and **CLXXXIII**) in which the nitroxide groups are separated by only one carbon atom. These dinitroxides show an enormous dipolar splitting constant, $2D = 1606$ G.

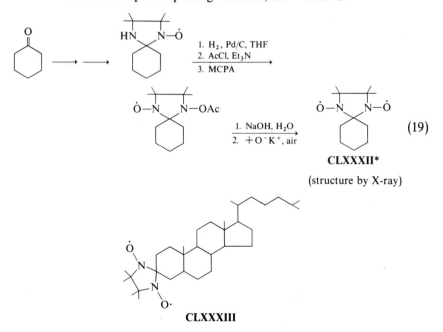

CLXXXII*

(structure by X-ray)

CLXXXIII

VIII. EXPERIMENTAL PROCEDURES FOR THE SYNTHESIS OF SELECTED PROXYL AND IMIDAZOLIDINE-DERIVED NITROXIDE SPIN LABELS

All reagents were routinely purified by recrystallization or distillation before use. Solvents were routinely dried and freshly distilled before use.

A. 2-Undecyl-5,5-dimethyl-Δ¹-pyrroline N-oxide(**CLXXXIV**) (Keana *et al.*, 1976)

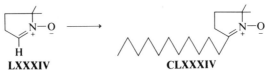

LXXXIV   CLXXXIV

A 1.0 *M* THF solution (46 ml) of the Grignard reagent prepared from undecyl chloride (K & K Chemical Co.) was added dropwise to a refluxing (N₂), stirred solution of freshly distilled nitrone **LXXIV** (Aldrich Co.) (3.52 gm) dissolved in THF (10 ml). After a 1-hr reflux period, the reaction was quenched with several drops of saturated aqueous NH₄Cl. The mixture was filtered, and the filtrate was evaporated. The crude *N*-hydroxy intermediate was taken up in MeOH–concentrated NH₄OH (15 : 2, 170 ml) containing cupric acid monohydrate (12.9 gm) and stirred under air for 1 hr. The resulting blue solution was concentrated under vacuum to 50 ml, and then chloroform (50 ml) was added. The solution was washed with saturated aqueous NaHCO₃ (50 ml) and dried (anhydrous K₂CO₃). Concentration followed by vacuum distillation afforded undecylnitrone **CLXXXIV** as a pale-yellow oil, bp 118°–121°C/0.005 mm (7.07 gm, 85%); *m/e* 267.257 (calculated, 267.256).

B. 7-Proxylstearyl alcohol(**CLXXVI**) (Keana *et al.*, 1976)

CLXXVI

6-Chlorohexanol (Aldrich Co., 11.4 gm) and 2,3-dihydropyran (9.59 gm) were combined (N₂) in dry CH₂Cl₂ (100 ml) containing 5 mg of *p*-toluenesulfonic acid monohydrate. The solution was stirred at 25°C for 1 hr and then washed with 1 *N* NaOH (50 ml), dried (anhydrous K₂CO₃), and then concentrated. Vacuum distillation gave the tetrahydropyranyl ether as a colorless oil, bp 100°–105°C/0.75 mm (18.22 gm, 94%).

A 1.0 *M* THF solution of the Grignard reagent was prepared from the above chloride ether. [Grignard formation proceeds best when the chloride is added to the refluxing THF–Mg mixture over several hours (Keana and

LaFleur, 1978) (see p. 158).] A 2.8-ml aliquot was added to a THF solution (10 ml) (N$_2$) of nitrone **CLXXXIV** (500 mg). After 18 hr at 25°C, the reaction was quenched at 0°C by dropwise addition of 15 ml of a solution of MeOH and 3 N HCl (2 : 1). After 1.5 hr at 0°C, the quenched mixture was diluted with ether (25 ml) and water (25 ml). The organic layer was washed with water (2 × 25 ml) and saturated NaHCO$_3$ (25 ml) and then evaporated. The resulting oil was taken up in MeOH (10 ml) containing cupric acetate monohydrate (5 mg) and stirred under air for 1 hr. This was diluted with ether (25 ml), washed with saturated NaHCO$_3$ (10 ml), dried (anhydrous K$_2$CO$_3$), and evaporated. Preparative thin-layer chromatography (TLC) (Analtech silica gel GF, elution with ether) gave 7-proxylstearyl alcohol as a yellow oil (154 mg, 22%); m/e 368.354 (calculated, 368.355); R$_f$ (ether) 0.42.

C. 7-Proxylstearyl alcohol methanesulfonate (**CLXXXV**) (Keana et al., 1976)

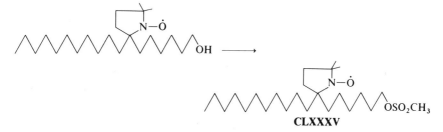

CLXXXV

The following procedure is essentially that of Crossland and Servis (1970). Proxyl alcohol **CLXXVI** (107 mg) and triethylamine (44 mg) were combined in dry CH$_2$Cl$_2$ (2.0 ml). The stirred solution (N$_2$) was cooled to −25°C, and then 0.32 ml of a 1.0 M methanesulfonyl chloride–CH$_2$Cl$_2$ solution was added. The cooling bath was removed and stirring was continued for 1 hr. The solution was diluted with CH$_2$Cl$_2$ (10 ml), washed with water (10 ml), dried (Na$_2$SO$_4$), and evaporated. Preparative TLC (Analtech silica gel GF, elution with hexane–ether, 2 : 1) gave methanesulfonate **CLXXXV** as a yellow oil (118 mg, 91%); R$_f$ (ether) 0.49. This substance was used immediately for the synthesis of salt **CLXXXVI**.

D. 7-Proxylstearyltrimethylammonium methanesulfonate (**CLXXXVI**) (Keana et al., 1976)

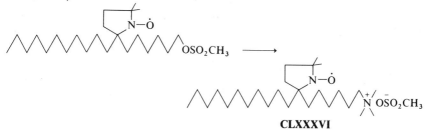

CLXXXVI

Proxyl methanesulfonate **CLXXXV** (54 mg) was dissolved in THF (0°C) (3.0 ml), and the solution was saturated with trimethylamine (gas). The solution then was placed in a sealed pressure reactor and heated at 110°C for 1 hr. The solution was evaporated, and the product was recrystallized from THF–ether, affording quaternary salt **CLXXXVI** as orange cubes (48 mg, 78%); mp 88.5°–89.5°C. *Analysis*: Calculated for $C_{27}H_{57}N_2O_4S \cdot \frac{1}{2}H_2O$: C, 62.99; H, 11.36; N, 5.44. Found: C, 63.10; H, 11.44; N, 5.23.

E. 7-Proxyloctadecanal(**CLXXVII**) (Keana *et al.*, 1976)

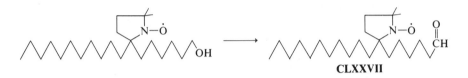

CLXXVII

The method of oxidation was that of Corey and Kim (1972). To a stirred mixture (N_2) of *N*-chlorosuccinimide (45 mg) in dry toluene (1.2 ml) was added at 0°C a 1.0 *M* solution of methyl sulfide in toluene (1.2 ml). This was cooled to $-25°C$, and then 7-proxylstearyl alcohol **CLXXVI** (83 mg) dissolved in toluene (1.0 ml) was added via syringe. After 3 hr at $-25°C$, triethylamine (34 mg) in toluene (1.0 ml) was added, and the solution was allowed to warm to 25°C. It then was diluted with ether (10 ml), washed with water (3 × 10 ml), dried (anhydrous K_2CO_3), and evaporated. Preparative TLC (silica gel, elution with ether–hexane, 1 : 1) gave aldehyde **CLXXVII** as a yellow oil (64 mg, 78%); *m/e* 366.338 (calculated, 366.337); R_f (ether) 0.63.

F. 7-Proxylstearic acid(**CLXXVIII**) (Keana *et al.*, 1976)

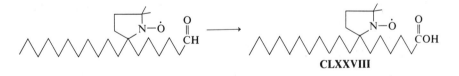

CLXXVIII

Proxyl aldehyde **CLXXVII** (54 mg) was dissolved in MeOH (2.0 ml). The solution was protected from light and stirred under air while Tollen's reagent (Roberts *et al.*, 1974) (0.5 ml) was added. After 3 hr at 25°C, ether (5 ml) and ice-cold 0.1 *N* HCl (5 ml) were added. The ether layer was washed with cold 0.1 *N* HCl (2 × 5 ml) and then water (2 × 5 ml), dried (MgSO₄), and evaporated. Preparative TLC (silica gel, elution with CHCl₃–MeOH, 95 : 5) gave acid **CLXXVIII** as a yellow oil (39 mg, 70%); *m/e* 382.330 (calculated, 382.332).

G. 2,5-Dimethyl-Δ¹-pyrroline N-oxide(**XCVII**) (T. D. Lee and J. F. W. Keana, unpublished results)

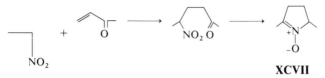

XCVII

To a solution of nitroethane (28.5 gm) in 150 ml of methanol was added a solution of sodium methoxide (made from 6.9 gm of sodium in 75 ml of methanol). To the resulting mixture was added dropwise with stirring over a 10-min period 21.1 gm of methyl vinyl ketone. After the flask had cooled to room temperature, the solution was acidified with 30 ml of acetic acid, and the solvent and excess nitroethane were evaporated. The residue was treated with a mixture of ether and water. The ether phase was separated, washed several times with 10% aqueous Na_2CO_3 solution and then brine, and dried over $MgSO_4$. The solvent was evaporated, and the resulting dark oil was distilled, affording 25.7 gm (59%) of the nitro ketone as a pale-yellow oil, bp 115°–116°C/12 mm.

The nitro ketone (25.7 gm), NH_4Cl (9.5 gm), and water (150 ml) were combined in a 300-ml three-necked flask fitted with an overhead stirrer and thermometer. Zinc dust (46.3 gm) was added in small portions over a period of 1.5 hr such that the temperature inside the flask did not exceed 10°C. The mixture was stirred an additional 0.5 hr and then was filtered. The filtrate was combined with several methanol washes of the filter cake. The solvent volume was reduced by evaporation to 50 ml. The solution was saturated with borax and then extracted several times with chloroform. The chloroform extracts were combined, dried over K_2CO_3, and evaporated. The resulting yellow oil was distilled to give 13.5 gm (68%) of nitrone **XCVII** as a colorless oil, bp 98°–100°C/5 mm.

H. 2,5-Dimethyl-5-nonyl-Δ¹-pyrroline N-oxide(**XCVIII**, R = C₉H₁₉) (Lee *et al.*, 1977)

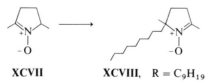

XCVII   **XCVIII**, R = C₉H₁₉

To 100 ml of a 1.0 M nonyl magnesium bromide solution in ether was added with stirring 5.65 gm (50.0 mM) of nitrone **XCVII** in 30 ml of ether at a rate sufficient to maintain gentle reflux. The solution was stirred an additional 30 min at 21°C and then treated with an amount of saturated aqueous

NH_4Cl sufficient to collect the precipitated aqueous salts in a mass at the bottom of the flask. The ether layer was decanted and combined with two ether washings of the aqueous residue. The solvent was evaporated to yield a yellow oil, which was taken up in 50 ml of CH_3OH and 5 ml of concentrated aqueous NH_4OH and stirred with 1 gm of $Cu(OAc)_2 \cdot H_2O$ under O_2 until the solution developed a deep-blue color. The solution was diluted with ether and H_2O. The ether layer was separated and combined with three ether extractions of the aqueous phase; washed with H_2O, saturated aqueous Na_2CO_3 and brine; dried over K_2CO_3; and evaporated to give a brown oil. Distillation at 0.005 mm gave 5.43 gm (45%) of **XCVIII** ($R = C_9H_{19}$): bp 100°–109°C; IR (CCl_4), 1595 cm^{-1} (C=N); NMR, δ 0.88 (3H, m, terminal Me), 1.38 (3H, s, ring Me), 2.02 (3H, t, $J = 2$ Hz, N=C—Me), 2.57 (2H, m, N=C—CH$_2$); mass spectrum, 239.225 (31) (calculated for $C_{15}H_{29}NO$, 239.255), 222 (34), 113 (68), 96 (75), 73 (27), 55 (28), 45 (53), 43 (100), 41 (36).

I. 8-Azethoxyloctadecanol tetrahydropyranyl ether [*mixture of cis and trans isomers* **CI** ($R = OTHP$) *and* **CII** ($R = OTHP$)] (Lee et al., 1977)

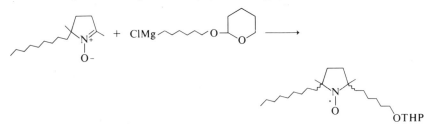

To 601 mg of magnesium turnings was added with stirring 5 ml of a 25-ml THF solution of 5.52 gm of 6-chlorohexanol tetrahydropyranyl ether. The mixture was heated to reflux, and 50 mg of 1,2-dibromoethane was added to start the reaction. The heat source was removed, and the remainder of the above THF solution was added dropwise with stirring over a period of 1 hr. Stirring was continued an additional 3 hr. The Grignard reagent was transferred via syringe to a clean, dry flask (N_2) and at 21°C with stirring was added dropwise a solution of 2.39 gm of nitrone **XCVIII** ($R = C_9H_{19}$) in 10 ml of THF. After the addition was complete ($\sim$ 15 min), saturated aqueous NH_4Cl was added. The ether layer was decanted and combined with an ether washing of the residue. The solvent was evaporated. The residue was taken up in 100 ml of CH_3OH and stirred vigorously with 30 mg of $Cu(OAc)_2 \cdot H_2O$ for 30 min, and the solvent was evaporated. Silica gel chromatography of the residue eluting with $CHCl_3$ gave 0.929 gm (22%) of the mixture of **CI** ($R = OTHP$) and **CII** ($R = OTHP$) as a yellow oil sufficiently pure for the next reaction. An analytical sample was prepared by preparative TLC on silica gel (ether, R_f 0.7); mass spectrum, 424.380 (6) (calculated for $C_{26}H_{50}NO_3$, 424.379), 352 (6), 340 (10), 282 (22), 240 (100), 224 (68), 214 (66), 198 (22), 85 (100), 73 (20), 69 (26), 57 (24), 55 (33), 43 (26), 41 (30).

J. *8-Azethoxyloctadecanol* [*mixture of cis and trans isomers* **CI** (*R = OH*) *and* **CII** (*R = OH*)] (Lee *et al.*, 1977)

To a 50-ml CH_3OH solution of 0.929 gm (2.19 mM) of the mixture of **CI** (R = OTHP) and **CII** (R = OTHP) was added 70 mg of *p*-toluenesulfonic acid. The solution was left standing for 2 hr at 21°C and then diluted with ether and H_2O. The ether phase was washed with H_2O and brine, dried over Na_2SO_4, filtered, and evaporated to give a yellow oil. This was taken up in $CHCl_3$ and put on a 2 × 20 cm dry silica gel column. The column was eluted with 200 ml of $CHCl_3$ followed by 200 ml of ether. The ether portion was evaporated to give 0.569 (76%) of the mixture of *cis*- and *trans*-alcohols **CI** (R = OH) and **CII** (R = OH), which showed a single spot on silica gel TLC (ether, R_f 0.4).

K. *cis- and trans-8-azethoxyl-1-iodooctadecane* (**CI**, R = I; *and* **CII**, R = I) (Lee *et al.*, 1977)

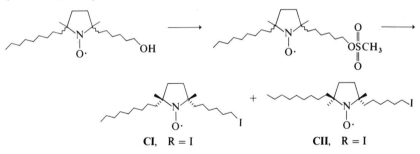

To a stirred solution of 320 mg (0.94 mM) of the mixture of *cis*- and *trans*-alcohols **CI** and **CII** (R = OH) and 162 mg (1.60 mM) of triethylamine in 5 ml of dry CH_2Cl_2 cooled to −20°C (N_2) was added 138 mg (1.20 mM) of methanesulfonyl chloride via a syringe. The mixture was allowed to warm to 25°C, diluted with cyclohexane (15 ml), and filtered. Evaporation of the solvent gave the crude methanesulfonyl esters, which were converted to a mixture of *cis*- and *trans*-iodides by treatment with 300 mg (2.0 mM) of sodium iodide in refluxing methyl ethyl ketone for

15 min. The reaction mixture was diluted with 30 ml of cyclohexane and filtered, and the solvent was evaporated, affording a brown oil. This was taken up in $CHCl_3$ and put on a 3 × 30 cm dry silica gel column and eluted with $CHCl_3$. The first 50 ml of eluant was discarded. The next 150 ml gave 249 mg of pure *trans*-iodide **CII** (R = I) as a yellow, waxy solid: Single spot by silica gel TLC ($CHCl_3$, R_f 0.4); mass spectrum, 450.221 (9) (calculated for $C_{21}H_{41}NOI$, 450.223), 324 (100), 308 (12), 240 (100), 224 (26), 196 (27), 96 (20), 69 (43), 57 (47), 55 (61), 43 (33), 41 (56). The next 100 ml yielded a mixture of cis and trans isomers. Evaporation of the next 100 ml gave 45 mg of cis isomer **CI** (R = I) as a yellow oil: Single spot by silica gel TLC ($CHCl_3$, R_f 0.3); mass spectrum, 450.221 (23) (calculated for $C_{21}H_{41}NOI$, 450.223), 324 (99), 308 (21), 294 (10), 254 (41), 240 (100), 223 (82), 198 (53), 196 (36), 182 (56), 127 (24), 96 (21), 85 (35), 71 (56), 69 (35), 55 (46), 43 (44), 41 (36). The mixture of isomers from the intermediate fraction was separated by preparative TLC on silica gel ($CHCl_3$). The total yield of trans isomer **CII** (R = I) was 313 mg (74%) and that of cis isomer **CI** (R = I) was 67.7 mg (16%).

L. *cis-* *and* *trans-oxazolines*(**CIII,** R = ⌐⟨⟩ ; *and* **CIV,** R = ⟨⟩)

(Lee *et al.*, 1977)

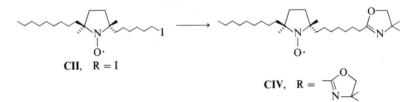

CII, R = I

CIV, R =

The procedure used was that of Meyers *et al.* (1974). To a stirred solution of 340 mg (3.00 m*M*) of 2,5,5-trimethyloxazoline in 2 ml of THF at −78°C (Dry Ice–acetone) under N_2 was added 1.6 ml of a 1.6 *M* solution of butyl lithium in hexane. After 3 min, 290.9 mg of the iodide **CII** (R = I) in 1.0 ml of THF was added. A white precipitate formed after 1 min, and the mixture was stirred an additional 20 min at −78°C before the bath was removed and the mixture allowed to warm to 0°C, during which time the precipitate dissolved. The solution was treated with saturated aqueous NH_4Cl, diluted with ether, washed with H_2O and brine, dried over K_2CO_3, and evaporated to give a yellow oil, which was chromatographed on silica gel to give

202.2 mg (72%) of **CIV** (R = ⟨⟩) as a yellow waxy solid: Single spot by TLC (ether, R_f 0.4); mass spectrum, 436 (72), 435 (68), 422 (9), 421 (8), 405 (17), 309 (75), 293 (66), 240 (100), 224 (68), 196 (34), 126 (40), 113 (68), 96 (18), 69 (16), 55 (25), 43 (12), 41 (16).

Similar treatment of the cis isomer **CI** (R = I) gave **CIII**

(R = 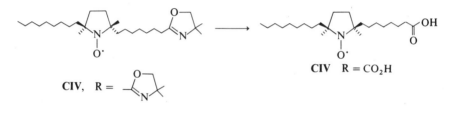) (59%) as a yellow oil: Single spot by TLC (ether, R_f 0.4);

mass spectrum, 436 (32), 435.396 (33) (calculated for $C_{27}H_{51}N_2O_2$, 435.395), 422 (5), 421 (6), 405 (7), 309 (64), 293 (21), 240 (100), 224 (27), 196 (30), 126 (44), 113 (56), 96 (29), 69 (38), 55 (77), 43 (39), 41 (53).

*M. cis-10-Azethoxyleicosanoic acid(**CIII**, R = CO$_2$H) and trans-10-azethoxyleicosanoic acid (**CIV**, R = CO$_2$H) (Lee et al., 1977)*

A solution of 196.8 mg (0.45 m*M*) of **CIV** (R =) in 3 ml of methyl iodide was left standing in the dark at 21°C for 14 hr. The methyl iodide was evaporated, and the residue was stirred with 8 ml of CH$_3$OH and 2 ml of 4 *N* NaOH for 20 hr. The solution was acidified with cold 1 *N* aqueous HCl and washed with ether. The ether solution was washed with H$_2$O and brine, dried over MgSO$_4$, and evaporated to give a yellow oil. Silica gel chromatography gave 145.7 mg (84%) of **CIV** (R = CO$_2$H) as a yellow waxy solid: Single spot by TLC (ether/0.5% HOAc, R_f 0.7); IR (CCl$_4$), 2800–3400 cm^{-1} (OH), 1710 cm^{-1} (acid carbonyl); mass spectrum, 382.332 (12) (calculated for $C_{23}H_{44}NO_3$, 382.332) 368 (2), 352 (2), 256 (71), 240 (100), 224 (10), 113 (10), 81 (10), 69 (10), 55 (19), 43 (11), 41 (13).

Similarly prepared from 23 mg of **CIII** (R =) was *cis*-azethoxyleicosanoic acid(**CIII**, R = CO$_2$H) (73%) as a yellow waxy solid: Single spot by TLC (ether/0.5% HOAc, R_f 0.7); IR (CCl$_4$), 2800–3400 cm^{-1} (OH), 1710 cm^{-1} (acid carbonyl); mass spectrum, 382 (21), 368 (4), 352 (3), 256 (81), 240 (100), 224 (17), 113 (5), 96 (8), 69 (10), 55 (13), 43 (10), 41 (14).

N. 1-Palmitoyl-2-(trans-10-azethoxyleicosoyl)glycerolphosphatidylcholine
(**CV**) (Lee *et al.*, 1977)

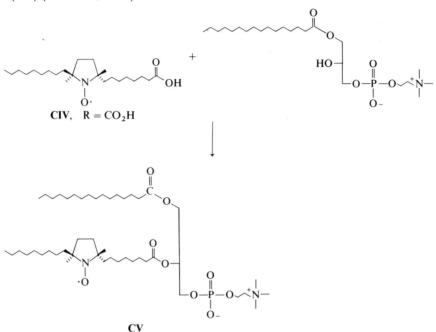

CV

The procedure followed was patterned after that of Boss *et al.* (1975). To a solution of 47.7 mg (0.122 m*M*) of **CIV** (R = CO$_2$H) in 0.4 ml of dry CHCl$_3$ was added 22.3 mg (0.137 mM) of carbonyldiimidazole. After a 20-min stir at 21°C, 35.2 mg (0.071 m*M*) of lysopalmitoylglycerolphosphatidylcholine (Sigma Co.) in 0.2 ml of CHCl$_3$ was added, and the mixture was heated to 50°–55° for 5 days, during which time most of the solvent had evaporated. The reaction was quenched with H$_2$O and taken up in CHCl$_3$. Azeotropic removal of the H$_2$O by several evaporations from CHCl$_3$ gave a yellow foam, which was put on a silica gel column prewashed with a 50 : 50 mixture of CH$_3$OH–CHCl$_3$ and then CHCl$_3$ only. The starting nitroxide acid and imidazole were eluted with CHCl$_3$–CH$_3$OH (85 : 15). Crude **CV** was eluted with CHCl$_3$–CH$_3$OH (50 : 50). Final purification on an 18 × 1650 mm Sephadex LH-20 column eluting with 95% EtOH gave 37.3 mg (61%) of **CV** as a yellow solid. *Analysis*: Calculated for C$_{47}$H$_{92}$N$_2$O$_8$P·3H$_2$O: C, 62.84; H, 10.10; N, 3.12. Found: C, 63.07; H, 10.44; N, 3.47.

It is likely that some of the molecules bear the spin label in the chain attached to C-1 of the glycerol backbone as a result of acyl migration before acylation with the spin-labeled acid. The extent to which acyl migration has occurred has not as yet been determined in our studies.

*O. 2,2-Pentamethylene-4,4,5,5-tetramethylimidazolidine(**CXII**) (Keana et al., 1978b)*

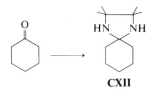

CXII

A 100-ml flask was fitted with a Dean–Stark water separator containing anhydrous K_2CO_3 and then was charged with 2,3-diamino-2,3-dimethyl-butane (4.43 gm), cyclohexanone (3.74 gm), toluenesulfonic acid monohydrate (15 mg), and benzene (60 ml). After a 48-hr reflux period fractional distillation afforded imidazolidine **CXII** as a colorless oil, bp 112°–116°C/26 mm (4.49 gm, 60%): NMR (CDCl$_3$), δ 1.10 (s, 12H, methyl groups), 1.56 (bm, 10H); 1.87 (bs, 2H, NH protons). *Analysis*: Calculated for $C_{12}H_{24}N_2$: C, 73.41; H, 12.32; N, 14.27. Found: C, 73.80; H, 12.48; N, 14.32.

*P. 2,2-Pentamethylene-4,4,5,5-tetramethylimidazolidine-1-oxyl(**CXIII**)* (Keana et al., 1978b)

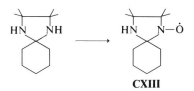

CXIII

To a stirred solution of **CXII** (4.49 gm) in ether (40 ml) at 0°C was added dropwise overnight a solution of MCPA (7.0 gm) dissolved in ether (70 ml). The yellow solution then was washed with 10% Na_2CO_3 (three times), dried (anhydrous K_2CO_3), and then evaporated. Chromatography of the resulting solid over silica gel (elution with CHCl$_3$) afforded a yellow solid, mp 96°–115°C (1.64 gm). Recrystallization from hexane afforded analytically pure nitroxide **CXIII**, mp 121°–123°C: ESR (4 : 1 MeOH–CHCl$_3$) three lines (a_N = 13.8 G); UV max (EtOH) 237 nm (2700); visible max 450 nm (6.5); m/e 211 (molecular ion). *Analysis*: Calculated for $C_{12}H_{23}N_2O$: C, 68.25; H, 10.90. Found: C, 68.19; H, 10.96. NMR spectrum (CDCl$_3$) after phenylhydrazine: δ 1.10 (s, 6H); 1.66 (s, 6H); 1.66 (bs, 10H).

Q. N-Acetoxy-2,2-pentamethylene-4,4,5,5-tetramethylimidazolidine
(**CLXXXVII**) (Keana *et al.*, 1978b)

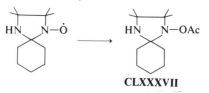

CLXXXVII

To a 25°C stirred mixture of 10% Pd/C (7 mg) and THF (3.0 ml) under an H_2 atmosphere was added dropwise a solution of nitroxide **CXIII** (328 mg) in THF (2.5 ml). Stirring was continued until H_2 uptake ceased (10 min). The mixture was filtered. The filtrate (colorless) was cooled to 0°C and treated with triethylamine (263 mg) dissolved in THF (0.3 ml) followed immediately by the addition of acetyl chloride (177 mg) dissolved in THF (0.3 ml). After a 1-hr stir, the white $Et_3N \cdot HCl$ was filtered out, and the filtrate was evaporated. The residual oil was treated with hexane, and this mixture was filtered. Evaporation of the filtrate afforded crude acetate **CLXXXVII** as a pale-yellow, hydrolytically unstable oil (288 mg, 73%); IR, 1775 cm^{-1} (N—OAc); NMR (CDCl$_3$), δ 1.06 (s, 6H); 1.16 (s, 6H); 1.63 (bs, 10H); 2.12 (s, 3H).

R. N-Acetoxy-2,2-pentamethylene-4,4,5,5-tetramethylimidazolidine-N′-oxyl
(**CLXXXVIII**) (Keana *et al.*, 1978b)

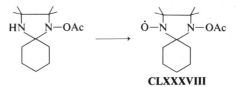

CLXXXVIII

The crude nitroxide was prepared in a manner similar to the preparation of **CXIII**. From 996 mg of **CLXXXVII** 930 mg of crude **CLXXXVIII** was obtained. Crude **CLXXXVIII** was purified by silica gel chromatography (elution with CHCl$_3$), producing 838 mg (80%) of **CLXXXVIII** as an orange oil: IR, 1775 cm^{-1} (N—OAc); NMR (CDCl$_3$) after phenylhydrazine reduction, δ 1.10 (s, 6H); 1.16 (s, 6H); 1.40–1.85 (b); 2.12 (s, 3H).

S. 1-Hydroxy-2,2-pentamethylene-4,4,5,5-tetramethylimidazolidine-3-oxyl
(**CLXXXIX**) (Keana *et al.*, 1978b)

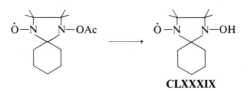

CLXXXIX

A solution of KOH (250 mg) dissolved in methanol (10 ml) was added dropwise to a stirred solution of acetate **CLXXXVIII** (385 mg) in MeOH (15 ml) at 0°C. After 1.5 hr at 25°C, the volatiles were removed at reduced pressure. The residue was leached with ether. The ether solution was washed with water (four times), dried ($MgSO_4$), and evaporated, giving nitroxide **CLXXXIX** as yellow crystals (242 mg, 75%). Recrystallization from hexane afforded 182 mg (56%) of pure nitroxide, mp 86°–87°C: UV max (EtOH) 245 nm (2430); m/e 227.176 (calculated, 227.174); ESR (CHCl₃), three lines (a_N = 15.5 G). *Analysis*: Calculated for $C_{12}H_{23}N_2O_2$: C, 63.40; H, 10.20; N, 12.32. Found: C, 63.23; H, 10.20; N, 12.30.

T. 2,2-Pentamethylene-4,4,5,5-tetramethylimidazolidine-1,3-dioxyl(**CLXXXII**) (Keana *et al.*, 1978b)

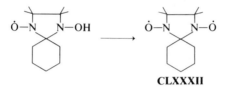

CLXXXII

To a solution of **CLXXXIX** (115 mg) in *tert*-butyl alcohol (5 ml) at 25°C was added 0.4 *M* potassium *tert*-butoxide in *tert*-butyl alcohol (2 ml). Oxygen gas was then bubbled through this solution for 3 min, giving an orange mixture. The solvent was removed at reduced pressure, and the residue was treated with several fresh portions of hexane. The combined hexane extracts were evaporated and the residue once again leached with hexane. Centrifugation followed by evaporation of the supernatant solution gave 97 mg of orange crystals. Low-temperature recrystallization from pentane gave pure dinitroxide as orange needles, mp 70.5°–71.5°C: UV max (*tert*-BuOH) 235 nm (4020); m/e 226.167 (calculated, 226.168). An X-ray crystallographic analysis confirmed the assigned structure.

ACKNOWLEDGEMENT

It is a pleasure to acknowledge support from the Public Health Service through research grant GM CA-24951 and through receipt of a Research Career Development Award (NS 70156). Expert assistance from Mr. Larry LaFleur with the literature searches also is gratefully acknowledged.

REFERENCES

(Supplemental references not discussed in the text may be found in the Appendix on page 347.)

Abdallah, M. A., J.-J. André, and J.-F. Biellmann (1977). A new spin-labelled analogue of nicotinamide adenine dinucleotide. *Bioorg. Chem.* **6**, 157–163.
Adackaparayil, M., and J. H. Smith (1977). Preparation and reactivity of a new spin label reagent. *J. Org. Chem.* **42**, 1655–1656.

Aurich, H. G., and P. Höhlein (1974). Aminyloxide. XIX. Bildung von Amidinyl-N-oxiden und Amidinyl-N,N-Dioxiden Durch Oxidative Kupplung. *Tetrahedron Lett.* pp. 279–282.

Aurich, H. G., and K. Stork (1975). Bildung von Aminyloxiden bei Reaktionen von Nitriloxiden mit Hydroxylaminen. *Chem. Ber.* **108**, 2764–2780.

Aurich, H. G., and J. Trösken (1973). Über die Darstellung von Acyl-Aminyloxiden und ihre Konformation. *Chem. Ber.* **106**, 3483–3494.

Aurich, H. G., and W. Weiss (1976). Verhalten von *N-tert*-Butyl(3-oxoindolin-2-yliden)amin-*N*-oxid und *N-tert*-Butyl(3-oxo-3H-indol-2-yl)aminyl-*N*-oxid gegenüber Nucleophilen. *Justus Liebigs Ann. Chem.*, **1976**, 432–439.

Aurich, H. G., K. Hahn, and K. Stork (1975). Vinylaminyloxide (Vinylnitroxide) mit grosser Spindichte in der Vinylgruppe. *Angew. Chem.* **87**, 590.

Balaban, A. T., and I. Pascaru (1972). Factors affecting stability and equilibria of free radicals. VI. Oxidation of cyclic hydroxamic acids to nitroxides. *J. Magn. Reson.* **7**, 241–246.

Baldassare, J. J., D. E. Robertson, A. G. McAfee, and C. Ho (1974). A spin-label study of energy-coupled active transport in *Escherichia coli* membrane vesicles. *Biochemistry* **13**, 5210–5214.

Baldry, P. J., A. R. Forrester, and R. H. Thomson (1976). Nitroxide radicals. Part XVIII. Further spectroscopic investigation of the Banfield and Kenyon nitroxide. *J. Chem. Soc., Perkin Trans.* 2 pp. 76–81.

Banks, R. E., D. J. Edge, J. Freear, and R. N. Haszeldine (1974). Nitroxide chemistry. Part VI. *N*-trifluoromethylsulphamate *N*-oxyl. *J. Chem. Soc., Perkin Trans.* 1 pp. 721–722.

Beereboom, J. J. (1966). The synthesis of piperitenone via mesityl oxide and methyl vinyl ketone. *J. Org. Chem.* **31**, 2026–2028.

Bennett, J. E., H. Sieper, and P. Tavs (1967). 2,2,6,6-Tetramethyl-piperidyl-1-thiyl. A stable new radical. *Tetrahedron* **23**, 1697–1699.

Berti, C., M. Colonna, L. Greci, and L. Marchetti (1975). Stable nitroxide radicals from phenylisatogen and arylimino-derivatives with organo-metallic compounds. *Tetrahedron* **31**, 1745–1753.

Berti, C., M. Colonna, and L. Greci (1976). Stable nitroxide radicals from 2-substituted quinoline-*N*-oxides with organometallic compounds. *Tetrahedron* **32**, 2147–2151.

Birrell, G. B., T. D. Lee, O. H. Griffith, and J. F. W. Keana (1978). Synthesis and properties of chlorophyll-derived nitroxide spin labels. *Bioorganic Chem.* **7**, 325–337.

Boocock, D. G. B., and E. F. Ullman (1968). Studies of stable radicals. III. A 1,3-Dioxy-2-imidazolidone Zwitterion and its stable nitronyl nitroxide radical anion. *J. Am. Chem. Soc.* **90**, 6873–6874.

Boocock, D. G. B., R. Darcy, and E. F. Ullman (1968). Studies of free radicals. II. Chemical properties of nitronyl nitroxides. A unique radical reaction. *J. Am. Chem. Soc.* **90**, 5945–5946.

Borch, R. F., M. D. Bernstein, and H. D. Durst (1971). The cyanohydridoborate anion as a selective reducing agent. *J. Am. Chem. Soc.* **93**, 2897–2904.

Boss, W. F., C. J. Kelley, and F. R. Landsberger (1975). A novel synthesis of spin labeled derivatives of phosphatidylcholine. *Anal. Biochem.* **64**, 289–292.

Bowman, D. F., T. Gillan, and K. U. Ingold (1971). Kinetic applications of electron paramagnetic resonance spectroscopy. III. Self-reactions of dialkyl nitroxide radicals. *J. Am. Chem. Soc.* **93**, 6555–6561.

Briere, R., and A. Rassat (1976). Synthese et étude cinétique de la decomposition du t-butyl isopropyl nitroxide. Effect isotopique. *Tetrahedron* **32**, 2891–2898.

Briere, R., H. Lemaire, A. Rassat, and J.-J. Dunand (1970). Nitroxides. XXXVI. Etude de l'interconversion d'un radical libre nitroxyde piperidinique, par RMN a 310 MHz. *Bull. Soc. Chim. Fr.* pp. 4220–4226.

Busby, S. J. W., M. A. Hemminga, G. K. Radda, W. E. Trommer, and H. Wenzel (1976). Spin-labeled AMP—an activator of phosphorylase. *Eur. J. Biochem.* **63**, 33–38.

Butterfield, D. A., A. L. Crumbliss, and D. B. Chesnut (1975). Radical decay kinetics in ferrocytochrome c model membranes. A spin label study. *J. Am. Chem. Soc.* **97**, 1388–1392.

Calder, A., and A. R. Forrester (1969). Nitroxide radicals. Part VI. Stability of meta- and para-alkyl substituted phenyl-*t*-butylnitroxides. *J. Chem. Soc. C* pp. 1459–1464.

Calder, A., A. R. Forrester, and S. P. Hepburn (1973). Nitroxide radicals. Part XII. Decomposition of ortho-, meta-, and para-halogenophenyl *t*-butyl nitroxides. *J. Chem. Soc., Perkin Trans. 1* pp. 456–465.

Calder, A., A. R. Forrester, and G. McConnachie (1974). Nitroxide radicals. Part XIV. Decomposition of 1- and 2-naphthyl *t*-butyl nitroxides. *J. Chem. Soc., Perkin Trans. 1* pp. 2198–2207.

Call, L., and E. F. Ullmann (1973). Stable free radicals. XI. Photochemistry of a nitronyl nitroxide. *Tetrahedron Lett.* pp. 961–964.

Carrington, A., and A. D. McLachlan (1967). "Introduction to Magnetic Resonance," Chapter 13. Harper, New York.

Cazianis, C. T., and D. R. Eaton (1974). Spin-labelled ligands. *Can. J. Chem.* **52**, 2454–2462.

Cella, J. A., J. A. Kelley, and E. F. Kenehan (1974). Unexpected oxidation of a nitroxide alcohol with *m*-chloroperbenzoic acid. *J. Chem. Soc., Chem. Commun.* p. 943.

Cella, J. A., J. A. Kelley, and E. F. Kenehan (1975a). Nitroxide-catalyzed oxidation of alcohols using *m*-chloroperbenzoic acid. A new method. *J. Org. Chem.* **40**, 1860–1862.

Cella, J. A., J. A. Kelley, and E. F. Kenehan (1975b). Oxidation of nitroxides by m-chloroperbenzoic acid. *Tetrahedron Lett.* pp. 2869–2872.

Chambaz, E., G. Defaye, A. Hadjian, P. Martin, R. Ramasseul, and A. Rassat (1971). Nitroxides. XLVIII. A study of the interaction between bovine serum albumin and a modified steroid by electron spin resonance. *FEBS Lett.* **19**, 55–59.

Chiarelli, R., and A. Rassat (1973). Synthese de radicaux nitroxydes deuteries. *Tetrahedron* **29**, 3639–3647.

Chou, S., J. A. Nelson, and T. A. Spencer (1974). Oxidation and mass spectra of 4,4-dimethyloxazolidine-N-oxyl (Doxyl) derivatives of ketones. *J. Org. Chem.* **39**, 2356–2361.

Comfurius, P., and R. F. A. Zwaal (1977). The enzymatic synthesis of phosphatidylserine and purification by CM-cellulose column chromatography. *Biochim. Biophys. Acta* **488**, 36–42.

Corey, E. J., and C. U. Kim (1972). A new and highly effective method for the oxidation of primary and secondary alcohols to carbonyl compounds. *J. Am. Chem. Soc.* **94**, 7586–7587.

Crossland, R. K., and K. L. Servis (1970). A facile synthesis of methanesulfonate esters. *J. Org. Chem.* **35**, 3195–3196.

Daniel, W. E., and M. Cohn (1976). Changes in tertiary structure accompanying a single base change in transfer RNA. Proton magnetic resonance and aminoacylation studies of *Escherichia coli* tRNA$_{f1}^{Met}$ and tRNA$_{f3}^{Met}$ and their spin-labeled (s^{4}U8) derivatives. *Biochemistry* **15**, 3917–3924.

Davies, A. P., A. Morrison, and M. D. Barratt (1974). Mass spectrometry of stable free radicals. II. Pyrrolidine nitroxides. *Org. Mass Spectrom.* **8**, 43–48.

Dupeyre, R.-M., and A. Rassat (1966). Nitroxides. XIX. Norpseudo-pelletierine-N-oxyl, a new, stable, unhindered free radical. *J. Am. Chem. Soc.* **88**, 3180–3181.

Dupeyre, R.-M., and A. Rassat (1973). Application de la réaction de Hoffman-Löffler-Freytag. Synthese de dérivés Diaza-2,6-Adamantane. *Tetrahedron Lett.* pp. 2699–2701.

Dupeyre, R.-M., and A. Rassat (1975). Nitroxydes. LXXIII. Oxydation d'amines secondaires et tertiaires par le permanganate de potassium en milieu basique. *Tetrahedron Lett.* pp. 1839–1840.

Dupeyre, R.-M., A. Rassat, and J. Ronzaud (1974). Nitroxides. LII. Synthesis and electron spin resonance studies of N,N'-Dioxy-2,6-diazaadamantane, a symmetrical ground state triplet. *J. Am. Chem. Soc.* **96**, 6559–6568.

Dvolaitzky, M., C. Taupin, and F. Poldy (1975). Nitroxydes piperidiniques-synthèse de nouvelles "sondes paramagnétiques." *Tetrahedron Lett.* pp. 1469–1472.

Flohr, K., R. M. Paton, and E. T. Kaiser (1975). Studies on the interaction of spin-labeled substrates with chymotrypsin and with cycloamyloses. *J. Am. Chem. Soc.* **97**, 1209–1218.

Forrester, A. R., and S. P. Hepburn (1969). *t*-Butylferrocenylnitroxide, a stable ferrocenyl radical. *Chem. Commun.* pp. 698–699.

Forrester, A. R., and S. P. Hepburn (1970). Nitroxide radicals. Part VIII. Stability of ortho-alkyl-substituted phenyl *t*-butyl nitroxides. *J. Chem. Soc. C* pp. 1277–1280.

Forrester, A. R., and S. P. Hepburn (1974). Nitroxide radicals. Part XV. *p*-methoxy- and *p*-phenoxy-phenyl *t*-butyl nitroxides. *J. Chem. Soc., Perkin Trans. 1* pp. 2208–2213.

Forrester, A. R., and R. Ramasseul (1971). Nitroxide radicals. Part IX. Preparation of [2,2]paracyclophane mono- and bis-nitroxides and [2,2]paracyclophanequinones. *J. Chem. Soc. B* pp. 1638–1644.

Forrester, A. R., and R. Ramasseul (1975). Nitroxide radicals. Part XVII. Transannular interactions in [2.2.]paracyclophenyl nitroxides. *J. Chem. Soc., Perkin Trans. 1* pp. 1753–1757.

Forrester, A. R., S. P. Hepburn, and G. McConnachie (1974). Nitroxide radicals. Part XVI. Unpaired electron distribution in para-substituted aryl *t*-butyl nitroxides and 2-naphthyl phenyl nitroxides. *J. Chem. Soc., Perkin Trans. 1* pp. 2213–2219.

Gaffney, B. J. (1976). The chemistry of spin labels. *In* "Spin Labeling: Theory and Applications" (L. J. Berliner, ed.), Vol. 1, Chapter 5. Academic Press, New York.

Ganem, B. (1975). Biological spin labels as organic reagents. Oxidation of alcohols to carbonyl compounds using nitroxyls. *J. Org. Chem.* **40**, 1998–1999.

Ghriofa, S. N., R. Darcy, and M. Conlon (1977). Synthesis and electron spin resonance of 3-oxy-1,3-diazacyclohexene-1-oxide (1,3-Nitrone-Nitroxide) radicals. *J. Chem. Soc., Perkin Trans. 1* pp. 651–653.

Giotta, G. J., and H. H. Wang (1972). Reduction of nitroxide free radicals by biological materials. *Biochem. Biophys. Res. Commun.* **46**, 1576–1580.

Griffith, O. H., J. F. W. Keana, D. L. Noall, and J. L. Ivey (1967). Nitroxide mixed carboxylic-carbonic acid anhydrides—a new class of versatile spin labels. *Biochim. Biophys. Acta* **148**, 583–585.

Grigor'ev, I. A., and L. B. Volodarskii (1975). The intramolecular oxidation of a 4-aminoalkyl group by a nitrone grouping in stable iminoxyl radicals derived from 3-imidazoline-3-oxide. *J. Org. Chem. USSR (Engl. Transl.)* **11**, 1313–1316.

Grigor'ev, I. A., A. G. Druganov, and L. B. Volodarskii (1976). Nitrosation of steric hindered 3-imidazolines. Formation of 4-nitromethylene-2,2,5,5-tetramethylimidazolidine-1-oxyl. *Izv. Akad. Nauk SSSR, Ser. Khim.* pp. 131–135.

Griller, D., and K. U. Ingold (1976). Persistent carbon-centered radicals. *Acc. Chem. Res.* **9**, 13–19.

Harcus, R., P. N. Preston, and J. S. Suffolk (1976). Synthesis of purine and pyrimidine substituted nitroxides. *Z. Naturforsch., Teil C* **31**, 101–102.

Hatch, G. F., and R. Kreilich (1971). ^{13}CMR spectra of nitroxide radicals. *Chem. Phys. Lett.* **10**, 490–492.

Hubbell, W. L., and H. M. McConnell (1969). Motion of steroid spin labels in membranes. *Proc. Natl. Acad. Sci. U.S.A.* **63**, 16–22.

Kavanagh-Caron, C., M. Caron, N. Brisson, and H. Dugas (1976). Evaluation of a spin-labeled inhibitor of acetylcholinesterase. *Can. J. Chem.* **54**, 3545–3547.

Keana, J. F. W. (1978). Newer aspects of the synthesis and chemistry of nitroxide spin labels. *Chem. Rev.* **78**, 37–64.

Keana, J. F. W., and F. Baitis (1968). Photolysis of the stable nitroxide, 3-carbamoyl-2,2,5,5-tetramethylpyrrolidine-1-oxyl. *Tetrahedron Lett.* pp. 365–368.

Keana, J. F. W., and R. J. Dinerstein (1971). A new highly anisotropic dinitroxide ketone spin label. A sensitive probe for membrane structure. *J. Am. Chem. Soc.* **93**, 2808–2810.

Keana, J. F. W., and T. D. Lee (1975). A versatile synthesis of doxyl spin labels bipassing the usual ketone precursors. *J. Am. Chem. Soc.* **97**, 1273–1274.

Keana, J. F. W., and L. E. LaFleur (1978). Saturated and unsaturated lipid spin labels with terminally located nitroxide groups. *Chem. Phys. Lipids*, in press.

Keana, J. F. W., S. B. Keana, and D. Beetham (1967). A new versatile ketone spin label. *J. Am. Chem. Soc.* **89**, 3055–3056.

Keana, J. F. W., R. J. Dinerstein, and F. Baitis (1971). Photolytic studies on 4-hydroxy-2,2,6,6-tetramethylpiperidine-1-oxyl, a stable nitroxide free radical. *J. Org. Chem.* **36**, 209–211.

Keana, J. F. W., T. D. Lee, and E. M. Bernard (1976). Side-chain substituted 2,2,5,5-tetramethylpyrrolidine-*N*-oxyl (proxyl) nitroxides. A new series of lipid spin labels showing improved properties for the study of biological membranes. *J. Am. Chem. Soc.* **98**, 3052–3053.

Keana, J. F. W., E. M. Bernard, and R. B. Roman (1978a). Selective reduction of doxyl and proxyl nitroxide carboxylic acids to the corresponding alcohols with borane methyl sulfide. *Synth. Commun.* **8**, 169–173.

Keana, J. F. W., R. S. Norton, M. Morello, D. Van Engen, and J. Clardy (1978b). Mono-nitroxides and proximate dinitroxides derived by oxidation of 2,2,4,4,5,5-hexasubstituted imidazolidines. A new series of nitroxide and dinitroxide spin labels. *J. Am. Chem. Soc.* **100**, 934–937.

Kornberg, R. D., and H. M. McConnell (1971). Inside-outside transitions of phospholipids in vesicle membranes. *Biochemistry* **10**, 1111–1120.

Kornblum, N., and H. W. Pinnick (1972). Reduction of nitroxides to amines by sodium sulfide. *J. Org. Chem.* **37**, 2050–2051.

Kreilich, R. W. (1968). The nuclear magnetic resonance spectrum of a phenoxy radical. Di-*t*-butyl nitroxide as a spin relaxer. *J. Am. Chem. Soc.* **90**, 2711–2713.

La Mar, G. N., W. D. Horrocks, Jr., and R. H. Holm, eds. (1973). "NMR of Paramagnetic Molecules." Academic Press, New York.

Lane, C. F. (1975). Reduction of organic functional groups with borane-methyl sulfide. *Aldrichimica Acta* **8**, 20–23.

Lee, T. D. (1977). Synthesis of spin labeled derivatives of lipids, diacylglycerophosphatidylcholine, and pyrochlorophyll-A via Grignard and alkyllithium additions to nitrones. Ph.D. Thesis, University of Oregon, Eugene.

Lee, T. D., and J. F. W. Keana (1975). In situ reduction of nitroxide spin labels with phenylhydrazine in deuteriochloroform solution. A convenient method for obtaining structural information on nitroxides using nuclear magnetic resonance spectroscopy. *J. Org. Chem.* **40**, 3145–3147.

Lee, T. D., and J. F. W. Keana (1976). Nitrones and nitroxides derived from oxazolines and dihydrooxazines. *J. Org. Chem.* **41**, 3237–3241.

Lee, T. D., and J. F. W. Keana (1978). Nitroxides derived from 3,4-dihydro-2,5-dimethyl-2*H*-pyrrole 1-oxide. A new series of minimum steric perturbation lipid spin labels. *J. Org. Chem.* **43**, 4226–4231.

Lee, T. D., G. B. Birrell, and J. F. W. Keana (1977). A new series of minimum steric perturbation nitroxide spin labels. *J. Am. Chem. Soc.*, **100**, 1618–1619.

Lee, T. D., G. B. Birrell, P. J. Bjorkman, and J. F. W. Keana (1978). Azethoxyl nitroxide spin labels. ESR studies involving thiourea crystals, model membrane systems and chromatophores, and chemical reduction with ascorbate and dithiothreitol. *Biochim. Biophys. Acta* (in press).

Lin, J. S., and H. S. Olcott (1975). Ethoxyquin nitroxide. *J. Agric. Food Chem.* **23**, 798–800.

Lin, J. S., T. C., Tom, and H. S. Olcott (1974). Proline nitroxide. *J. Agric. Food Chem.* **22**, 526–528.

Luckhurst, G. R. (1975). Biradicals as spin probes. *In* "Spin Labeling: Theory and Applications" (L. J. Berliner, ed.), Vol. 1, Chapter 4. Academic Press, New York.

Marai, L., J. J. Myher, A. Kuksis, L. Stuhne-Sekalec, and N. Z. Stanacev (1976). Identification of isomeric doxyl stearic acids by gas-liquid chromatography—mass spectrometry. *Chem. Phys. Lipids* **17**, 213–221.

Martinie, J., J. Michon, and A. Rassat (1975). Nitroxides. LXX. Electron spin resonance study of cyclodextrin inclusion compounds. *J. Am. Chem. Soc.* **97**, 1818–1823.

Martinie-Hombrouck, J., and A. Rassat (1974). Nitroxides. LX. Isolement and autodecomposition du *N*-methyl *N*- t ri-*t*-butyl-2,4,6 phenyl nitroxide. *Tetrahedron* **30**, 433–436.

Maruyama, K., and S. Ohnishi (1974). A spin-label study of the photosynthetic bacterium *Rhodospirillum rubrum*. Reduction and regeneration of nitroxide spin-labels. *J. Biochem. (Tokyo)* **75**, 1153–1164.

Mendenhall, G. D., and K. U. Ingold (1973a). Reversible dimerization and some solid-state properties of two bicyclic nitroxides. *J. Am. Chem. Soc.* **95**, 6390–6394.

Mendenhall, G. D., and K. U. Ingold (1973b). Reactions of bicyclic nitroxides involving reduction of the NO group. *J. Am. Chem. Soc.* **95**, 6395–6400.

Metzner, E. K., L. J. Libertini, and M. Calvin (1977). Electron spin exchange in rigid biradicals. *J. Am. Chem. Soc.* **99**, 4500–4502.

Meyers, A. I., D. L. Temple, R. L. Nolen, and E. D. Mihelich (1974). Oxazolines. IX. Synthesis of homologated acetic acids and esters. *J. Org. Chem.* **39**, 2778–2783.

Michon, J., and A. Rassat (1974). Nitroxides. LIX. Rotational correlation time determination of nitroxide biradical application to solvation studies. *J. Am. Chem. Soc.* **96**, 335–337.

Michon, P., and A. Rassat (1974). Nitroxides. LVIII. Structure of steroidal spin labels. *J. Org. Chem.* **39**, 2121–2124.

Morat, C., and A. Rassat (1972). Nitroxides. XLVI. Determination de la fréquence d'élongation NO dans des radicaux libres nitroxydes piperidiniques. *Tetrahedron* **28**, 735–740.

Morrisett, J. D. (1975). The use of spin labels for studying the structure and function of enzymes. *In* "Spin Labeling: Theory and Applications" (L. J. Berliner, ed.), Chapter 8. Academic Press, New York.

Morrison, A., and A. P. Davies (1970). Mass spectrometry of piperidine nitroxides—a class of stable free radicals. *Org. Mass Spectrom.* **3**, 353–366.

Moutin, M., and A. Rassat (1976). Nitroxydes. LXXIV. Mise en évidence de deux oxazolidines nitroxydes dérivées du norcamphre. *J. Mol. Struct.* **31**, 275–282.

Murphy, P. A., J. S. Lin, and H. S. Olcott (1974). Peroxide oxidation of tris to a free radical. *Arch. Biochem. Biophys.* **164**, 776–777.

Neely, J. W., G. F. Hatch, and R. W. Kreilich (1974). Electron-carbon couplings of aryl nitronyl nitroxide radicals. *J. Am. Chem. Chem. Soc.* **96**, 652–656.

Nelson, J. A., S. Chou, and T. A. Spencer (1971). Reaction of 4,4-dimethyloxazolidine-*N*-oxyl (Doxyl) derivatives with nitrogen dioxide: A novel and efficient reconversion into the parent ketone. *Chem. Commun.* p. 1580.

Nelson, J. A., S. Chou, and T. A. Spencer (1975). Oxidative demethylation at C-4 of a steroid via nitroxide photolysis. *J. Am. Chem. Soc.* **97**, 648–649.

Osiecki, J. H., and E. F. Ullman (1968). Studies of free radicals. I. α-nitronyl nitroxides, a new class of stable radicals. *J. Am. Chem. Soc.* **90**, 1078–1079.

Paleos, C. M., and P. Dais (1977). Ready reduction of some nitroxide free radicals with ascorbic acid. *J. Chem. Soc., Chem. Commun.* pp. 345–346.

Quon, H. H., T. Tezuka, and Y. L. Chow (1974). A stable nitroxide derived from α-pinene. *J. Chem. Soc., Chem. Commun.* pp. 428–429.

Ramasseul, R., and A. Rassat (1970). Nitroxides. XXXIII. Radicaux: Nitroxydes pyrroliques encombres. Un pyrryloxyle stable. *Bull. Soc. Chim. Fr.* pp. 4330–4341.

Ramasseul, R., and A. Rassat (1971). Nitroxides. XLIX. Steroidal nitroxides. *Tetrahedron Lett.* pp. 4623–4624.

Ramasseul, R., A. Rassat, and P. Rey (1975). Nitroxydes. LXXII. Preparation d'amines et d'ammonium quaternaires nitroxydes curarisants potentiels. *Tetrahedron Lett.* pp. 839–841.

Ramasseul, R., A. Rassat, and P. Rey (1976). A useful protecting group in the preparation of amino-nitroxides. *J. Chem. Soc., Chem. Commun.* pp. 83–84.

Rassat, A., and P. Rey (1971). Nitroxides: Photochemical synthesis of trimethylisoquinuclidine-*N*-oxyl. *Chem. Commun.* pp. 1161–1162.

Rassat, A., and P. Rey (1972). Nitroxydes. XLVIII. Nitroxydes isoquinuclidiniques. *Tetrahedron* **28**, 741–750.

Rassat, A., and P. Rey (1974a). Nitroxydes. LXII. Nitroxydes oxaziniques: Synthese et etude conformationelle. *Tetrahedron* **30**, 3315–3325.

Rassat, A., and P. Rey (1974b). Nitroxydes. LXIII. Oxidation de nitroxydes isoquinuclidiniques par l'oxide d'argent humide. Preparation de nitroxydes Aza-6 bicyclo[3.2.1]octeniques. *Tetrahedron* **30**, 3597–3604.

Rassat, A., and J. Ronzaud (1976). Nitroxides. LXXI. Synthèse de dérives nitroxydes du tropane. Etude par RPE et RMN des interactions hyperfines a longue distance dans ces radicaux. *Tetrahedron* **32**, 239–244.

Rassat, A., and H. U. Sieveking (1972). Ein Stabiles Aromatisches Diradical mit Starker Dipolar Elektronenwechselwirkung. *Angew. Chem.* **84**, 353–354.

Rauchman, E. J., and G. M. Rosen (1976). Synthesis of spin labeled probes: Esterification and reduction. *Synth. Commun.* **6**, 325–329.

Rauchman, E. J., G. M. Rosen, and M. B. Abou-Donia (1975). Improved methods for the oxidation of secondary amines to nitroxides. *Synth. Commun.* **5**, 409–413.

Rauchman, E. J., G. M. Rosen, and M. B. Abou-Donia (1976a). Synthesis of a useful spin labeled probe, 1-oxyl-4-carboxyl-2,2,6,6-tetramethylpiperidine. *J. Org. Chem.* **41**, 564–565.

Rauchman, E. J., G. M. Rosen, and R. J. Lefkowitz (1976b). Pharmacological activity of nitroxide analogues of dichloroisoproterenol and propranolol. *J. Med. Chem.* **19**, 1254–1256.

Rauchman, E. J., G. M. Rosen, and M. B. Abou-Donia (1976c). The use of trimethylsulfonium iodide in the synthesis of biologically active nitroxides. *Org. Prep. Proced. Int.* **8**, 159–161.

Rawson, G., and J. B. F. N. Engberts (1970). Mannich-type condensation products of sulfinic acids with aldehydes and hydroxylamines or hydroxamic acids. *Tetrahedron* **26**, 5653–5664.

Roberts, R. M., J. C. Gilbert, L. B. Rodewald, and A. S. Wingrove (1974). "An Introduction to Modern Experimental Organic Chemistry." Holt, New York.

Rosen, G. M. (1974). Use of sodium cyanoborohydride in the preparation of biologically active nitroxides. *J. Med. Chem.* **17**, 358–360.

Rosen, G. M., and M. B. Abou-Donia (1975). The synthesis of spin labelled acetylcholine analogs. *Synth. Commun.* **5**, 415–422.

Rozantsev, E. G. (1970). "Free Nitroxyl Radicals" (transl. by B. J. Hazzard). Plenum, New York.

Rozantsev, E. G., and V. D. Sholle (1969). Products from the oxidation of cocaine and other narcotics of the tropane series. *Dokl. Akad. Nauk SSSR* **187**, 1319–1321.

Rozantsev, E. G., and V. D. Sholle (1971). Synthesis and reactions of stable nitroxyl radicals. I. Synthesis. *Synthesis* pp. 190–202.

Schlude, H. (1973). Oxidation of hydroxylamines to nitroxyl radicals with Fremy's salt. *Tetrahedron* **29**, 4007–4011.

Schlude, H. (1975). New phosphonium salts for the Wittig synthesis of aldehydes from ketones. *Tetrahedron* **31**, 89–92.

Schlude, H. (1976). A new reagent for the spin labeling of aldehydes and ketones. *Tetrahedron Lett.* pp. 2179–2182.

Setaka, M., and C. Lagercrantz (1975). Orientation of some nitroxide spin labels in the lamellar mesophases of aerosol-OT-water and decanol-decanoate-water systems. *J. Am. Chem. Soc.* **97**, 6013–6018.

Sevastyanova, T. K., and L. B. Volodarskii (1972). Preparation of stable iminoxyl radicals of 3-imidazolines. *Izv. Akad. Nauk SSSR, Ser. Khim.* pp. 2339–2341.

Sholle, V. D., V. A. Golubev, and E. G. Rozantsev (1971). Interaction of nitroxyl radicals with Grignard reagents. *Dokl. Akad. Nauk SSSR* **200**, 137–139.

Sinha, B., and C. F. Chignell (1975). Synthesis and biological activity of spin-labeled analogues of biotin, hexamethonium, decamethonium, dichloroisoproterenol, and propranolol. *J. Med. Chem.* **18**, 669–673.

Sosnovsky, G., and M. Konieczny (1976). Preparation of 4-hydroxy-2,2,6,6-

tetramethylpiperidine-1-oxyl. A reinvestigation of methods using hydrogen peroxide in the presence of catalysts. Z. Naturforsch., Teil B **31**, 1376–1378.

Sosnovsky, G., and M. Konieczny (1977a). An improved method for the preparation of spin-labeled phosphates using imidazole as transfer agent. Z. Naturforsch., Teil B **32**, 82–86.

Sosnovsky, G., and M. Konieczny (1977b). Preparation of spin-labeled phosphoramidates in high yield using imidazole as the transfer agent. Z. Naturforsch., Teil B **32**, 321–327.

Sosnovsky, G., and M. Konieczny (1977c). Preparation of triacetoneamine. II. Z. Naturforsch., Teil B **32**, 338–346.

Sümmerman, W., and U. Deffner (1975). Die electrochemische oxidation aliphatischer nitroxyl radikale. Tetrahedron **31**, 593–596.

Torsell, K., J. Goldman, and T. E. Petersen (1973). Spindelokalisierungen bei Elementen der Gruppe IV B in organischen Verbindungen. Justus Liebigs Ann. Chem. **1973**, 231–240.

Ullman, E. F., L. Call, and J. H. Osiecki (1970). Stable free radicals. VIII. New imino, amido and carbamoyl nitroxides. J. Org. Chem. **35**, 3623–3631.

Ullman, E. F., J. H. Osiecki, D. G. B. Boocock, and R. Darcy (1972). Studies of stable free radicals. X. Nitronyl nitroxide monoradicals and biradicals as possible small molecule spin labels. J. Am. Chem. Soc. **94**, 7049–7059.

Volkamer, K., and H. W. Zimmerman (1970). Über N-Oxide von Imidazolylen. Chem. Ber. **103**, 296–306.

Volodarskii, L. B., G. A. Kutikova, V. S. Kobrin, R. Z. Sagdeev, and Y. N. Molin (1971). Synthesis and spectral studies of 2,2,5,5-tetramethyl-3-imidazoline-1-oxil-4-carboxylic acid and its amide, salt and Cu(11)-complex. Izv. Sib. Otd. Akad. Nauk SSSR, Ser. Khim. Nauk pp. 101–103.

Volodarskii, L. B., I. A. Grigor'ev, and G. A. Kutikova (1973). Reaction of 1-hydroxy-2,2,4,5,5-pentamethyl-Δ^3-imidazoline 3-oxide with aldehydes, bromine, amylnitrite and nitrosobenzene in the presence of bases. J. Org. Chem. USSR (Engl. Transl.) **9**, 1990–1994.

Volodarskii, L. B., I. A. Grigorev, G. A. Kutikova, and T. K. Sevastyanova (1974). USSR Patent 412,193; Chem. Abstr. **80**, 120956q (1974).

Wagner, B. E., J. W. Linowski, J. A. Potenza, R. D. Bates, Jr., J. N. Helbert, and E. H. Poindexter (1976). Dynamic nuclear polarization studies of labile complex formation between lithium ion and nitronyl nitroxide or imidazoline-1-oxyl radical ligands. J. Am. Chem. Soc. **98**, 4405–4409.

Weinkam, R. J., and E. C. Jorgensen (1971a). Free radical analogues of histidine J. Am. Chem. Soc. **93**, 7028–7033.

Weinkam, R. J., and E. C. Jorgensen (1971b). Angiotensin II analogues. VIII. The use of free radical containing peptides to indicate the conformation of the carboxyl terminal region of angiotensin II. J. Am. Chem. Soc. **93**, 7033–7038.

Wenzel, H. R., G. Pfleiderer, W. E. Trommer, K. Paschenda, and A. Redhardt (1977). The synthesis of spin-labeled derivatives of NAD$^+$ and its structural components and their binding to lactate dehydrogenase. Biochim. Biophys. Acta **452**, 292–301.

Wetherington, J. B., S. S. Ament, and J. W. Moncrief (1974). The structure and absolute configuration of the spin-label R-(+)-3-carboxy-2,2,5,5-tetramethyl-1-pyrrolidinyloxy. Acta Crystallogr., Sect. B **30**, 568–573.

Whitesides, G. M., and T. L. Newirth (1975). Reaction of n-butyllithium and 2,2,6,6-tetramethylpiperidine nitroxide. J. Org. Chem. **40**, 3448–3450.

Williams, J. C., R. Mehlhorn, and A. D. Keith (1971). Synthesis and novel uses of nitroxide motion probes. Chem. Phys. Lipids **7**, 207–230.

Wong, L. T. L., R. Schwenk, and J. C. Hsia (1974). New synthesis of nitroxyl radicals of the piperidine and tetrahydropyridine series. Can. J. Chem. **54**, 3381–3383.

Yu, K.-Y., J. J. Baldassare, and C. Ho (1974). Physical-chemical studies of phospholipids and poly (amino acids) interactions. Biochemistry **13**, 4375–4381.

4

Spin-Labeled Synthetic Polymers

WILMER G. MILLER

DEPARTMENT OF CHEMISTRY
UNIVERSITY OF MINNESOTA
MINNEAPOLIS, MINNESOTA

I.	Introduction	173
II.	Sample Preparation	174
	A. Spin-Labeled Polymers	174
	B. Doping Solid Polymers with Spin Probes	184
	C. Labeling Complex Polymeric Materials	185
III.	Analysis in Polymeric Materials	185
	A. Motionally Slowed Spectra	187
	B. Motionally Narrowed Spectra	191
IV.	Motion in Bulk Polymers	195
	A. Comparison with Relaxations Determined by Other Techniques	195
	B. Glass Transition	202
	C. Other Studies with Solid Polymers	203
V.	Effect of Diluents	204
	A. Dilute Polymer Solutions	204
	B. Concentrated Polymer Solutions	209
	C. Effect of Nonsolvents and Mixed Solvent Systems	212
VI.	Conformation of Polymers Adsorbed at Solid–Liquid Interfaces	214
VII.	Conclusion	216
	References	217

I. INTRODUCTION

The use of spin labels and spin probes in biologically related systems has been the thrust of numerous investigations, as many chapters in these volumes testify. The application of spin labels to synthetic polymer systems is in its infancy, even though a synthetic polymer was labeled in one of the earliest

173

nitroxide studies (Stone *et al.*, 1965). The use of spin probes has received much more attention, possibly due to the ease of sample preparation. Russian workers have been foremost in the application of spin probes. Much of their work has been reviewed in Russian (Buchachenko *et al.*, 1973) as well as in English translation (Buchachenko *et al.*, 1976). This chapter deals primarily with spin-labeled polymers, with occasional reference to spin-probe studies wherever it seems appropriate. Bullock, Cameron and co-workers in England and Tormala and co-workers in Finland have made sustained efforts to explore the application of nitroxide studies to synthetic polymer problems. Each group has reviewed particular aspects of the spin-probe and spin-label polymer literature (Tormala and Lindberg, 1976; Bullock and Cameron, 1976). In this chapter, the major emphasis is on the types of problems for which the spin-label method does or may yield important information. There is a critical discussion of that which has been done, with the interpretation frequently colored by the author's own experience in the area.

The most important use of nitroxide spin labels is as a monitor of motion. To the polymer scientist, molecular motion is of interest over an extremely wide range of conditions, from internal rotation in dilute polymer solution to motion in crystalline and amorphous bulk material. A serious potential limitation of the spin-label technique is that one knows for certain only that the electron spin resonance (ESR) spectrum is a monitor of the motion of the nitroxide. Whether this reflects a property of the polymeric material must be established, if possible. Fortunately, molecular motion in polymers can and has been studied by many techniques. Among those giving direct dynamic information are nuclear magnetic resonance, fluorescence depolarization, quasi-elastic light scattering, dielectric relaxation, the torsion pendulum, lumped resonators, and other dynamic mechanical methods. In addition, techniques such as scanning calorimetry and dilatometry as well as other equilibrium measurements give indirect information concerning motion. The techniques vary as to the frequency range of usefulness, due to fundamental or practical limitations. The nitroxide ESR spectrum, being responsive to almost a 10^5 range in rotational correlation time, seems potentially well suited for exploring a variety of polymer problems.†

II. SAMPLE PREPARATION

A. Spin-Labeled Polymers

Approaches to spin labeling biological macromolecules start with the preformed polymer. Strategies with synthetic polymers may include the formation of the polymer itself. Thus, methods and mechanisms of polymer

---

† *Editor's note*: See Chapter 1 by Hyde and Dalton for applications to even slower motions.

synthesis as well as polymer modification are of interest. The increased breadth of labeling approaches is counterbalanced somewhat by the large number of identical functional groups generally present in a synthetic polymer. Consequently, in many cases the labeling is at random rather than at one or a very few specific sites in the macromolecule. The approaches we will consider are (1) copolymerization using a small amount of a nitroxide-containing monomer, (2) copolymerization with a small amount of a particular comonomer that may be modified after polymerization to contain a nitroxide, (3) labeling of a preformed polymer, and (4) initiating or terminating a polymerization by use of a nitroxide or nitroxide precursor. The use of each of these approaches can be found in the literature. Examples of spin-labeled synthetic polymers that have been prepared are shown in Figs. 1 and 3, and examples of spin labels are shown in Figs. 2 and 3.

Methods 1 and 2 are presumed to give randomly labeled polymers. The monomer reactivity ratios describing the copolymerization are frequently unknown. Consequently, it is not known if the distribution is indeed random. In most cases, the nitroxide is present in low percentage. This at least ensures that the nitroxide-bearing monomeric units will not be closely spaced except under the most adverse monomer reactivity ratios (Flory, 1953).

1. COPOLYMERIZATION WITH A NITROXIDE MONOMER

Many of the polymers that have been studied are randomly labeled vinyl addition polymers. The polymerization of vinyl monomers typically follows a free-radical mechanism if a free-radical initiator is utilized or an ionic mechanism if an ionic initiator or catalyst is employed. During a free-radical-initiated polymerization, the presence of a nitroxide free radical is inhibitory. There are no examples in the literature of spin-labeled polymers prepared in this manner.

When the polymerization mechanism does not involve free-radical intermediates, copolymerization with a nitroxide-containing comonomer appears to be feasible, PMMA II was prepared by copolymerization of methyl methacrylate with 4-methylacryloxyl-2,2,6,6-tetramethylpiperidine (100 : 1) using an anionic initiator (Shiotani and Sohma, 1977). The resulting polymer was of rather low molecular weight (7800), containing less than 80 monomeric units. There is no information as to whether the molecular weight was affected by the presence of the nitroxide monomer. One would expect that nitroxide-containing vinyl monomers also could be copolymerized using cation catalysts.

Other polymerizations that do not involve free-radical intermediates include most condensation polymers. Although it is clear that labeled monomers may be used in the preparation of polyamides and polyesters (Section

Polyethylene

$$-CH_2-CH_2-\underset{\underset{R'}{|}}{\overset{\overset{R}{|}}{C}}-CH_2-CH_2-CH_2-$$

PE **IV, V, VI**

Polystyrene

$$-\underset{\underset{\bigcirc}{|}}{\overset{\overset{H}{|}}{C}}-CH_2-\underset{\underset{R'}{|}}{\overset{\overset{R}{|}}{C}}-CH_2-\underset{\underset{\bigcirc}{|}}{\overset{\overset{H}{|}}{C}}-CH_2-$$

PS **II, III, V, VIII, IX, X, XI, XIV, XV**

Poly(vinyl acetate)

$$-\underset{\underset{\underset{CH_3}{|}}{\underset{C=O}{|}}}{\overset{\overset{H}{|}}{C}}-CH_2-\underset{\underset{R'}{|}}{\overset{\overset{R}{|}}{C}}-CH_2-\underset{\underset{\underset{CH_3}{|}}{\underset{C=O}{|}}}{\overset{\overset{H}{|}}{C}}-CH_2-$$

PVAc **VII, XVIII**

Poly(methyl acrylate)

$$-\underset{\underset{\underset{CH_3}{|}}{\underset{O}{|}}}{\overset{\overset{H}{|}}{\underset{C=O}{C}}}-CH_2-\underset{\underset{R'}{|}}{\overset{\overset{R}{|}}{C}}-CH_2-\underset{\underset{\underset{CH_3}{|}}{\underset{O}{|}}}{\overset{\overset{H}{|}}{\underset{C=O}{C}}}-CH_2-$$

PMA **I**

Poly(methyl methacrylate)

$$-\underset{\underset{\underset{CH_3}{|}}{\underset{O}{|}}}{\overset{\overset{CH_3}{|}}{\underset{C=O}{C}}}-CH_2-\underset{\underset{R'}{|}}{\overset{\overset{R}{|}}{C}}-CH_2-\underset{\underset{\underset{CH_3}{|}}{\underset{O}{|}}}{\overset{\overset{CH_3}{|}}{\underset{C=O}{C}}}-CH_2-$$

PMMA **II, III, XIV, XV**

Poly(n-butyl methacrylate)

$$-\underset{\underset{\underset{n-Bu}{|}}{\underset{O}{|}}}{\overset{\overset{CH_3}{|}}{\underset{C=O}{C}}}-CH_2-\underset{\underset{R'}{|}}{\overset{\overset{R}{|}}{C}}-CH_2-\underset{\underset{\underset{n-Bu}{|}}{\underset{O}{|}}}{\overset{\overset{CH_3}{|}}{\underset{C=O}{C}}}-CH_2-$$

PBMA **II**

Fig. 1. Examples of synthetic polymers that have been randomly labeled. The labels (**I–XVIII**) are shown in Fig. 2. References for the labeling procedures are as follows: PE **IV, V**. Bullock *et al.* (1976a). PE **VI**. Bullock *et al.* (1975a). PS **II**. Kurosaki *et al.* (1972). PS **III**. Kurosaki *et al.* (1972), Veksli and Miller (1975), Veksli *et al.* (1976). PS **V**. Bullock *et al.* (1975b). PS **VIII**. Regen (1974). PS **IX**. Drefahl *et al.* (1968), Bullock *et al.* (1971). PS **X**. Bullock *et al.* (1972). PS **XI**. Bullock *et al.* (1973c). PS **XIV, XV**. Veksli and Miller (1975). PVAc **VII**. Rudolph (1976). PVAc **XVIII**. Wassermann *et al.* (1976). PMA **I**. Bullock *et al.* (1976b). PMMA **II**. Bullock *et al.* (1976b), Shiotani and Sohma (1977). PMMA **III**. Kurosaki *et al.* (1972), Veksli and Miller (1975), Veksli *et al.* (1976). PMMA **XIV, XV**. Veksli and Miller (1975). PBMA **XI**. Bullock *et al.* (1976b). PVP **XII**. Fox *et al.* (1974). PHPMA **III, XIII**. Labský *et al.* (1977). PL **XVI**. Stone *et al.* (1965). PBLG **XVII**. Wu (1973).

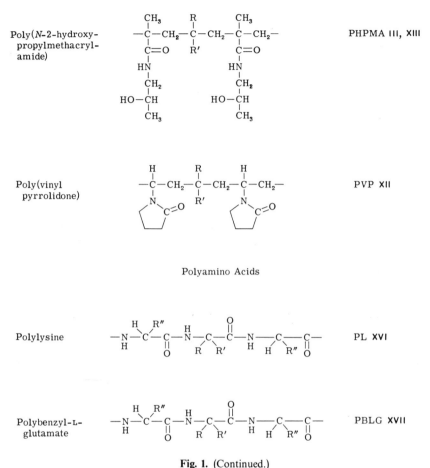

Poly(N-2-hydroxy-
propylmethacryl-
amide)   PHPMA III, XIII

Poly(vinyl
pyrrolidone)   PVP XII

Polyamino Acids

Polylysine   PL XVI

Polybenzyl-L-
glutamate   PBLG XVII

Fig. 1. (Continued.)

II,A,4), no randomly labeled condensation polymers have been prepared in this manner.

The polymerization of N-carboxyanhydrides proceeds by a nonradical mechanism. It is also evident (Section II,A,4) that the presence of a nitroxide does not restrict polymerization, although no randomly labeled polymers have been prepared.

The use of nitroxide-containing monomers to prepare randomly labeled polymers is attractive in that chemical modification of the polymer, leading to potential alteration in the polymer molecular weight, molecular weight distribution, and structure, is avoided. Although the feasibility of many such polymerizations is evident, this synthetic approach has not been explored.

Spin Label	R	R′

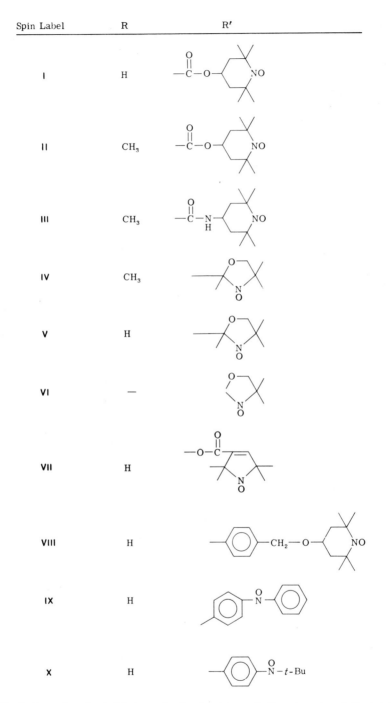

Fig. 2. Examples of spin labels randomly placed in synthetic polymers as indicated in Fig. 1.

XI H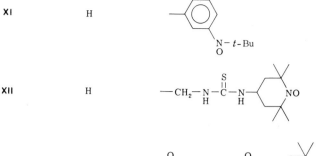

XII H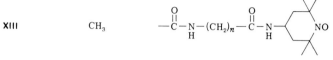

XIII CH$_3$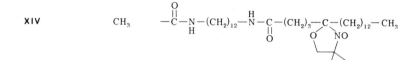

$$n = 1, 2, 4, 5, 6, 7, 11$$

XIV CH$_3$

XV CH$_3$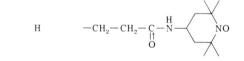

XVI H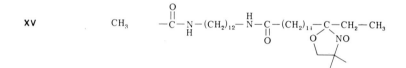

XVII H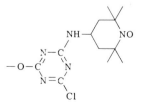

XVIII H

Fig. 2. (Continued.)

179

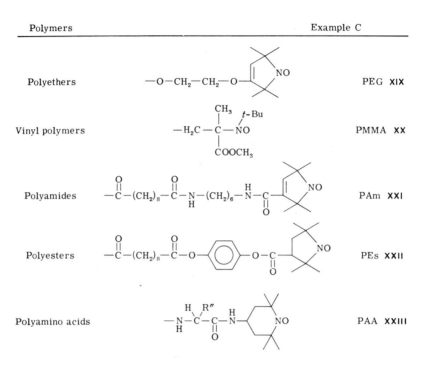

Fig. 3. Examples of end-labeled synthetic polymers. PEG XIX. Tormala *et al.* (1970, 1973). PMMA XX. Forrester and Hepburn (1971), Bullock *et al.* (1974a). PAm XXI, PEs XXII. Tormala *et al.* (1971). PAA XXIII. Sanson and Ptak (1970), Wee (1971), Santee (1972), Wu (1973), Wee and Miller (1973), Miller *et al.* (1974).

2. COPOLYMERIZATION WITH A NITROXIDE PRECURSOR

This approach has been effectively employed to prepare a variety of randomly labeled polymers, especially by Bullock, Cameron, and co-workers. Thus, low-density, spin-labeled polyethylene was prepared by high-pressure, free-radical copolymerization of ethylene with carbon monoxide (200 : 1), methyl vinyl ketone (50 : 1), or methyl isopropenyl ketone (50 : 1), which, upon modification through the ketone group after polymerization, led to PE VI (Bullock *et al.*, 1975a), PE IV, or PE V (Bullock *et al.*, 1976a), respectively. Assuming that the copolymer composition was the same as the monomer feed, the percentage of the ketone groups modified to become nitroxides was about 0.5% with PE VI and less than 0.1% with PE IV and PE V. Labeled polystyrene was prepared by free-radical copolymerization (4 : 1) of styrene with 4-methacryloylamino-2,2,6,6-tetramethylpiperidine, 4-methacryloyloxy-2,2,6,6-tetramethylpiperidine (4 : 1), methyl methacrylate (50 : 1), ethylene glycol dimethacrylate (100 : 1), or methyl vinyl ketone (ca. 15 : 1), which, upon modification of the nonstyrene component, led to

PS **III** and PS **II** (Kurosaki *et al.*, 1972), PS **III** (Veksli and Miller, 1975), and PS **V** (Bullock *et al.*, 1975b), respectively. The labeled polymers prepared by Kurosaki *et al.* (1972) were highly labeled. The other preparations had less than 1% of the minor component labeled. Labeled poly-(methyl acrylate) was prepared by a free-radical copolymerization of methyl acrylate with its acid chloride followed by reaction with 4-hydroxy-2,2,6,6-tetramethylpiperidino-1-oxy to give the nitroxide-labeled polymer PMA **I** (Bullock *et al.*, 1976b). Approximately 1% of the acid chloride units were labeled, and the remainder was converted to the methyl ester. PMMA **III** was prepared by Kurosaki *et al.* (1972) by a procedure similar to that used to prepare PS **III**, again with a very high nitroxide content. PMMA **II** and PBMA **II** were prepared in a manner analogous to PMA **I** (Bullock *et al.*, 1976b), with about 0.1% of the acid chlorides being labeled in PMMA **II**, and 0.5% in PBMA **II**. A series of labeled poly(*N*-2-hydroxypropylmetha-crylamides) were prepared by precipitation copolymerization of *N*-(2-hydroxypropyl)methacrylamide with 4-nitrophenyl esters of methacry-loylated ω-amino acids, followed by aminolysis of the copolymers with 4-amino-2,2,6,6-tetramethylpiperidino-1-oxy to yield PHPMA **III** and **XIII** (Labský *et al.*, 1977). The copolymers, which contained 3 mol% of the nitrophenyl ester, were reported to have an aminolysis yield of about 80% after spin labeling. The molecular weights of the polymers formed were generally high, 30,000 to over 500,000, and in any case can be controlled by the usual procedures employed in polymer synthesis for a given type of polymerization.

In all cases cited except the PHPMA's and the preparation of Kurosaki *et al.* (1972), the percentage of the minor component labeled was very low. Inasmuch as one wants to avoid spin exchange effects by keeping the spin concentration low, the low yields may be by design. However, the resulting polymer is a terpolymer containing major component, unlabeled minor component, and labeled minor component. An exception to this occurs with the ester–acid chloride polymers, in which the unlabeled acid chloride was converted to the ester (Bullock *et al.*, 1976b).

3. LABELING A PREFORMED SYNTHETIC POLYMER

Unlike methods 1, 2, and 4, in which a knowledge of polymer synthesis is relevant and sometimes essential, random labeling of a preformed polymer is little different from labeling of a small molecule. One need only avoid a labeling density per polymer chain that is too high; otherwise, spin exchange will become significant, which cannot be eliminated by adding a diluent (Drefahl *et al.*, 1968; Kurosaki *et al.*, 1972, 1974; Tormala *et al.*, 1971). On the other hand, the large number of identical functional groups allows a very low yield reaction to be a successful one. The molecular weight may range from oligomers to very high molecular weight material and may even be cross-linked if the cross-linking density is sufficiently low to allow solvent swelling (Veksli and Miller, 1975; Regen, 1974). A few examples will be given.

Regen (1974) prepared **PS VIII** by reaction of 2% cross-linked chloro-methylated polystyrene with 2,2,6,6-tetramethyl-4-piperidinol-1-oxyl; Drefahl *et al.* (1968) and Bullock *et al.* (1971) prepared **PS IX** by mercura-tion, followed by treatment with nitrosyl chloride, phenyl magnesium brom-ide, and oxidation; and Bullock *et al.* (1972) prepared **PS X** and **XI** by a procedure similar to that for **PS IX**. PVAc **VII** was prepared by acid–ester interchange (Rudolph, 1976; Rudolph and Miller, 1978), and PVAc **XVIII** was prepared by reaction of partially hydrolyzed PVAc with 2,2,6,6-tetra-methyl-4-aminodichlorotriazinepiperidine-1-oxyl (Wasserman *et al.,* 1976). Aminolysis with a spin-labeled amine was used to produce PMMA **III** (Veksli and Miller, 1975) and PBLG **XVII** (Wu, 1973; Miller *et al.*, 1974). Stone *et al.* (1965) prepared PL **XVI** by reaction of polylysine with 2,2,5,5-tetramethyl-3-isocyanotopyrrolidin-1-oxyl. PMVK **IV** (Bullock *et al.,* 1975b) was prepared by oxidation of 4,4-dimethyloxazolidin-N-oxyl reacted with poly(methyl vinyl ketone).

4. END-LABELED POLYMERS

The terminal monomeric units on a polymer chain frequently possess a functional group found in no other part of the chain. An end-labeled poly-mer is produced if a nitroxide is attached to this unique functional group. Although this procedure could be placed in the previous section, it seems more appropriate here. Thus, Tormala *et al.* (1973) labeled the terminal hydroxyls in poly(ethylene glycols) from 200 to 22,000 molecular weight to give PEG **XIX**. The success in end labeling as a function of molecular weight is shown in Fig. 4. The poor labeling yield at low molecular weight was thought to be controlled by residual water, and at high molecular weight by a "molecular weight effect." There is no *a priori* reason why a "high molecu-

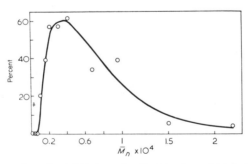

Fig. 4. The percentage of poly(ethylene glycol) molecules end labeled as a function of the number average molecular weight (Tormala *et al.*, 1973).

lar weight effect " should exist, and under different labeling conditions it may not be present. The method has considerable potential for end labeling. Even free-radical-polymerized vinyl polymers could be end labeled if a free-radical initiator containing an appropriate functional group for coupling after polymerization were employed. The polymers could be labeled at both ends if termination were by coupling or at one end if other termination mechanisms predominated.

Other methods of end labeling depend on a knowledge of the mechanism of polymer synthesis. A general method for vinyl polymers uses anionic polymerization followed by terminating the "living polymer" with 2-methyl-2-nitrosopropane or N-benzylidene-*tert*-butylamine N-oxide (Chalfent *et al.*, 1968; Forrester and Hepburn, 1971; Bullock *et al.*, 1974a), giving, for example, PMMA **XX**. Polystyrene, poly(α-methylstyrene), and poly-(vinyl pyridines) have also been end labeled, and the method appears to be applicable to most monomers that can be polymerized anionically. Competing reactions occur under some conditions, leading to more than one type of radical in the labeled polymer (Forrester and Hepburn, 1971; Bullock *et al.*, 1974a).

Condensation or stepwise polymerizations generally do not involve free-radical mechanisms. As stated in Section II,A,1, polymerization with a nitroxide-containing monomer should be feasible. Tormala *et al.* (1971) prepared end-labeled polyesters and polyamides by just this procedure, by copolymerizing two bifunctional, nonlabeled monomers together with a monofunctional nitroxide-containing monomer. Incorporation of the monofunctional monomer into the polymer stops the chain from further growth at that end, thus leading to end-labeled polymers such as PAm **XXI** and PEs **XXII**. Whether the chain is labeled at one or at both ends depends on the stoichiometric ratio of the mono- and bifunctional monomers. The end-labeled polymers produced by Tormala *et al.* (1971) are all of low molecular weight (2600–9100). This is not likely to be due to an inhibitory effect of the nitroxide, particularly with the polyamides, in which a large imbalance in functional groups was employed. In such a polymerization, the limiting number average chain length is given by $(1 + r)/(1 - r)$, where r is the ratio of the two types of functional groups present (Flory, 1953). Using their stated monomer ratios, an average chain length of 2 is expected for the polyamides, and 41 is expected for the polyesters. It thus is surprising that Tormala *et al.* were able to obtain polyamides of as high molecular weight as they did. The method, however, should be generally useful for producing end-labeled condensation polymers and devoid of the multiple-radical problems encountered in the anionic polymerization scheme described above.

Polyamino acids frequently are obtained by polymerization of the N-carboxyanhydride of the amino acid. If the polymerization is initiated by a primary amine, the mechanism does not involve radicals and will incorporate the initiator at the end of the chain. Thus, if the polymerization is initiated with a nitroxide amine, the resulting polymer is end labeled. A variety of end-labeled polyamino acids, e.g., polybenzyl glutamate, polybenzyl asparate, polycarbobenzoxylysine, polyalanine, and polyleucine, have been prepared by this approach (Sanson and Ptak, 1970; Wee, 1971; Santee, 1972; Wu, 1973; Wee and Miller, 1973; Miller *et al.*, 1974). The molecular weight is controlled by the monomer–initiator ratio, with molecular weights up to 180,000 having been prepared.

Many of the major types of synthetic polymers have been spin labeled successfully. An overall view of the labeling strategies has been given, with few detailed accounts of the labeling procedure. In fact, many of the original references themselves have a paucity of detail regarding the synthesis procedure.

The spin labels vary greatly as to how tightly they are coupled to the polymer main chain, ranging from the tightly coupled PE **VI** to PHPMA **XIII**, in which the spin label is at the end of a long tether.

B. Doping Solid Polymers with Spin Probes

The doping of a spin probe into a solid polymer is so much simpler than attaching a label that it would be the method of choice as long as comparable information were attainable. Although the emphasis in this chapter is on spin-labeled polymers, some results of spin-probe studies will be discussed. Many spin-probe studies have been carried out using a variety of polymers and probes (Buchachenko *et al.*, 1973, 1976; Rabold, 1969a,b; Gross, 1971; Bullock *et al.*, 1975a; Wasserman *et al.*, 1976; Kusumoto *et al.*, 1974; Kumler and Boyer, 1976; Lenk, 1974; Tormala *et al.*, 1973, 1974; Savolainen and Tormala, 1974; Tormala and Tulikoura, 1974; Tsutsumi *et al.*, 1976; Catoire and Hagege, 1973; Hagege and Catoire, 1976; Catoire *et al.*, 1977; Veksli and Miller, 1977b). In most cases, the doping is done by removing the solvent from a solution of the nitroxide and polymer. It also has been done by adsorption of a nitroxide vapor into the polymer matrix (Wasserman *et al.*, 1976) and by direct dissolution of the nitroxide into a molten polymer (Kumler and Boyer, 1976). A potential problem with any method is the inhomogeneous distribution of the nitroxide. This includes partitioning between amorphous and crystalline regions in semicrystalline polymers, partitioning between different monomeric units in copolymers, particularly block copolymers, and changes in partitioning as a function of temperature. The partitioning problem is not always easy to identify, avoid, or interpret.

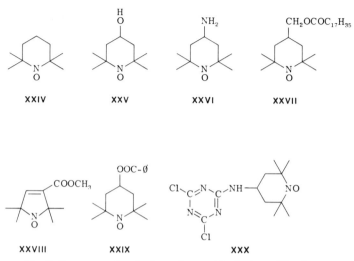

Fig. 5. Some examples of spin probes used in doping solid polymers.

Examples of spin probes doped into solid synthetic polymers are shown in Fig. 5. The concentration is kept sufficiently low ($< 10^{-3}$ M) that spin exchange is not a problem.

C. Labeling Complex Polymeric Materials

Most spin-labeled polymer work to date has involved methods of synthesis and exploring the types of information that can be obtained with simple polymeric materials. Applications involving complex polymeric materials have considerable potential. As an example, latex spheres were surface labeled (Veksli and Miller, 1975; Miller and Veksli, 1975; Veksli *et al.*, 1976) with nitroxides, e.g., PMMA **III**, **XIV**, and **XV**, in an attempt to monitor molecular motion at the polymer–diluent interface when the latex spheres were dispersed in nonsolvents. One can envision other surface-labeling applications, e.g., to study laminates and composites, and the effect of vapors and solvents on films and coatings.

III. ANALYSIS IN POLYMERIC MATERIALS

The control of the concentration of spin label or probe in synthetic polymeric systems is in the hands of the investigator, as we have seen. Signal intensity generally is not a problem, except with high molecular weight end-labeled polymers or with studies on adsorbed monolayers. The difficulty in interpreting ESR spectra from polymer systems depends on the degree of detail desired and on the type of problem one is investigating. Thus, one may be interested only in monitoring the sudden change in the outer hyperfine extrema as a measure of the glass transition (Kumler and Boyer, 1976), or in determining accurately the temperature dependence of the rotational correlation time in order to compare activation energies with those determined by other dynamic measurements. In a single study, the sample may range from a glassy or crystalline material to a highly viscous fluid to a dilute solution of polymer in solvent. An example of this is shown in Figs. 6 and 7. The spectra are seen to range from rigid-limit or glass-type spectra to isotropic, motionally narrowed spectra. In analyzing the spectra to determine correlation times, one must consider motional anisotropy and the possibility of superposition of noninterconverting spin populations of differing correlation times. In addition, there is always the problem of uniqueness. The theoretical aspects of nitroxide spectra have been dealt with extensively and expertly in other chapters in these volumes. The remainder of this section is

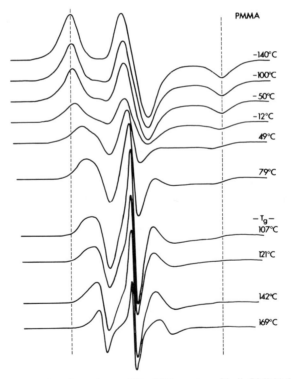

Fig. 6. The temperature dependence of the ESR spectra of bulk PMMA III (Veksli and Miller, 1977a). The dashed lines are the positions of the outer hyperfine entrema at $-140°C$. The glass transition (T_g) occurs at $110° \pm 5°C$.

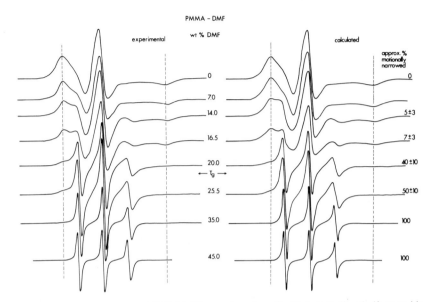

Fig. 7. The ESR spectra of PMMA III as a function of added N,N-dimethylformamide (DMF) at 25°C. The dashed lines indicate the outer hyperfine extrema in the dry polymer. The solvent composition at which the glass transition reaches 25°C is indicated (22 wt % by scanning calorimetry). The calculated spectra are explained in the text (Section V) (Veksli and Miller, 1977a).

colored significantly by the author's experience with spectral interpretation in polymeric systems, with particular emphasis on obtaining and interpreting rotational correlation times. In some cases, it may be at variance with literature interpretations.

Theoretical treatments exist which allow rotational correlation times to be determined from motionally narrowed as well as from motionally slowed spectra. For accurate determinations, the components of the **g** and **A** tensors must be known. In most cases, these are not known accurately for spin-labeled polymers. In this circumstance, one uses published values for a small molecular weight analogue as close in chemical structure as possible. Although the **g** and **A** components do not vary greatly from one spin label to another, the correlation time calculated from experimental spectra depends on the values assumed.

A. Motionally Slowed Spectra

All the following interpretations are based on the treatment by Freed (Freed *et al.*, 1971; Freed, 1976). Although the theory predicts the entire spectral line shape, polymer data generally are interpreted, not from a complete line shape analysis, but from approximate relationships deduced by Freed and co-workers.

1. CORRELATION TIMES CALCULATED FROM LIFETIME BROADENING OF THE OUTER HYPERFINE EXTREMA

The ESR spectra are insensitive to motion when the nitroxide rotational correlation time is more than a few microseconds. An increase in motion brought about, for example, by an increase in temperature can both broaden the outer hyperfine extrema as well as decrease their separation. Mason and Freed (1974) found the broadening to be more sensitive to motion than changes in separation in this time range. On the basis of computer simulations, Mason and Freed have given approximate expressions for the rotational correlation time as a function of the extrema broadening for *isotropic* rotational reorientation, and Brownian as well as jump diffusion.

Shown in Fig. 8 is the temperature dependence of the half-width at half-height of some low-field extrema (Δ_l). The linewidth of the extrema is very large and decreases with increasing temperature. In our experience, this is frequently observed in solid synthetic polymers. It has been observed by others (Bullock *et al.*, 1975a). Over this temperature interval the extrema separation decreases by 3 G in the PMMA samples and by a very small amount in the end-labeled PBLG. In the randomly labeled, amorphous, glassy PMMA samples, Δ_l appears to depend on sample preparation, e.g., on the thermal history of the sample. This method does not appear to be useful for synthetic polymers. The very large, rigid-limit linewidths are probably due to quite heterogeneous environments. As motion sets in, the observed linewidth decreases as a result of both averaging of the environmental hetero-

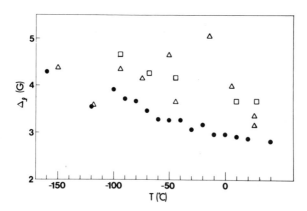

Fig. 8. The half-width at half-height of the lower hyperfine extrema (Δ_l) as a function of temperature for solid PMMA III, (△) PMMA XXVII (□), and PBLG XXIII (180,000) (●).

geneities and the intermolecular inhomogeneous broadening. These narrowing effects apparently are larger than the lifetime broadening contribution. In addition, the motion in the solid polymer may well be anisotropic. Approximate expressions from theory for anisotropic motion have not been given.

2. CORRELATION TIMES CALCULATED FROM A CHANGE IN THE SEPARATION OF THE OUTER HYPERFINE EXTREMA

For *isotropic* motion of the spin label, the correlation time τ may be estimated from the separation of the extrema (Goldman *et al.*, 1972),

$$\tau = a(1 - S)^b \tag{1}$$

where S is the ratio of the extrema separation $(2A_z')$ to its rigid limit $(2A_z)$. The constants a and b depend primarily on the peak-to-peak derivative Lorentzian linewidth and the diffusion model (Brownian, free or jump). An alternate calculation assuming isotropic motion can be made using the formulation of McCalley *et al.* (1972).

For *anisotropic* motion the situation is considerably more complex. Mason *et al.* (1974) computed the τ dependence of the separation for spin label **III** (Fig. 2) with fast motion about the C–N bond. Even assuming axially symmetric rotation, the $\tau - A_z'$ dependence will be different for other labels. No compilation has been made.

Correlations for spectra in Fig. 6 up to 49°C could be calculated from Eq. (1). If the motion of the label is anisotropic, which seems highly likely in a solid polymer, the correlation time calculated from Eq. (1) assuming isotropic motion will be in error. This is a serious problem, which is discussed in Section IV.

3. ANISOTROPIC SLOW MOTION VERSUS SUPERPOSITION OF NONINTERCONVERTING MOTIONALLY SLOWED AND MOTIONALLY NARROWED SPECTRA

A problem can arise that is not always simple to resolve. In Fig. 9 are shown spectra for PBLG **XXIII**. The structure of this labeled polymer is shown in Fig. 10. Mason *et al.* (1974) showed convincingly that the nitroxide motion in this system is highly anisotropic, dominated by a fast motion about the C–N bond. For the 16.5 wt % DMF spectrum in Fig. 7, Veksli and Miller (1977a) have argued that this spectrum corresponds to a superposition of a motionally narrowed and a motionally slowed spectrum. In spectra B–E in Fig. 9 and in the 16.5 wt % DMF spectra in Fig. 7, there are both low-field and high-field doublets. Spectrum F in Fig. 9 and the 7 and 14 wt % DMF spectra in Fig. 7 are similar, in that the low-field extrema have a shoulder and there is a high-field doublet. In Fig. 7 the conclusion of

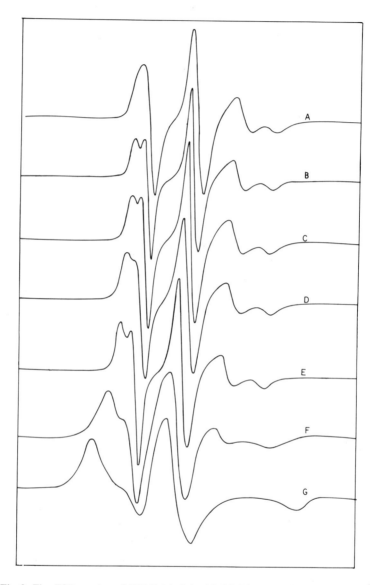

Fig. 9. The ESR spectra of PBLG labeled with 2,2,6,6-tetramethyl-4-aminopiperidin-1-oxyl in DMF solutions of various concentrations at room temperature. The polymer concentration (volume fraction) was (A) 0.008, (B) 0.0917, (C) 0.128, (D) 0.148, (E) 0.200, (F) 0.42 (0.5 weight fraction), and (G) 1.0 (solid polymer). The molecular weight was 122,000 (Wee and Miller, 1973).

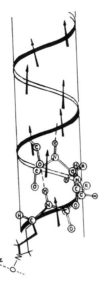

Fig. 10. Conformation of PBLG in the solid state or in helicogenic solvents such as DMF. In end-labeled PBLG **XXIII** the spin label is as shown (Wee and Miller, 1973).

superposition was based on the invariance of A_z' for the slow component, on the magnitude of the separation of the low-field "doublet," and on the analysis of the motion in the absence of solvent. In many instances, less information is available, and distinguishing between these two situations is not always possible.

When there is a superposition of motionally narrowed and motionally slowed spectra, an additional problem exists. The signal from nitroxides that are motionally narrowed may be at least two orders of magnitude more intense than that from the same number of motionally slowed nitroxides (see, for example, the 25.5 wt % DMF spectrum in Fig. 7). Thus, with a slightly noisy signal one could miss entirely the presence of the slow component.

B. Motionally Narrowed Spectra

When the nitroxide motion is sufficiently fast that a three-line spectrum is observed, correlation times are calculated at three levels of refinement. These can be stated as (1) isotropic motion assuming no proton hyperfine interaction, (2) isotropic motion with proton hyperfine interaction, and (3) anisotropic motion.

1. Isotropic Motion Assuming No Proton Hyperfine Interaction

The most frequently used analysis is based on the Kivelson theory (Kivelson, 1960), in which the linewidth $[T_2(m_N)]^{-1}$ is given by

$$[T_2(m_N)]^{-1} = \left[\frac{3b^2}{20} + \frac{4}{45}(\Delta\gamma\, H_0)^2 + \frac{b^2 m_N^{\,2}}{8} - \frac{4}{15} b\, \Delta\gamma\, H_0 m_N \right] \tau + X \quad (2)$$

Here, m_N is the nitrogen nuclear spin state, H_0 the applied magnetic field, τ the rotational correlation time, and X the contributions to the broadening not dependent on m_N. Also, b is given by

$$b = (4\pi/3)[A_{zz} - \tfrac{1}{2}(A_{xx} + A_{yy})] \quad (3)$$

and $\Delta\gamma$ by

$$\Delta\gamma = \frac{|\beta|}{h}[g_{zz} - \tfrac{1}{2}(g_{xx} + g_{yy})] \quad (4)$$

Equation (2) can be modified to eliminate X (Stone et al., 1965); thus,

$$\frac{[T_2(m_N)]^{-1}}{[T_2(0)]^{-1}} = 1 - [\tfrac{4}{15}b\, \Delta\gamma\, H_0 T_2(0)\tau]m_N + [\tfrac{1}{8}b^2 T_2(0)\tau]m_N^{\,2} \quad (5)$$

or

$$\frac{[T_2(m_N)]^{-1}}{T_2(0)^{-1}} = 1 + Bm_N + Cm_N^{\,2} \quad (6)$$

By using the ratio of the widths of all three lines, τ can be calculated from either the linear or quadratic term in Eq. (6). Inasmuch as the peak-to-peak line height $h(m_N)$ of the first derivative of a Lorentzian line varies with the inverse square of the width, the correlation time is calculated typically from the three peak-to-peak line heights and the linewidth of the central line $[T_2(0)]^{-1}$; thus,

$$\tau = 4\left[\left(\frac{h(0)}{h(1)} \right)^{1/2} + \left(\frac{h(0)}{h(-1)} \right)^{1/2} - 2 \right] b^{-2}[T_2(0)]^{-1} \quad (7)$$

2. Isotropic Motion with Proton Hyperfine Interaction

The proton hyperfine coupling with the electron is a source of inhomogeneous broadening which, if not taken into account, will lead to erroneous correlation times (Poggi and Johnson, 1970). The proton couplings are sometimes very clearly observable (Bullock et al., 1973a; Murakami et al., 1976). In particular, Bullock, Cameron, and co-workers have made an effort to correct the peak-to-peak intensities and linewidths before using the Kivelson theory to calculate a correlation time. The procedure is to take the

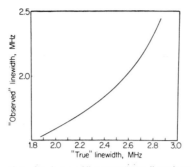

Fig. 11. Calibration plot of "observed" versus "true" peak-to-peak linewidth (Bullock *et al.*, 1973a).

center multiplet from an arbitrarily chosen experimental spectrum, simulate it with arbitrarily chosen proton hyperfine coupling constants until the computed and experimental lines fit, and then generate "true" versus "observed" intensity ratios and linewidths. An example is shown in Fig. 11 (Bullock *et al.*, 1973a). The "true" or corrected values are then used in Eq. (7). The correlation times corrected for proton hyperfine coupling may differ in some cases by only a few percent from the uncorrected ones (Bullock *et al.*, 1975a). In other cases there is a much larger effect on the correlation time.

Alternatively, Kuznetsov *et al.* (1974) showed that Eq. (7) can be used without correction for the broadening due to the proton–electron interaction if there is a suitable additional source of broadening such as oxygen. Labský *et al.* (1977) showed that correlation times calculated from Eq. (7) with nondegassed samples of PHPMA **XIII** are in good agreement with those calculated from degassed samples, taking into account the proton hyperfine interaction.

3. ANISOTROPIC MOTION

In most spin-labeled polymers the nitroxide is pendant to the main chain and connected to it by one or more single bonds. Rotational relaxation about pendant single bonds in solution is typically 10^{-10}–10^{-12} sec. Motion about main-chain bonds must be correlated with motion about adjacent main-chain bonds and is typically slower. Thus, it is not surprising that nitroxide motion in labeled polymers may be anisotropic irrespective of whether the spectra are motionally narrowed or motionally slowed. When the slowest motion is sufficiently fast that the **g** and **A** tensors are averaged, the spectrum is insensitive to motional anisotropy. If the motion about the terminal bond is held constant and other sources of motion are made progressively slower, the spectrum will be affected. A particularly clear example

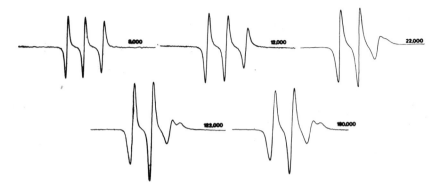

Fig. 12. The ESR spectra for 1 wt % solutions of helical PBLG **XXIII** in DMF at 60°C as a function of molecular weight (Wu, 1973).

of this is shown in Fig. 12, in which the molecular weight of end-labeled helical PBLG **XXIII** is made progressively larger. The fast rotation (Mason *et al.*, 1974) about the terminal N–C bond (Fig. 10) should be independent of molecular weight, whereas the rotational motion of the rodlike α helix is strongly dependent on molecular weight. The two lowest molecular weights are motionally narrowed. One can calculate an erroneous correlation time using Eq. (7), with or without proton hyperfine interaction corrections.

Alternatively, one can analyze in terms of anisotropic motion. The simplest approach is to assume axially symmetric rotary diffusion, whereby the motion is described by two rotary diffusion coefficients and two correlation times. Buchachenko *et al.* (1973, 1976) and Freed (1976) have reviewed the analysis of motionally narrowed spectra assuming axially symmetric diffusion. On the other hand, the Freed "slow-motion" theory using an asymmetric diffusion tensor can be used (Freed, 1976). However, to even contemplate such an analysis one would have to have a solid basis for the motions involved, as, for example, with the end-labeled helical polypeptides. Thus, in most cases analysis in terms of a pseudo-axially symmetric rotation is the most feasible approach.

There are several ways of estimating whether anisotropic motion is significant. The commonly used nitroxides have **g** and **A** components such that in X-band measurements the central line is never more than slightly broader than the low-field line. Thus, the spectra for the two lowest molecular weights in Fig. 12 clearly indicate anisotropic motion. Another clue is to calculate the correlation time from Eq. (6) using the linear term and the quadratic term. If the values do not agree, this can be taken as good evidence of significant motional anisotropy (see, for example, Bullock *et al.*, 1976b; Labský *et al.*, 1977).

IV. MOTION IN BULK POLYMERS

A. Comparison with Relaxations Determined by Other Techniques

Motion in bulk polymers is related to their mechanical properties and has been studied extensively (Ferry, 1970). Experimentally, the most frequently used techniques are dynamic mechanical spectroscopy, dielectric relaxation, fluorescence depolarization, and nuclear magnetic resonance (NMR) spectroscopy. From these, relaxation frequencies ($\nu = 1/2\pi\tau$) as a function of temperature are determined and typically presented as an Arrhenius plot. The results for poly(methyl methacrylate) are shown in Fig. 13, in which four distinct relaxations are observed. The molecular motion corresponding to a particular relaxation must be deduced by comparison with other measurements and other polymers. Relaxation plots for many polymers have been determined (McCall, 1969, 1971). With dynamic mechanical data, the relaxation in the absence of molecular identification are labeled α, β, γ, etc. in order of descending temperature. Thus, the dynamic mechanical data in Fig. 13 would be designated as α (in all amorphous polymers related to the glass transition as deduced by volume or calorimetric data), β (labeled the "ester dipole"), and γ (labeled "chain methyls") relaxations.

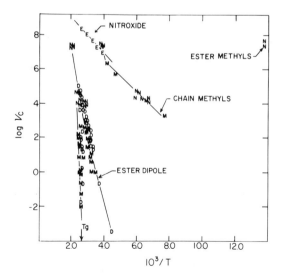

Fig. 13. Temperature dependence of the relaxation frequencies in PMMA as determined from dielectric (D), dynamic mechanical (M), and NMR (N) data. Also shown is the correlation frequency for nitroxide motion about the terminal N–C bond in PMMA III (E, ESR nitroxide) (Veksli and Miller, 1977a).

The temperature dependence of bulk, spin-labeled PMMA, PS, PVAc, PMA, PBMA, PE, and PEG (see Figs. 1 and 3) have been studied in some detail. In all cases, the separation of the hyperfine extrema decreases as the temperature increases (see, e.g., Fig. 6). Veksli and Miller (1977a) analyzed the ESR spectra in terms of anisotropic motion. All other studies in which a correlation time was deduced assumed isotropic motion.

1. Poly(Methyl Methacrylate), Poly(Methyl Acrylate), and Poly(n-Butyl Methacrylate)

PMMA II has been studied by two groups (Shiotani and Sohna, 1977; Bullock *et al.*, 1976b), and PMMA III has been studied by Veksli and Miller (1977a). The structural difference between the two spin labels is minor (see Fig. 2). The extrema separation for PMMA II is shown in Fig. 14. Its temperature dependence is sigmoidal, approaching the rigid limit at low temperature and the isotropic value at high temperature, significantly above the glass transition ($T_g \approx 100°C$). Shiotani and Sohna calculated correlation times assuming isotropic motion, as shown in Fig. 15. The apparent activation energy ranges from nearly zero below 80°C to greater than 10 kcal/mole at high temperature. The nearly zero activation energy from 0° to 80°C results from the small temperature variation in the extrema. However, the data of Bullock *et al.* show the extrema to be temperature dependent to at least −200°C. If their extrapolated rigid limit is used, the correlation times will be up to an order of magnitude smaller and show a temperature dependence of the correlation time in the 0°–80° range.

Using PMMA III, Veksli and Miller (1977a) calculated correlation times from the extrema separation assuming anisotropic motion, in particular, motion about the terminal N–C bond (see Fig. 2). The values are shown in

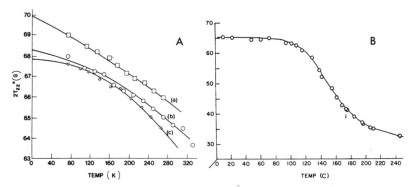

Fig. 14. Extrema separation for PMMA II as determined (A,b) by Bullock *et al.* (1976b) or (B) by Shiotani and Sohma (1977). The values for PMA II (A,a) and PBMA II (A,c) are also shown.

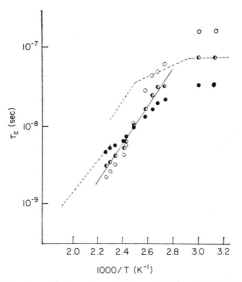

Fig. 15. Correlation times (Shiotani and Sohma, 1977) calculated from data in Fig. 14B assuming isotropic motion and Eq. (1) with Brownian (○), moderate-jump (◑), or strong-jump (●) diffusion. Dashed lines calculated from slow-motion formulation of Shiotani and Sohma.

Fig. 13. If isotropic motion were assumed, and correlation times were calculated according to Eq. (1), the values are smaller by about a factor of 5 and are close to the values calculated for PMMA II using the rigid-limit value of Bullock *et al.* (Fig. 14A). Veksli and Miller argue that, in order to have isotropic motion, motion at least about the ester group would be necessary, in addition to motion about the terminal N–C (in PMMA III) or O–C (in PMMA II) bond. But in Fig. 13 one sees that at room temperature the ester group motion is orders of magnitude slower than the nitroxide motion irrespective of the method of calculation and so can hardly contribute to the nitroxide motion. Only at higher temperature, particularly above T_g, does other motion become comparable to the terminal bond motion. Veksli and Miller found the motion to be still significantly anisotropic even at 169°C.

The extrema separation for PMA II and PBMA II is shown in Fig. 14A at temperatures less than T_g. Qualitatively, one can see that the nitroxide motion is similar for all three polymers in the glassy state. The β relaxation (ester rotation) is very slow in this region and similar for all three polymers (McCall, 1969). If the dominant motion in this temperature region is about the terminal O–C bond, similar motion for the three polymers is not surprising.

It seems very reasonable to the author that the motion of a covalently attached nitroxide in a bulk polymer below its glass transition most likely would be highly anisotropic in many polymers. It should be clear from the discussion above that to identify molecular processes and determine the associated activation energy with any certainty is difficult.

2. POLYSTYRENE

The relaxation map of polystyrene, shown in Fig. 16, is complex. The molecular motion corresponding to each relaxation is not well established. Yano and Wada (1971) believe that the β relaxation represents a local backbone oscillation, the γ a phenyl rotation, and the δ a lattice defect motion. Others believe that neither the δ nor the γ relaxation involves phenyl rotation (Tonelli, 1973). Bullock *et al.* (1973b) studied bulk PS **X** in the glass state ($T_g \approx 100°C$). The spectra were analyzed in terms of isotropic motion. Arrhenius plots are nonlinear, with low (1–2 kcal/mole) activation energies (see Fig. 16). The assumption of isotropic motion coupled with the complex relaxation map for polystyrene makes identification of the molecular basis of the motion difficult.

PS **X** and end-labeled PS **XX** have been examined by electron–electron double resonance by Dorio and Chien (1975). They observed a fast nitroxide motion down to $-150°C$ with only a slight temperature dependence. The motion was suggested to be a low activation energy torsional oscillation. The nitroxide in the end-labeled polymer had a much longer relaxation time and a higher activation energy.

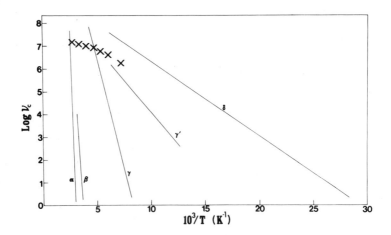

Fig. 16. Relaxation map for polystyrene [based on the composite map of Yano and Wada (1971)] and the nitroxide relaxation in PS **X** (Bullock *et al.*, 1973b) based on Eq. (1) and moderate-jump diffusion.

PS **X** has also been studied by pulsed ESR spectroscopy (Brown, 1976). A minimum was observed in the phase memory time at $-123°C$, which was interpreted as possibly due to the onset of molecular motion associated with phenyl rotation.

3. POLYETHYLENE

Linear polyethylene and branched polyethylene have surprisingly similar relaxation maps (McCall, 1969), with transitions associated with the crystalline as well as the amorphous state. Branched polyethylene PE **VI** was studied by Bullock *et al.* (1975a) from $-196°$ to $+116°C$, ranging from the glassy state through the glass transition, variously reported from $-123°$ to $-80°C$ (Brandrup and Immergut, 1975), and on up through the melting temperature ($\sim 110°C$ for low-density PE). The sample investigated had an estimated crystallinity of about 60%. The spin label is coupled directly to the chain backbone, with no intermediary single bonds. Therefore, nitroxide motion requires backbone motion. The ESR spectra showed virtually no change in the hyperfine extrema up through the glass transition. A gradual decrease in separation occurred above T_g, with a rapid decrease about midway between T_g and T_m. Bullock *et al.* calculated correlation times assuming isotropic motion. They associated the motion between $-50°$ and $75°C$ with the β relaxation observed by other techniques. An alternate explanation is possible. The spectra above room temperature are shown in Fig. 17. It is clear to this author that the spectra in Fig. 17 below T_m ($110°C$) are composite spectra, consisting of a motionally narrowed and a motionally slowed population of spins. This is based on the observation that the nitrogen hyperfine splitting of the fast component is seen to be close to that of the splitting in the melt or in the solution spectra, and that the separation in both the low- and high-field "doublets" is much larger than with anisotropic motion (Mason *et al.*, 1974). A logical guess as to the double population of nitroxides is that the fast component is in the amorphous regions and that the slow is in the crystalline regions. As the temperature is increased, the correlation time of the fast component decreases. The intensity of the fast component therefore will increase at the apparent expense of the slow component, or possibly the number of spins in amorphous regions as well as their motion will increase with increasing temperature as the melting point is approached. Bullock *et al.* did consider superposition but felt the spectra represented anisotropic motion. One additionally should be able to differentiate between superposition and anisotropic motion using samples of different crystallinity by appropriate annealing. The annealed sample should show less "fast" component than the less crystalline ones. Although Bullock *et al.* did anneal, no spectra were reported.

The temperature dependence of bulk PE **IV** and PE **V** has been studied

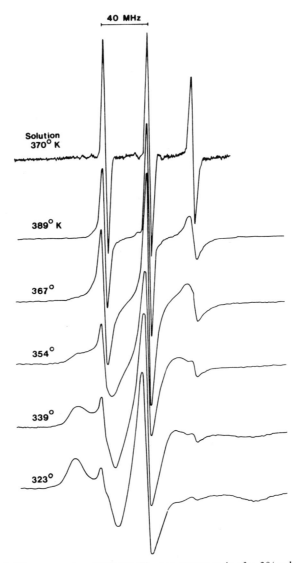

Fig. 17. Solid-state spectra of PE **VI**. The top spectrum is of a 3% solution in xylene (Bullock *et al.*, 1975a).

(Bullock *et al.*, 1976a). These compounds contain one single bond between the label and the polymer backbone. The extrema separation is qualitatively similar to that of PE **VI**. No spectra were reported, except at 299° and 321°K for PE **V**. The 299°K spectra is strongly indicative of a composite spectrum, although the authors believe that it is due to anisotropic motion.

4. POLYETHYLENE GLYCOL

The temperature dependence of end-labeled PEG **XIX** from 200 to 22,000 molecular weight (MW) has been studied (Tormala *et al.*, 1973) using highly crystalline samples (> 80%). A rapid narrowing of the hyperfine extrema occurs between T_g and T_m, analogous to the results with randomly labeled PS, PE, and PMMA. The published spectra ($M_n = 4000$) at temperatures slightly below T_m suggest a composite spectrum, but not as dramatic as in polyethylene. The correlation times determined on the assumption of isotropic motion are shown in Fig. 18. At high temperatures, above T_m, and at low temperatures, but above T_g, the correlation times show only a small molecular weight dependence except at very low molecular weight. In the melt, the apparent activation energy (~ 6 kcal/mole) has little molecular weight dependence, and the correlation times obviously have little relationship to the bulk viscosity. Further considerations by Tormala (1974a) suggest that the nitroxide relaxation between T_m and T_g correlates with the γ relaxation.

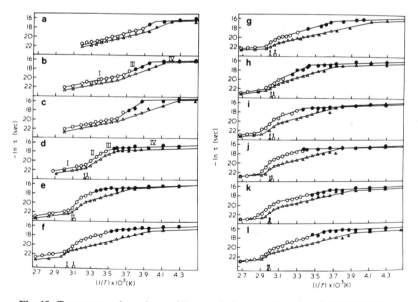

Fig. 18. Temperature dependence of the correlation time of end-labeled PEG **XIX** (○) and probe **XXV** in PEG (△) as a function of molecular weight. All values were calculated assuming motionally narrowed (open symbols) or motionally slowed (shaded symbols) isotropic motion. Melting points are indicated by arrows, and all temperatures are above T_g. The molecular weights were (a) 200, (b) 400, (c) 600, (d) 1000, (e) 1550, (f) 2050, (g) 3000, (h) 4000, (i) 67,000, (j) 9500, (k) 15,000, (l) 22,000 (Tormala *et al.*, 1973).

5. Poly(Vinyl Acetate)

PVAc **XVIII** of MW 10^5 together with the spin probes PVAc **XXIV** and PVAc **XXX** was studied by Wasserman *et al.* (1976). The spectrum of the labeled polymer is not motionally narrowed even 100°C above T_g. Analogous to the previous discussions of other labeled polymers, the separation of the hyperfine extrema changes much more rapidly above than below T_g. Correlation times, assuming isotropic motion, were calculated and found to have much lower activation energies below than above T_g (~ 30°C). Comparison with NMR data above T_g indicated that the nitroxide motion depends not only on polymer segmental motion but on additional relaxation mechanisms as well. Motion about the N–C bond, as in PMMA **III**, would seem likely. Motion of the spin probe **XXX** was similar to the labeled polymer except at high temperature, whereas the probe **XXIV** showed faster motion at all temperatures.

6. Polyurethane Networks

Ward and Books (1974) studied the temperature dependence of two labeled polyurethane networks, one containing a very low "hard block" content, and the other a very high one. The network with low "hard block" content clearly shows composite spectra over a range in temperature above the glass transition, with no significant change during applied strains. In contrast, the sample containing a high "hard block" content showed little temperature dependence of the spectral line shape. As would be expected, swelling of the network shifted the onset of rapid motion to lower temperature.

B. Glass Transition

The glass transition, which refers to the narrow range in temperature during which correlated, segmental backbone motion becomes frozen in, is usually measured by dilatometry or calorimetry. It also can be related to free volume (Ferry, 1970). The α relaxation in amorphous polymers generally correlates well with calorimetric or dilatometric data as, e.g., in Fig. 13. In the previous section, the rotational correlation time of spin-labeled polymers was seen to show no dramatic effects at T_g, although it may rapidly decrease between T_g and T_m. The separation of the hyperfine extrema, from which the correlation times are generally calculated, changes rather rapidly from a value near the rigid limit (60–65 G) to the isotropic value (30–35 G) somewhere between T_g and T_m, but not at T_g.

Spin probes, not being covalently bound to the polymer, typically show much more motional freedom near T_g than spin labels. Tormala (1974b), extending earlier work (Fig. 18, and Tormala *et al.*, 1973) on polyethylene glycol, determined the correlation time of probe **XXVIII**, which was located primarily in the amorphous phase (Tormala and Tulikoura, 1974). An Arrhenius plot of the correlation time showed a break at $-60°C$, similar to the T_g determined by other dynamic measurements (McCall, 1969). In this temperature range, the spectra are all motionally slowed and far from the temperature when the extrema separation is changing rapidly. Tormala concluded that the rotation of the probes at low temperature is a volume-controlled process.

Arrhenius plots of the isotropic correlation time of probes **XXIV** and **XXX** in PVAc (Wasserman et al., 1976) consist of two linear segments. The temperature of intersection of the linear segments corresponds well to the T_g of PVAc measured by other methods, even though the smaller radical was found to move by jump diffusion and the larger by Brownian diffusion.

Motion of spin probes above T_g has been studied and reviewed extensively (Buchachenko et al., 1973, 1976). The motion depends on the size of the probe and can be correlated well with segmental motion of the polymer matrix as determined by other techniques.

Early spin-probe work found a correlation between extrema separation and T_g (Rabold, 1969b; Buchachenko et al., 1973). An extrema separation of 50 G is approximately midway between the rigid limit and the isotropic separation. As an empirical parameter, the temperature ($T_{50\,G}$) at which the extrema separation was 50 G was chosen arbitrarily to correlate to T_g. A linear relationship using probe **XXIX** initially was suggested (Boyer, 1973). Further extension to polymers with low T_g values showed the relationship to be nonlinear (Kumler and Boyer, 1976) but describable by

$$T_{50\,G} = \frac{T_g}{1 - 0.03 T_g / \Delta H_a} \tag{8}$$

where ΔH_a is the apparent activation enthalpy observed in low-frequency mechanical or dielectric relaxation measurements. Extension of this work to the molecular weight dependence of T_g for polystyrene has given impressive results (Kumler et al., 1977). The utility of this correlation is in estimating T_g using a spin probe in complex polymeric material or in polymers that yield ambiguous results by other techniques. Inasmuch as a superposition of two populations of spins of different motional properties is sometimes observed even in homopolymers (Veksli and Miller, 1977b; Bullock et al., 1975a; Kusumoto et al., 1974; Ward and Books, 1974), analysis in complex materials must be approached with caution.

C. Other Studies with Solid Polymers

Some interesting applications have been made of spin probes with solid polymers for which comparable studies have not been made with labeled polymers. The decreased motion resulting from covalent bonding may, in fact, make the spin-label method much less useful than the spin-probe method in these cases. Much of the work has been done by the Russian group. The kinetics of the crystallization of poly(ethylene terephthalate) using probe **XXV** was found to follow the Avrami equation,† with activation energies comparable to those determined from other techniques. Morphological studies have been made, including the effect of spherulite size on motion in the amorphous region and the effect of stretching on motion in crystalline as well as noncrystalline polymers. The reviews by Buchachenko *et al.* (1973, 1976) should be consulted for the original Russian references.

V. EFFECT OF DILUENTS

A. Dilute Polymer Solutions

Small spin probes in low-viscosity solvents have rotational correlation times of the order of hundredths of microseconds. Attaching the nitroxide by one or more single bonds would be expected to have an effect on the nitroxide motion. Most synthetic polymers, whether amorphous or semicrystalline, assume random conformations with much internal motion when in solution. Dilute polymer solutions give a high-resolution NMR spectrum, although broadened somewhat over their low molecular weight analogues (Bovey and Tiers, 1963). The motion on the NMR time scale is clearly dominated by local mode motion. One would anticipate that nitroxides attached to random-coil polymers in solution would show motionally narrowed spectra but would be motionally slowed when attached to rigid-chain polymers. This is realized experimentally, as can be seen from Figs. 7 and 12. When attached to flexible polymers, the nitroxide motion may be anisotropic, since correlated local mode motion may be slower than single-bond rotation. However, all motions may be sufficiently fast or the spectroscopic parameters known insufficiently to make a clear analysis for motional anisotropy. It is difficult generally to assess this anisotropy from the literature. Very few dilute-solution spectra are published. Instead, a correlation time is reported, generally assuming isotropic motion. When motional anisotropy is investigated, it is usually done by comparing correlation times calculated from the B and C terms in Eq. (6).

The correlation times for several spin-labeled polymers in relatively dilute solution are given in Table I. The polymer concentration, molecular weight, temperature, and method of analysis differ somewhat from one example to another. Consequently, comparisons with other than a homolo-

---

† This is the frequently used equation to describe the kinetics of phase formation.

TABLE I

CORRELATION TIMES IN DILUTE SOLUTION AT ROOM TEMPERATURE[a]

Polymer	Solvent side chain	τ (nsec)[b]	Reference
PE IV, V	Xylene	0.36	Bullock et al. (1976a)
PS III[c]	DMF	~ 0.1	Veksli and Miller (1977a)
	CHCl₃	0.4	Veksli and Miller (1977a)
	Benzene	0.3 ± 0.1	Veksli and Miller (1977a)
PS VIII[c,d]	Benzene	0.2	Regen (1974)
	CH₂Cl₂	0.4	Regen (1974)
	CCl₄	0.5 ± 0.1	Regen (1974, 1975b)
	Tetrahydrofuran	0.3	Regen (1975b)
PS IX	Toluene	0.55	Bullock et al. (1971)
PS X	Toluene	0.52	Bullock et al. (1974b)
PS XI	Toluene	0.65 (196,000)	Bullock et al. (1973a)
		0.65 (97,200)	Bullock et al. (1973a)
		0.55 (50,000)	Bullock et al. (1973a)
		0.5 (19,700)	Bullock et al. (1973a)
		0.4 (10,000)	Bullock et al. (1973a)
		0.35 (3550)	Bullock et al. (1973a)
		0.3 (2025)	Bullock et al. (1973a)
PVAc VII	CHCl₃	0.04	Rudolph (1976)
	Benzene	0.05	Rudolph (1976)
	Toluene	0.06	Rudolph (1976)
	CCl₄	0.20	Rudolph (1976)
	C₂H₅OH	0.07	Rudolph (1976)
PVAc XVIII	CH₃OH	0.63	Wasserman et al. (1976)
PMMA II	Benzene	0.13–0.17	Murakami et al. (1976)
	Toluene	0.36	Bullock et al. (1976b)
PMMA III	DMF	0.07	Veksli and Miller (1977a)
	CHCl₃	0.05	Veksli and Miller (1977a)
PMMA III[c]	Benzene	0.30	Veksli and Miller (1977a)
PHPMA III	CH₃OH	1.7	Labský et al. (1977)
PHPMA XIII (n = 11)	CH₃OH	0.17	Labský et al. (1977)
PL XVI	H₂O	0.4 (pH 8)	Stone et al. (1965)
PBLG XVII	DMF	0.2	Miller et al. (1974)
End labeled PMMA XX[e]	CH₃COOC₂H₅	0.33, 0.22 (82,000)	Bullock et al. (1976b)
		0.36, 0.12 (25,000)	Bullock et al. (1976b)
		0.34, 0.12 (2300)	Bullock et al. (1976b)
PAm XXI	n-Cresol	5.	Tormala et al. (1971)
PEs XXII	CHCl₃	0.1	Tormala et al. (1971)

(*Continued*)

TABLE I (Continued)

Polymer	Solvent side chain	τ (nsec)[b]	Reference
Probes			
XXIV	$CHCl_3$	~ 0.01	Veksli and Miller (1977a,b)
XXVI	$CHCl_3$	0.02	Veksli and Miller (1977a,b)
	DMF	0.02	Veksli and Miller (1977a,b)
	CH_3OH	0.02	Veksli and Miller (1977a,b)
XXX	CH_3OH	0.15	Wasserman et al. (1976)

[a] Concentration generally less than 10 wt %, and temperature 20°–30°C.

[b] Numbers in parentheses indicate molecular weight.

[c] Lightly cross-linked polymer (1–2%).

[d] Solvents listed swell the cross-linked polymer to at least three times the dry volume.

[e] Anisotropic motion reported. Values calculated from Eq. (6) using C (first value) or B (second value) terms.

gous series from the same laboratory should be restricted to a single significant figure at best. As a gross generalization, the correlation times fall in the range of tenths of nanoseconds, and the corresponding spin probe 5–10 times smaller.

1. EFFECT OF MOLECULAR WEIGHT

PMMA II of 33,000 and 550,000 MW in toluene exhibited identical correlation times within experimental error over the entire temperature range studied (Bullock et al., 1976b). PVAc VII of 6×10^4 to 6×10^5 MW in several solvents at room temperature showed no perceptible molecular weight dependence of the correlation time (Rudolph, 1976). PS XI in toluene (Bullock et al., 1973a) had a measurable decrease in τ as the molecular weight was decreased (Table I), being particularly noticeable with molecular weights less than 20,000. End-labeled PMMA XX in ethyl acetate has little molecular weight dependence at room temperature (Table I) but a noticeable dependence at very low molecular weight at lower temperatures. The independence of molecular weight above approximately 10^4 is indicative of main-chain contributions to the nitroxide relaxation being dominated by local mode motions. This is quite consistent with the NMR data on flexible polymers. Only at low molecular weight do highly correlated, multiple-bond backbone modes become fast enough to contribute to the nitroxide relaxation (Liu and Ullman, 1968; Carlstrom et al., 1973). At very low molecular weights the nitroxide motion has been sensibly described in terms of independent contributions from internal motion and end-over-end tumbling (Bullock et al., 1973a, 1974b).

2. EFFECT OF SOLVENT

The rotation of a solid body in a continuum as given, for example, by the Stokes–Einstein equation for spheres or the Perrin equations for ellipsoids of revolution, etc., depends on T/η, where η is the solvent viscosity. Many internal rotations, determined by various techniques, also show a T/η dependence. Additional factors may be important. As the flexible polymer is placed in a thermodynamically poorer and poorer solvent, the polymer chain dimensions decrease (Flory, 1953). In a sufficiently poor solvent, the polymer has a tendency to collapse, excluding solvent from its domain (Ptitsyn et al., 1968). The rate of bond rotations may thus depend on polymer–solvent interaction parameters as well as on the solvent viscosity. In addition, the hydrodynamic behavior of short chains has been shown to be very different from that of long chains and to depend on local solvent–solute interactions rather than on a model assuming continuum hydrodynamics (Devan et al., 1971). The applicability of these results to the motion of short chains attached to a polymer backbone is possible, although not well studied.

The effect of solvent viscosity is evident in Table I. The 5-nsec correlation time for PAm **XXI** is unusually large. When the high viscosity (20 cP at 20°C) of the solvent is taken into account, the value is quite in line with other correlation times for flexible-chain polymers. PMMA **II** and PMMA **XX**, when corrected for solvent viscosity, are very close, although the motional anisotropy in PMMA **XX** somewhat clouds the comparison.

The solvents $CHCl_3$, benzene, toluene, and CCl_4, with 25°C viscosities of 0.54, 0.60, 0.55, and 0.89 cP, respectively, are progressively worse solvents for PVAc, ranging from an athermal solvent to an almost θ solvent. The correlation times for PVAc **VII**, after correction for viscosity differences, still show a clear increase on going to progressively worse solvents. The PMMA data show a similar trend. Thus, a change in solvent can have effects in addition to a simple viscosity effect.

3. EFFECT OF POLYMER AND SPIN-LABEL STRUCTURE

The data in Table I suggest that spin label as well as polymer structure effect the observed correlation times. Thus, PVAc **XVIII** in methanol has a correlation time larger by at least a factor of 10 than that of PVAc **VII** in solvents of similar solvent power and viscosity. Even though there must be some internal rotation in the two rings of **XVIII**, the size of the attached nitroxide has an effect. This is supported by comparing the corresponding spin probes. However, after considering the benzene and toluene data for PS **III**, **VIII**, **IX**, **X**, and **XI**, it is not yet possible to separate spin-label size and internal motion and backbone motion into separate contributions. The results on PHPMA make separation of motion into various factors even more difficult. PHPMA **III** has a factor of 10 larger correlation time than PHPMA **XIII** ($n = 11$). Label **XIII** may be looked on as label **III** attached to the polymer backbone through a series of single bonds. PHPMA **XIII** with n values less than 11 is reported to have progressively less motion, approaching PHPMA **III** as n goes to 0, thus suggesting an additive contribution per single bond. But the order of magnitude difference between PHPMA **III** and PMMA **III** requires other considerations. Both polymers have identical groups coming off the backbone, and the spin labels themselves are identical. The conformational space available to the two polymers should be similar and quite restricted (Sundararsjan and Flory, 1974). The large difference may be a solvent effect, since methacrylate polymers in polar solvents can exhibit very unusual conformational effects (Priel and Silberberg, 1969). Without further characterization of the polymer–solvent systems, it is quite difficult to assign the molecular basis of the effect of side-chain length.

A comparison of PE **IV**, **V**, and **VI** in xylene is instructive since PE **VI** is attached to the chain backbone as a ring, whereas PE **IV** and **V** are attached through single bonds. In addition, the polyethylene chain has much more conformational freedom than mono- and disubstituted vinyl polymers. At 60°C, PE **VI** has a correlation time of 0.25 msec, whereas that of PE **IV** and **V** is 0.1 nsec (Bullock *et al.*, 1975a, 1976a). The presence of internal rotation in the pendant label seems evident, although the motional freedom of the unsubstituted backbone is a dominant factor.

Finally, it should be noted that lightly cross-linking the polymer has no more than a small effect in a highly swollen network, at least in PS **III**, PS **VIII**, and PMMA **III**. This is consistent with NMR data (Liu and Burlant, 1967).

4. Comparison with Other Methods

Rotational relaxation times determined from fluorescence depolarization experiments in dilute solution depend on the size of the fluorescent label as well as on polymer motion (North and Soutar, 1972). With anthracence and fluorescein labels, relaxation times of nanoseconds are observed with many of the polymers listed in Table I. The order of magnitude slower motion is somewhat related to the larger size of the fluorescent label compared to the spin labels generally employed.

Dielectric relaxation of a dilute, atactic PMMA solution in benzene at 30°C is 1.4 nsec (Pohl *et al.*, 1960). Numerous studies of dielectric relaxation in PMMA solutions suggest that the mode of relaxation involves cooperative rotation about backbone as well as side-chain bonds (Block and North, 1970). The faster relaxation in PMMA **II** and PMMA **III** may come from additional motion about the single bonds past the carbonyl, although this is likely to be too simplistic an interpretation. The dielectric relaxation frequency for PMA in toluene at room temperature is reported to be 1.8 GHz or a relaxation time of 0.1 nsec, with PVAc showing remarkably similar results (Block and North, 1970). The increased motion compared to PMMA is thought to be a result of less steric hindrance. The spin-label data show a similar trend.

The correlation time for PS IX corresponds closely with the proton NMR result for polystyrene (Bullock et al., 1971). Correlation times were calculated by Bullock et al. (1973a) from ^{13}C T_1 measurements of Allerhand and Hailstone (1972) for polystyrene in tetrachloroethylene and compared to spin-label correlation times of PS X in toluene. The correspondence was excellent and taken as evidence of the absence of significant single-bond rotation in the side-chain-attached label. More recent NMR studies cast some doubt on this interpretation. A very extensive ^{13}C NMR study of polystyrene (Heatley and Begum, 1976) reveals a distribution of correlation times, as shown by several models, with a mean value of 4 nsec at 30°C. This is an order of magnitude slower than the spin-label result, suggesting internal motion in the nitroxide as well as backbone segmental motion. The ^{13}C NMR studies by Heatley and Begum on PMMA also indicated a distribution of relaxation times, with a mean relaxation time of 1–2 nsec at 30°C. Again, the effect of internal side-chain motion seems evident.

In much of the dilute-solution spin-label work, activation parameters have been determined from Arrhenius plots. These generally are similar to those obtained by other means. They have not been emphasized here due to the difficulty in molecular interpretation resulting from internal side-chain as well as backbone segmental motion.

B. Concentrated Polymer Solutions

The effect of concentration has been systematically studied in only a few cases, primarily with polystyrene (Bullock et al., 1974b; Regen, 1974, 1975a,b; Veksli and Miller, 1975, 1977a,b), with poly(methyl methacrylate) (Veksli and Miller, 1975, 1977a,b; Murakami et al., 1976), and with the rigid, α-helical polypeptides (Wee and Miller, 1973; Miller et al., 1974; Wu, 1973). In random-coil polymers, starting at infinite dilution, there is little effect of concentration up to about 10 wt % polymer. As the polymer concentration is further increased, the correlation time increases but generally by less than a factor of 5, even at 30–40% polymer. At these concentrations, using a high molecular weight polymer, the bulk viscosity of the solution is several orders of magnitude higher than in dilute solution. This further emphasizes the control of the nitroxide relaxation by internal rotation and local segmental backbone motion. This motion depends on the local or microscopic viscosity rather than on the bulk viscosity. A measure of this can be obtained from the concentration dependence of the solvent translational diffusion coefficient, which is weakly concentration dependent until high polymer compositions are reached and which can be correlated with the monomeric frictional coefficient of the polymer (Ferry, 1970; Crank and Park, 1968; Boss et al., 1967). The Fujita–Doolittle free volume theory (Fujita, 1961) successfully correlates the concentration dependence of the solvent translational diffusion and the solvent dependence of the monomeric friction coefficient. Thus, Veksli and Miller (1977a,b), by analogy, applied the Fujita–Doolittle equation to the concentration dependence of the nitroxide motion, as shown by the solid lines in Fig. 19. The fit was taken to mean that the backbone segmental motion dominated the nitroxide relaxation, whereas the solvent primarily affected the monomeric frictional coefficient. An alternate explanation may be that both side-chain and backbone segmental motion are important, but in the concentration range both types of motion are affected similarly.

Veksli and Miller (1977a,b) studied the concentration dependence of good solvents on nitroxide motion in PMMA III and PS III starting with the dry polymer. The results for PMMA III can be summarized by referring to Figs. 7 and 19. No nitroxide motion was observed until sufficient solvent was present to nearly plasticize (lower the glass transition to the temperature of measurement) the polymer. Just before plasticization, some nitroxides began to show a much shorter correlation time, with the bulk of the nitroxides unaffected by the presence of solvent at least to a first approximation. Further addition of solvent after plasticization resulted in a continuing bimodal distribution of correlation times. The slow component was approximately the same as in the dry polymer, and the correlation time of the other component showed a decreasing correlation time while constituting an increasing fraction of the spins. At some concentration, depending on the solvent, the slow component was no longer evident. The fast component could be modeled by the Fujita–Doolittle equation as described above. With spin label III in polystyrene, the results appear to be similar to those in Fig. 19 for PMMA, with an important exception. With PS III in benzene and chloroform, the appearance of the fast component was not evident until after the polystyrene was plasticized and corresponded more closely to the solvent concentration needed to plasticize PMMA. Since in both polymers the labeled monomeric unit was a methacrylate, this was taken to mean that the backbone motion contributing to the nitroxide motion is indeed a very local motion involving the labeled monomeric unit and few additional backbone bonds. The origin of the bimodal distribution was suggested to be related to differences in the rate of conformational interconversion among the various stereosequences. This suggestion is highly speculative, although evidence does exist that isotactic and syndiotactic units have different mo-

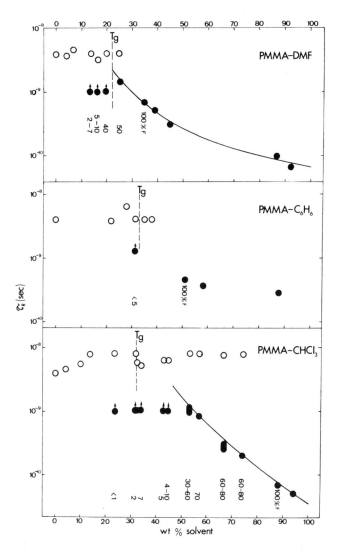

Fig. 19. The concentration dependence of the rotational correlation times of nitroxide-labeled PMMA III at room temperature (Veksli and Miller, 1977a). The slow-motion component (○) was calculated assuming anisotropic motion, whereas the fast component was either calculated assuming isotropic rotation (●) or observed to be present but a correlation time was not calculable (↑). The vertical dashed lines indicate the solvent composition at which T_g reaches the temperature of measurement. The percentage of the spin labels undergoing fast motion as determined by computer simulation is indicated. The solid curves through the fast-motion component were calculated from a Fujita–Doolittle free volume diffusion equation (Fujita, 1961). The DMF and CHCl$_3$ data were obtained using homogeneously labeled PMMA (linear), and the benzene data were obtained with surface-labeled PMMA (latex) (Veksli and Miller, 1977a).

tional freedom (Higgins *et al.*, 1972; Lyerla *et al.*, 1977). In **PMMA III** and **PS III** the effect of solvent appears to be different from the temperature dependence of the bulk polymer in that a bimodal distribution of correlation times was not observed in the latter. However, spin probes **XXV** and **XXVI**, hydrogen bonded to the **PMMA** ester groups, have a bimodal distribution of correlation times in some temperature ranges (Veksli and Miller, 1977b).

The solvent dependence of a spin label attached to a long, rigid polymer depends markedly on the number of single bonds between the rigid backbone and the nitroxide. Thus, end-labeled PBLG **XXIII** (Fig. 10), attached to the rigid polymer through only one single bond, exhibits anisotropic, motionally slowed spectra even at infinite dilution (Fig. 9). On the other hand, side-chain-labeled PBLG **XVI** has a motionally narrowed spectrum and a correlation time of 2 nsec with only 11 wt % solvent present. The correlation time is within a factor of 10 of the infinite dilution value (Miller *et al.*, 1974; Wu, 1973). PBLG **XVI**, having four single bonds between the rigid backbone and the nitroxide, with probably no more than three of these being effective for motion, illustrates vividly the contribution of single-bond motion to nitroxide relaxation in polymeric systems.

C. Effect of Nonsolvents and Mixed Solvent Systems

Polymer latex spheres of 1000 Å nominal diameter were prepared by emulsion polymerization, surface labeled, and studied when dispersed in nonsolvents for the polymer (Miller and Veksli, 1975; Veksli and Miller, 1975; Veksli *et al.*, 1976) in an effort to observe motion at the polymer–nonsolvent interface, as shown schematically in Fig. 20. Spheres were homogeneously labeled as a control. Many nonsolvents were found to give motion to both the surface and the homogeneously labeled spheres, a result that was surprising and confounding. Some examples are shown in Fig. 21. It was shown eventually that many fluids classified as nonsolvents penetrate

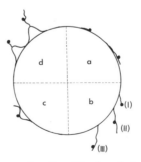

Fig. 20. Representation of types of motional freedom that might be expected with surface-labeled latexes dispersed in a nonsolvent.

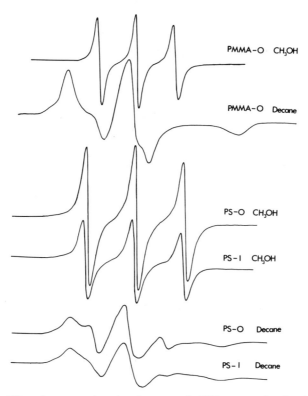

PMMA-O   CH₃OH

PMMA-O   Decane

PS-O   CH₃OH

PS-I   CH₃OH

PS-O   Decane

PS-I   Decane

Fig. 21. Effect of excess methanol or decane on the ESR spectra of surface-labeled, cross-linked, poly(methyl methacrylate) (PMMA O) and polystyrene (PS O) latex spheres and on homogeneously labeled, cross-linked polystyrene latex (PS I). Measurements at room temperature (Miller and Veksli, 1975).

a few hundred angstroms into the polymers in a relatively short time period. The spectra frequently show a bimodal distribution of correlation times, as was observed with the polymers when nearly plasticized with good solvents (Section V,B). Thus, many of the questions represented in Fig. 20 cannot be answered clearly because of the solvent penetration. Instead, this work suggests the use of spin labels to monitor the penetration of solvents and nonsolvents into a polymeric material.

The effect of fluids on homogeneously labeled, cross-linked polystyrene resin beads (**PS VIII**) of 200–400 mesh has been investigated (Regen, 1974, 1975a,b). The nitroxide motion as a function of degree of swelling with solvents and mixed solvents (Regen, 1974) was similar to the fast component curves in Fig. 19. The nitroxide motion was unaffected by nonsolvents such as alcohol in a 24-hr exposure, unlike the results of Veksli et al. (1976) with latex spheres. However, the penetration of nonsolvents into glassy polymers can be exceedingly slow and non-Fickian (Crank and Park, 1968). In our experience, both by nitroxide and by calorimetric measurements (Coon, 1978), particles of the order of tens of microns (200–400 mesh) may take weeks to equilibrate with nonsolvents, whereas 1000 Å particles may take only minutes to hours. Thus, the discrepancy is probably a result of equilibration time and particle size difference. Furthermore, it is our observation that in mixed solvent systems a partitioning of the solvents may occur. Thus, the nitroxide in a homogeneously labeled PMMA III latex sphere dispersed in 0.1 M acetic acid became very mobile over a period of a few days as the acetic acid, a moderate solvent, was extracted from the aqueous solution (Z. Veksli and W. G. Miller, unpublished) without significant destruction of the spin label.

Regen (1975a) found the degree of cross-linking in the polystyrene resin to significantly affect the nitroxide motion. Inasmuch as the solvent uptake is also a function of the cross-linking, such an effect is expected. In a comparison study of the motion of a spin probe, analogous to label VIII, in the polystyrene resins significantly more motion was found with the probe than with the label. The results were interpreted in terms of motion in solvent channels. There is no other evidence that solvent channels exist in a polymer plasticized with a good solvent. The results can be comprehended without solvent channels from the discussion in the preceding section on the effect of solvent concentration and the effect of covalent versus noncovalent attachment (Table I and Veksli and Miller, 1977b).

VI. CONFORMATION OF POLYMERS ADSORBED AT SOLID–LIQUID INTERFACES

The adsorption of polymers onto solid surfaces is of much importance when considering coatings, adhesives, filters, reinforcers, etc. The conformation of a polymer molecule adsorbed at a solid–liquid interface has received much attention both theoretically and experimentally (Lipatov and Sergeeva, 1974; Mittal, 1975). Theoretically the size of loops, trains, and tails is predicted as a function of solvent–surface–polymer interaction and chain length. Experimentally a mean surface thickness is inferred by ellipsometry, viscosity, or isotherm measurements. The fraction of monomeric units attached to the surface occasionally is determined spectroscopically. At the molecular level, knowledge of the polymer conformation is difficult to obtain. From our previous discussions of solvent effects on nitroxide motion in labeled polymers, the spin-labeling technique would seem well suited to differentiate between loops and tails from surface-adsorbed units. Little work in this area has been reported, however.

Fox *et al.* (1974), Clark *et al.* (1976), and Robb and Smith (1974; 1977) prepared labeled poly(vinyl pyrrolidone), PVP **XII**, of 18,000 and 40,000 MW and studied its adsorption from aqueous or chloroform solutions onto aerosil silica. Spectra composed of two populations of spins of widely different correlation times similar to spectra found in Fig. 7 were observed. The spectra were resolved into relative amounts of the two populations by comparison with the spectrum of the polymer in aqueous glycerol at various temperatures. The fast component was assumed to come from loops, and the slow component from trains. Several significant conclusions were made. At low surface covering approximately 90% of the spins were in the slow component, indicating a rather flattened conformation. The fraction of the slow component decreased with increasing surface coverage. The higher molecular weight had a lower percent slow component than did the lower molecular weight at the same surface coverage. With a saturated surface no desorption was observed in 24 hr, yet the polymer was able to change its conformation reversibly and rapidly (<2 min) while adsorbed. The concentration dependence of the slow component is consistent with the concentration dependence of the fraction of bound monomeric units deduced from other techniques, whereas the fractions of bound monomeric units at low coverage was higher than that generally found by other methods.

Labeled poly(vinyl acetate), PVAc **VII**, of 61,000, 194,000, and 603,000 MW was adsorbed on several oxide surfaces (TiO_2, Al_2O_3, silica) from four solvents ranging from relatively good (chloroform, nearly an athermal solvent) to relatively poor (CCl_4, nearly a θ solvent), using concentrations corresponding to saturation surface coverage (Rudolph, 1976). With Al_2O_3 only a glass-type spectrum was observed, indicating that the chain was very tightly held to the surface. Composite spectra were observed with TiO_2 and silica surfaces, similar to PVAc **XII**. With any solvent and either TiO_2 or silica, a distinct molecular weight dependence was observed in that the fraction of fast spins decreased significantly with increasing molecular weight. At fixed molecular weight and surface, the fraction of fast component decreased with decreasing solvent power. Further studies have shown that paramagnetic impurities in the high surface area oxides can significantly effect the fast component intensity, and that one must approach surface adsorption studies with caution (Miller *et al.*, 1979). The molecular weight dependence with a given surface and solvent seemed consistent only with the fast component coming from tails rather than loops. This interpretation results from loop size and fraction of units in loops becoming independent of molecular weight at high molecular weight, whereas the fraction of units in tails vanishes at infinite molecular weight. These results indicate that loop size in these systems is small. Most theoretical treatments consider only the infinite molecular weight. The calculation of Lal and co-workers (Mittal, 1975) suggests that tails may be significant and loop size small, although the calculations are not strictly applicable to the conditions studied.

VII. CONCLUSION

Procedures are currently available to either randomly label or end label virtually any synthetic polymer, with the spin concentration controlled such that there is good signal intensity but no spin–spin interaction. Generally one or more single bonds connect the nitroxide to the polymer backbone. In many applications, it would be desirable to eliminate this source of internal rotation—for example, in studies with dilute polymer solution. Convenient methods to attach the label rigidly to the backbone without the possibility of rotation should be investigated. Inasmuch as attaching the nitroxide to the polymer increases the correlation time, there are some applications in which the use of spin probes seems better suited than spin labels, e.g., when glass transition temperatures are being determined.

Studies on bulk polymers and highly concentrated solutions currently are hampered by lack of definitive means to distinguish anisotropic from isotropic motionally slowed spectra and to differentiate anisotropic motionally slowed spectra from composite motionally slowed and motionally narrowed spectra. Further theoretical calculations in this area would be useful. It may well turn out that a problem in spectral uniqueness exists, and auxiliary data may be needed. On a qualitative basis changes due to annealing, etc., can be studied without a detailed understanding of the motion.

Studies in dilute polymer solutions give results that can be sensibly correlated with results from other types of measurements. In both dilute and concentrated solution, the spin-label method can provide information on motion at the molecular level conveniently and with much less effort than nuclear magnetic resonance.

The use of spin labels to monitor the conformation of a polymer adsorbed at a solid–liquid interface has received little attention but appears to be very promising. It is restricted to solids that will not interfere with ESR measurements on the spin label. It can provide information not easily obtainable by other techniques.

The application of the spin-label method thus far has emphasized studies on homopolymers. As more experience is gained in labeling and in interpreting spectra, the method will undoubtedly be applied to more complex materials such as composites, laminates, and phase-separating systems.

ACKNOWLEDGMENTS

Support of the experimental work by grants from the U.S. Public Health Service (GM-16922), by the Education Committee of the Rubber Division, American Chemical Society, and by the Petroleum Research Fund (8749-AC5, 6), administered by the American Chemical Society, is gratefully acknowledged.

REFERENCES

Allerhand, A., and R. K. Hailstone (1972). Carbon-13 Fourier transform nuclear magnetic resonance spectra. X. Effect of molecular weight in ^{13}C spin-lattice relaxation times of polystyrene in solution. *J. Chem. Phys.* **56**, 3718–3720.

Block, H., and A. M. North (1970). Dielectric relaxation in polymer solutions. *Adv. Mol. Relaxation Process.* **1**, 309–374.

Boss, B. D., E. O. Stejskal, and J. D. Ferry (1967). Self-diffusion in high molecular weight polyisobutylene-benzene mixtures determined by the pulsed-gradient, spin-echo method. *J. Phys. Chem.* **71**, 1501–1506.

Bovey, F. A., and G. V. D. Tiers (1963). The high resolution nuclear magnetic resonance spectroscopy of polymers. *Fortschr. Hochpolym.-Forsch.* **3**, 139–195.

Boyer, R. F. (1973). Glass temperatures of polyethylene. *Macromolecules* **6**, 288–299.

Brandrup, J., and E. H. Immergut, eds. (1975). "Polymer Handbook," 2nd ed. Wiley, New York.

Brown, I. M. (1976). Molecular oxygen effects on the electron spin relaxation in a spin-labeled polystyrene. *J. Chem. Phys.* **65**, 630–638.

Buchachenko, A. L., A. L. Kovarskii, and A. M. Wasserman (1973). Study of polymers by the paramagnetic probe method. *Usp. Khim. Fiz. Polim.* pp. 31–63.

Buchachenko, A. L., A. L. Kovarskii, and A. M. Wassermann (1976). Study of polymers by the paramagnetic probe method. *In* "Advances in Polymer Science" (Z. A. Rogovin, ed.), pp. 26–57. Halsted Press, New York.

Bullock, A. T., and G. G. Cameron (1976). E.S.R. studies of spin-labeled synthetic polymers. *In* "Structural Studies of Macromolecules by Spectroscopic Methods" (K. J. Ivin, ed.), pp. 273–316. Wiley, New York.

Bullock, A. T., J. H. Butterworth, and G. G. Cameron (1971). Electron spin resonance studies of spin-labeled polymers. I. Segmental relaxation of polystyrene in solution. *Eur. Polym. J.* **7**, 445–451.

Bullock, A. T., G. G. Cameron, and P. Smith (1972). Electron spin resonance studies of spin-labeled polymers. II. Preparation and characterization of spin-labeled polystyrene. *Polymer* **13**, 89–90.

Bullock, A. T., G. G. Cameron, and P. M. Smith (1973a). Electron spin resonance studies of spin-labeled polymers. III. The molecular weight dependence of segmental rotational correlation times of polystyrene in dilute solution. *J. Phys. Chem.* **77**, 1635–1639.

Bullock, A. T., G. G. Cameron, and P. M. Smith (1973b). Electron spin resonance studies of spin-labeled polymers. IV. Slow-motional effects in solid polystyrene. *J. Polym. Sci., Polym. Phys. Ed.* **11**, 1263–1269.

Bullock, A. T., G. G. Cameron, and P. M. Smith (1973c). Electron spin resonance studies of spin-labeled polymers. V. Synthesis and characterization of meta-labeled polystyrene. *Polymer* **14**, 525–526.

Bullock, A. T., G. G. Cameron, and J. M. Elsom (1974a). Electron spin resonance studies of spin-labeled polymers. VI. End-labeled poly(methyl methacrylate). *Polymer* **15**, 74–76.

Bullock, A. T., G. G. Cameron, and P. M. Smith (1974b). Electron spin resonance studies of spin-labeled polymers. VII. Dependence of rotational correlation times on solvent properties and polymer concentration. *J. Chem. Soc., Faraday Trans.* 2 **70**, 1202–1210.

Bullock, A. T., G. G. Cameron, and P. M. Smith (1975a). Electron spin resonance studies of spin-labeled polymers. VIII. Relaxation processes in low density polyethylene. *Eur. Polym. J.* **11**, 617–624.

Bullock, A. T., G. G. Cameron, and P. M. Smith (1975b). Electron spin resonance studies of spin-labeled polymers. IX. Polymers of methyl vinyl ketone. *Makromol. Chem.* **176**, 2153–2157.

Bullock, A. T., G. G. Cameron, and P. M. Smith (1976a). Electron spin resonance studies of spin-labeled polymers. X. Polyethylene containing ketone side groups. *Macromolecules* **9**, 650–653.

Bullock, A. T., G. G. Cameron, and V. Krajewski (1976b). Electron spin resonance studies of spin-labeled polymers. XI. Segmental and end-group mobility of some acrylic ester polymers. *J. Phys. Chem.* **80**, 1792–1797.

Carlstrom, D. E., W. G. Miller, and R. G. Bryant (1973). Molecular weight dependence of the chlorine-35 nuclear magnetic resonance line width in polypeptides. *J. Phys. Chem.* **77**, 2759–2770.

Catoire, B., and R. Hagege (1973). Mise en évidence de la pénétration dans les fibres de solvents avec "marqueur de spin" par resonance paramagnétique électronique (RPE). *Bull. Sci. Inst. Text. Fr.* **2**, No. 7, 209–219.

Catoire, B., P. Bouriot, and R. Hagege (1977). Etude en spectroscopie de resonance paramagnétique électronique (RPE) des variations de porosité et de structure de fibres acryliques. *Bull. Sci. Inst. Text. Fr.* **6**, No. 21, 1–8.

Chalfent, G. R., M. J. Perkins, and A. Horsfield (1968). A probe for homolytic reactions in solution. II. The polymerization of styrene. *J. Am. Chem. Soc.* **90**, 7141–7142.

Clark, A. T., I. D. Robb, and R. Smith (1970). Influence of solvent on the conformation of polymers adsorbed at the solid/liquid interface. *J. Chem. Soc. Faraday Trans. I.* **72**, 1489–1494.

Coon, D. L. (1978). Relaxation in the glassy state of amorphous polystyrene: Temperature and diluent effects. M.S. Thesis, University of Minnesota, Minneapolis.

Crank, J., G. S. Park, eds. (1968). "Diffusion in Polymers." Academic Press, New York.

Devan, R. K., V. A. Bloomfield, and P. B. Berget (1971). Intrinsic viscosity of short-chain normal alkanes. *J. Phys. Chem.* **75**, 3120–3124.

Dorio, M. M., and J. C. W. Chien (1975). A study of molecular motion in polymeric solids by electron-electron double resonance. *Macromolecules* **8**, 734–739.

Drefahl, G., H.-H. Horhold, and K. D. Hofmann (1968). Ein makromolekulaus polyradikal vom diphenylstickstoffoxide-typ. *J. Prakt. Chem.* **37**, 137–142.

Ferry, J. D. (1970). "Viscoelastic Properties of Polymers." Wiley, New York.

Flory, P. J. (1953). "Principles of Polymer Chemistry." Cornell Univ. Press, Ithaca, New York.

Forrester, A. R., and S. P. Hepburn (1971). Spin traps. A cautionary note. *J. Chem. Soc. C* p. 701.

Fox, K. K., I. D. Robb, and R. Smith (1974). Electron paramagnetic resonance study of the conformation of macromolecules adsorbed at the solid of liquid interface. *J. Chem. Soc., Faraday Trans. I* **70**, 1186–1190.

Freed, J. H. (1976). Theory of slow tumbling esr spectra for nitroxides. *In* "Spin Labeling: Theory and Applications" (L. J. Berliner, ed.), Vol. 1, pp. 53–132. Academic Press, New York.

Freed, J. H., G. V. Bruno, and C. F. Polnaszek (1971). Electron spin resonance line shapes and saturation in the slow motion region. *J. Phys. Chem.* **75**, 3385–3399.

Fujita, H. (1961). Diffusion in polymer-diluent systems. *Fortschr. Hochpolym.-Forsch.* **3**, 1–47.

Goldman, S. A., G. V. Bruno, and J. H. Freed (1972). Estimating slow-motional rotational correlation times for nitroxides by electron spin resonance. *J. Phys. Chem.* **76**, 1858–1860.

Gross, S. (1971). Rotational mobility of nitroxyl radicals in polyesters. *J. Polym. Sci., Part A-1* **9**, 3327-3335.

Hagege, R., and B. Catoire (1976). Thermofixage des colorants plastosolubles sur polyester (PETP) après impregnation in milieux aqueux on non aqueux. *Teintex* **5**, 249-259.

Heatley, F., and A. Begum (1976). Molecular motion of poly(methyl methacrylate), polystyrene and poly(propylene oxide) in solution studied by ^{13}C n.m.r. spin-lattice relaxation measurements: Effects due to distribution of correlation times. *Polymer* **17**, 399-408.

Higgins, J. S., G. Allen, and P. N. Brier (1972). Methyl group motion in poly(propylene oxide), polypropylene and poly(methyl methacrylate). *Polymer* **13**, 157-163.

Kivelson, D. (1960). Theory of ESR linewidths of free radicals. *J. Chem. Phys.* **33**, 1094-1106.

Kumler, P. L., and R. F. Boyer (1976). ESR studies of polymer transition. I. *Macromolecules* **9**, 903-910.

Kumler, P. L., S. E. Keinath, and R. F. Boyer (1977). ESR studies of polymer transitions. III. Effect of molecular weight and molecular weight distribution on T_g's of polystyrene as determined by esr spin-probe studies. *J. Macromol. Sci. Phys.* **B13**, 631-646.

Kurosaki, T., K. W. Lee, and M. Okawara (1972). Polymers having stable radicals. I. Synthesis of nitroxyl polymers from 4-methyacryloxyl derivatives of 2,2,6,6-tetramethylpiperidine. *J. Polym. Sci., Polym. Chem. Ed.* **10**, 3295-3310.

Kurosaki, T., O. Takahashi, and M. Okawara (1974). Polymers having stable radicals. II. Synthesis of nitroxyl polymers from 4-methacryloxyl derivatives of 1-hydroxy-2,2,6,6-tetramethylpiperidine. *J. Polym. Sci., Polym. Chem. Ed.* **12**, 1407-1420.

Kusumoto, N., M. Yonezawa, and Y. Motozato (1974). Studies on the surface of as-grown and annealed polyethylene single crystals by the spin-probe method. *Polymer* **15**, 793-798.

Kuznetsov, A. N., and B. Ebert (1974). The dependence of the character of rotational motion of nitroxyl radicals in their molecular sizes. *Chem. Phys. Lett.* **25**, 342-345.

Kuznetsov, A. N., A. Y. Volkov, V. A. Livshits, and A. T. Mirzoian (1974). A method of studying the anisotropic rotation of organic nitroxyl radicals. *Chem. Phys. Lett.* **26**, 369-372.

Labský, J., J. Pilăr, and J. Kálal (1977). Spin-label study of poly(methacrylamide)-type copolymers in solution. *Macromolecules* **10**, 1153-1157.

Lenk, R. (1974). Application d'une sonde radicalaire aux études des macromolecules, des cristaux plastiques et des cristaux liquides. *Chimia* **28**, 51-55.

Lipatov, Yu. S., and L. M. Sergeeva (1974). "Adsorption of Polymers." Wiley, New York.

Liu, K. J., and W. Burlant (1967). High-resolution nmr of crosslinked polymers: Effects of crosslinked density and solvent interaction. *J. Polym. Sci., Part A-1* **5**, 1407-1413.

Liu, K. J., and R. Ullman (1968). Proton magnetic relaxation in polyethylene oxide solutions. *J. Chem. Phys.* **48**, 1158-1168.

Lyerla, J. R., T. T. Horikawa, and D. E. Johnson (1977). Carbon-13 relaxation study of stereoregular poly(methyl methacrylate) in solution. *J. Am. Chem. Soc.* **99**, 2463-2467.

McCall, D. W. (1969). Relaxation in solid polymers. *Natl. Bur. Stand. (U.S.), Spec. Publ.* **301**, 475-537.

McCall, D. W. (1971). Nuclear magnetic resonance studies of molecular relaxation mechanisms in polymers. *Acc. Chem. Res.* **4**, 223-232.

McCalley, R. C., E. J. Shimshick, and H. M. McConnell (1972). The effect of slow rotational motion on paramagnetic resonance spectra. *Chem. Phys. Lett.* **13**, 115-119.

Mason, R. P., and J. H. Freed (1974). Estimating microsecond rotational correlation times from lifetime broadening of nitroxide electron spin resonance spectra near the rigid limit. *J. Phys. Chem.* **78**, 1321-1323.

Mason, R. P., C. F. Polnaszek, and J. H. Freed (1974). Comments on the interpretation of electron spin resonance spectra of spin labels undergoing very anisotropic rotational reorientation. *J. Phys. Chem.* **78**, 1324-1329.

Miller, W. G., and Z. Veksli (1975). The use of stable free radicals to determine motion in polymeric systems. *Rubber Chem. Technol.* **48**, 1078–1089.

Miller, W. G., C. C. Wu, E. L. Wee, G. L. Santee, J. H. Rai, and K. D. Goebel (1974). Thermodynamics and dynamics of polypeptide liquid crystals. *Pure Appl. Chem.* **38**, 37–58.

Miller, W. G., W. T. Rudolph, Z. Veksli, D. L. Coon, C. C. Wu, and T. M. Liang (1979). Spin label studies of polymer motion at or near an interface. *In*, "Application of Nitroxides to Synthetic Polymers" (R. Boyer, ed.), in press. Gordon and Breech, New York.

Mittal, K. L., ed. (1975). "Adsorption at Interfaces." Am. Chem. Soc., Washington, D.C.

Murakami, K., M. Shiotani, and J. Sohma (1976). ESR study of molecular motion of spin-labeled PMMA in benzene solution. *Rep. Prog. Polym. Phys. Jpn.* **19**, 445–446.

North, A. M., and I. Soutar (1972). Fluorescence dipolarization measurements of polymer segmental motion. *J. Chem. Soc., Faraday Trans. 1* pp. 1101–1116.

Poggi, G., and C. S. Johnson (1970). Factors involved in the determination of rotational correlation times for spin labels. *J. Magn. Reson.* **3**, 436–445.

Pohl, H. A., R. Backsai, and W. P. Purcell (1960). Steric order and dielectric behavior in polymethylmethacrylates. *J. Phys. Chem.* **64**, 1701–1705.

Priel, Z., and A. Silberberg (1969). Conformation of poly(methacrylic acid) in alcohol-water mixtures. *J. Polym. Sci., Part A-2* **8**, 689–726.

Ptitsyn, O. B., A. K. Kron, and Y. Y. Eizner (1968). The models of the denaturation of globular proteins. I. Theory of globular-coil transitions in macromolecules. *J. Polym. Sci., Part C* **16**, 3509–3517.

Rabold, G. P. (1969a). Spin-probe studies. I. Applications to latexes and micelle characterization. *J. Polym. Sci., Part A-1* **7**, 1187–1201.

Rabold, G. P. (1969b). Spin-probe studies. II. Applications to polymer characterization. *J. Polym. Sci., Part A-1* **7**, 1203–1223.

Regen, S. L. (1974). Influence of solvent on the mobility of molecules covalently bound to polystyrene matrices. *J. Am. Chem. Soc.* **96**, 5275–5276.

Regen, S. L. (1975a). Motion of molecules within solvent channels of polystyrene matrices. *Macromolecules* **8**, 689–690.

Regen, S. L. (1975b). Influence of solvent on the motion of molecules "immobilized" on polystyrene matrices and on glass surfaces. *J. Am. Chem. Soc.* **97**, 3108–3111.

Robb, I. D., and R. Smith (1974). The adsorption of a copolymer of vinyl pyrrolidone and allylamine at the silica-solution interface. *Eur. Polym. J.* **10**, 1005–1010.

Robb, I. D., and R. Smith (1977). Adsorption of polymers at the solid–liquid interface; a comparison of the EPR and IR techniques. *Polymer* **18**, 500–504.

Rudolph, W. T. (1976). The conformation of synthetic polymers adsorbed at a solid, liquid interface, M.S. Thesis, University of Minnesota, Minneapolis.

Sanson, A., and M. Ptak (1970). Résonance paramagnetique electronique d'un radical nitroxyde fixé sur l'extrémité carboxylique d'un polypeptide. *C. R. Hebd. Seances Acad. Sci., Ser. D* **271**, 1319–1322.

Santee, G. L. (1972). Temperature-composition phase-equilibria in polymer liquid crystals. M.S. Thesis, University of Minnesota, Minneapolis.

Savolainen, A., and P. Tormala (1974). Spin probe studies on molecular motions of 2,6-disubstituted poly(phenylene oxides) near glass transition. *J. Polym. Sci., Polym. Phys. Ed.* **12**, 1251–1254.

Shiotani, M., and J. Sohma (1977). ESR studies of molecular motion of spin labeled poly(methyl methacrylate). *Polym. J.* **9**, 283–291.

Stone, T. J., T. Buckman, P. L. Nordio, and H. M. McConnell (1965). Spin labeled biomolecules. *Proc. Natl. Acad. Sci. U.S.A.* **54**, 1010–1017.

Sundararsjan, P. R., and P. J. Flory (1974). Configurational characteristics of poly(methyl methacrylate). *J. Am. Chem. Soc.* **96**, 5025–5031.

Tonelli, A. E. (1973). Phenyl group rotation in polystyrene. *Macromolecules* **6**, 682–883.
Tormala, P. (1974a). On the mechanism of motions of nitroxyl radicals in polymers. *Polymer* **15**, 124–125.
Tormala, P. (1974b). Determination of glass transition temperature of poly(ethylene glycol) by spin probe technique. *Eur. Polym. J.* **10**, 519–521.
Tormala, P., and J. J. Lindberg (1976). Spin labels and probes in dynamic and structural studies of synthetic and modified polymers. *In* " Structural Studies of Macromolecules by Spectroscopic Methods" (K. J. Ivin, ed.), pp. 255–272. Wiley, New York.
Tormala, P., and J. Tulikoura (1974). Effect of end-groups on the motion of free nitroxyl radicals in poly(ethylene glycol). *Polymer* **15**, 248–250.
Tormala, P., J. Martinmaa, K. Silvennoinen, and K. Vaahtera (1970). Spin labeling of poly-(ethylene glycols). *Acta Chem. Scand.* **24**, 3066–3067.
Tormala, P., K. Silvennoinen, and J. J. Lindberg (1971). Spin-labeling investigations of polyamides and polyesters. *Acta Chem. Scand.* **25**, 2659–2665.
Tormala, P., H. Lattila, and J. J. Lindberg (1973). Solid and liquid state relaxations in spin-labeled poly(ethylene glycol) at high temperatures $(T > T_g)$. *Polymer* **14**, 481–487.
Tsutsumi, A., K. Hikichi, and M. Kaneko (1976). Spin probe studies of poly(γ-benzyl L-glutamate). *Polym. J.* **8**, 511–515.
Veksli, Z., and W. G. Miller (1975). Motion of nitroxide spin-labels covalently attached to synthetic polymers. *Macromolecules* **8**, 248–250.
Veksli, Z., and W. G. Miller (1977a). The effect of good solvents on molecular motion of nitroxide free radicals in covalently labeled polystyrene and poly(methyl methacrylate). *Macromolecules* **10**, 686–692.
Veksli, Z., and W. G. Miller (1977b). The effect of solvents on molecular motion of nitroxide free radicals dipped in polystyrene and poly(methyl methacrylate). *Macromolecules* **10**, 1245–1250.
Veksli, Z., W. G. Miller, and E. L. Thomas (1976). Penetration of non-solvents into glassy, amorphous polymers. *J. Polym. Sci., Polym. Symp.* **54**, 299–313.
Ward, T. C., and J. T. Books (1974). Spin-labeled investigation of polyurethane networks. *Macromolecules* **7**, 207–212.
Wasserman, A. M., T. A. Alexandrova, and A. L. Buchachenko (1976). The study of rotational mobility of stable nitroxyl radicals in polyvinylacetate. *Eur. Polym. J.* **12**, 691–695.
Wee, E. L. (1971). Studies on the isotropic and the liquid crystal phases of the system poly-γ-benzyl-α,L-glutamate-*N*,*N*-dimethylformamide. Ph.D. Thesis, University of Minnesota, Minneapolis.
Wee, E. L., and W. G. Miller (1973). Studies on nitroxide spin-labeled poly-γ-benzyl-α,L-glutamate. *J. Phys. Chem.* **77**, 182–189.
Wu, C. C. (1973). Studies on nitroxide-labeled polypeptides. Ph.D. Thesis, University of Minnesota, Minneapolis.
Yano, O., and Y. Wada (1971). Dynamic mechanical and dielectric relaxation of polystyrene below the glass temperatures. *J. Polym. Sci., Part A-2* **9**, 669–686.

5

Spin Labeling in Pharmacology

COLIN F. CHIGNELL

LABORATORY OF ENVIRONMENTAL BIOPHYSICS
NATIONAL INSTITUTE OF ENVIRONMENTAL HEALTH SCIENCES
RESEARCH TRIANGLE PARK, NORTH CAROLINA

I.	Introduction	223
II.	Drug Absorption, Distribution, and Excretion	224
	A. Spin Immunoassay	224
	B. Drug Binding to Plasma Proteins	225
	C. Drug Distribution	226
III.	Drug Metabolism	227
	A. Cytochrome *P*-450	227
	B. Spin Trapping of Reactive Intermediates of Drug Metabolism	229
IV.	Pharmacological Effects of Drugs	231
	A. Effects Mediated via Receptors	231
	B. Effects Not Mediated via Receptors	240
V.	Future of Spin-Labeling Studies in Pharmacology	243
	References	243

I. INTRODUCTION

Pharmacology has been defined as the science of drugs, their chemical constitution, their biological action, and their therapeutic application in man (Goldstein *et al.*, 1974). In this chapter, the impact that the technique of spin labeling has had on pharmacology will be assessed. However, before such an evaluation is made, it should be pointed out that the first spin-labeling experiment, carried out by Ohnishi and McConnell (1965), involved the interaction of a drug, the chlorpromazine cation radical, with a putative biological target, namely, DNA. Although structural considerations as well as lack of stability made the chlorpromazine radical a poor choice for more

223

general spin-labeling applications, it did serve to emphasize the usefulness of the technique.

The aim of this chapter is to demonstrate the usefulness of the spin-labeling technique in pharmacology. No attempt has been made to include all of the published pharmacological applications of the technique. Instead, some general principles and guidelines will be presented, together with selected examples that demonstrate the kinds of pharmacological problems that can be solved with the aid of spin labels.

The biological fate of a drug can be summarized briefly as follows. After administration, the drug usually enters the blood fairly rapidly unless it was applied topically to achieve a local effect. Once present in the blood, the drug is distributed throughout the body and can be removed from the circulation by one or more of the following mechanisms: metabolism, excretion, and accumulation in the tissues. When the drug reaches the target tissue, it may combine with a cellular macromolecule or receptor to initiate a series of events that results in the expression of pharmacological activity. Spin labeling can be used to study the biological fate of a drug at each of these levels of increasing complexity.

II. DRUG ABSORPTION, DISTRIBUTION, AND EXCRETION

A. Spin Immunoassay

The blood level of a drug provides a useful indication of whether a sufficient amount of the agent is being absorbed and delivered to the site of action. Blood levels that are too low may result in insufficient drug reaching the target tissue, whereas blood levels that are too high may elicit toxic symptoms. Many analytical techniques are now available for the determination of plasma drug concentration. The spin-immunoassay procedure, introduced by Leute et al. (1972), is of considerable interest, since it provided the first application of spin labeling to clinical pharmacology. The technique is analogous to radioimmunoassay in that it involves the use of antibodies directed against a drug molecule. However, in the spin-immunoassay procedure, a spin-labeled drug analogue instead of a radioactive drug molecule is employed. On addition of the antibody, the spin-labeled drug binds to the active site, with a resultant decrease in mobility. When plasma samples containing the drug are added to the spin-probe–antibody complex, the spin-labeled drug is displaced, and its free concentration is estimated from the amplitude of the low-field line. The concentration of the added drug then can be estimated with the aid of a standard curve. The main advantage of the technique is its rapidity. The disadvantages of the procedure include lack of sensitivity and susceptibility to the presence of interfering substances, e.g., ascorbic acid, in the plasma sample. The spin immunoassay is discussed more fully in Chapter 6, by Piette and Hsia.

B. Drug Binding to Plasma Proteins

Many drugs are quite highly bound by albumin or other plasma proteins (Chignell, 1976). Drug binding decreases the free concentration of drug in plasma, which may in turn lead to an alteration in the rate of metabolism, excretion, and passage into the tissues (Chignell, 1974). There are many methods for studying drug binding, including equilibrium dialysis, ultrafiltration, ultracentrifugation, and gel chromatography (Chignell, 1977). One major disadvantage of all these techniques is that they are rather time-consuming. Montgomery and Holtzman (1974) and Montgomery *et al.* (1975) used spin-labeled analogues of morphine(**I**) and diphenylhydantoin(**II**) to estimate the plasma binding of these two drugs. The technique

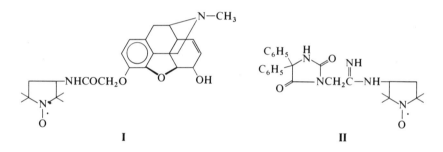

I   **II**

consists of adding the spin-labeled drug to plasma and estimating the concentration of free spin label from the amplitude of its low-field line. The entire procedure can be completed within a few minutes. The morphine analogue (**I**) was 30.5% bound in human serum, a value which compared well with the 35.4% bound by ultrafiltration for the radiolabeled parent drug. In contrast, 61% of the diphenylhydantoin spin label was bound, whereas ultrafiltration with the radiolabeled parent compound gave a value of 88.7% bound. Undoubtedly, the introduction of the pyrrolidine moiety into diphenylhydantoin caused a major perturbation of the molecular structure that in turn led to an alteration in binding characteristics. Thus, a spin-labeled drug cannot always be used with certainty to predict the binding of the parent molecule in plasma. The binding of a spin-labeled drug to plasma proteins is, however, an important aspect of the spin-immunoassay technique as applied to plasma samples. If such binding does occur, it may be necessary to modify the experimental procedure to take account of the fraction of spin-labeled drug that is bound to the plasma proteins and not available for binding to the antibody.

C. Drug Distribution

Radioisotopes have long been used in pharmacological studies to determine the biological fate of drugs and other compounds. The radiation hazard posed by such compounds often precludes their use in man, and so alternative procedures, e.g., the use of stable isotopes, have been sought. There are two reported uses of spin labels as biological tracers. Sherman and Landgraf (1967) used albumin covalently labeled with a piperidine maleimide spin label (III) to determine the plasma volume of dogs. The method

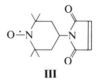

III

gave values very close to those determined by a standard procedure employing radioactive iodinated serum albumin. These workers took no precautions to determine whether the nitroxide group had undergone reduction *in vivo*. However, the close agreement between the values obtained by means of the spin-probe and radiotracer techniques suggests that reduction of the labeled albumin did not occur.

The second reported use of a spin-labeled drug as a radiotracer was described by Emanuel *et al.* (1976). These workers synthesized two spin-labeled analogues (IV and V) of the antineoplastic agent triethylenethiophosphoramide (Thio-TEPA, VI). Spin probe IV was found to have greater

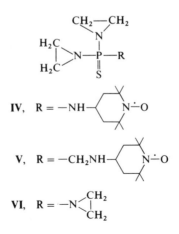

antitumor activity than either **V** or **VI**. Guerin carcinoma, Walker carcinoma, and Schwetz erythromyelosis carried in mice underwent complete regression under the action of **IV**. The levels of **IV** in the tumor, liver, and blood of mice bearing adenocarcinoma 755 were measured by means of electron spin resonance (ESR). All tissues were treated with alkaline H_2O_2 to reoxidize any spin probe that had been reduced *in vivo*. The level of **IV** in the tumor reached a maximum 2 hr after injection and then declined rapidly. In contrast, the level of **IV** in liver and blood reached a maximum only 45 min after administration. It is of interest that there was a second peak in blood level at 5 hr, which probably was due to the release of metabolites of **IV** from the tumor.

This study is important for several reasons. First, it represents the only reported use of a spin-labeled drug in pharmacokinetic studies. Second, it illustrates the necessity of demonstrating pharmacological activity of the spin-labeled drug before attempting the spin-label studies themselves. Finally, it emphasizes one of the potential frustrations of this approach. Label **IV** differs considerably in structure from the parent molecule **VI**, and yet it is more active. In contrast, label **V** differs from **IV** in that it has an extra methylene group, but it is inactive.

III. DRUG METABOLISM

A. Cytochrome *P*-450

The microsomal fraction from liver and other tissues contains an NADPH-dependent mixed-function oxidase that is responsible for the oxidation of drugs and other xenobiotics. The cytochrome responsible for these oxidations has been termed *P*-450 because in its reduced form it is capable of binding carbon monoxide to yield a pigment with a characteristic absorption peak of 450 nm (Gillette, 1963).

When a nitroxide group comes in close proximity to another paramagnetic center, the magnetic interaction between them may be sufficient to produce measurable characteristic changes in both of their ESR spectra (Leigh, 1970). Since *P*-450 contains a paramagnetic iron atom, it follows that if a spin-labeled substrate binds to this cytochrome it may be possible to detect an interaction between these two magnetic species. This phenomenon has been used by three groups to monitor the binding of two different substrates to *P*-450. Reichmann *et al.* (1972) synthesized an isocyanide spin probe (**VII**) and studied its interaction with rat liver microsomal *P*-450. Metyrapone spin probes **VIII** and **IX** subsequently were used by other

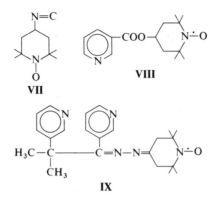

VII

VIII

IX

groups to examine substrate binding to cytochrome P-450 from *Pseudomonas putida* (Griffin *et al.*, 1974) and rat liver microsomes (Ruf and Nastainczyk, 1976). By measuring the decrease in intensity of **IX** complexed to liver microsomal P-450, Ruf and Nastainczyk estimated that the heme–nitroxide distance was 11 Å. This distance is compatible with a model in which one of the pyridine nitrogens is a ligand of the heme iron. The bound spin label showed virtually no interaction with added ferricyanide, suggesting that the active site of cytochrome P-450 is a hydrophobic pocket not accessible to the aqueous phase. Similar conclusions were reached by Griffin and co-workers (1974), who worked with a purified cytochrome P-450 from *Pseudomonas putida*.

These studies illustrate the usefulness of spin-labeled substrates for examining the topography of the active site of cytochrome P-450. Experiments with the liver microsomal system suffer from the disadvantage that it is impure. Under these conditions, it often is difficult to differentiate an interaction with the cytochrome from nonspecific binding to other membrane components such as proteins and phospholipids. Reichman and co-workers (1972) in their studies with the piperidine isocyanide **VII** did not demonstrate that it was binding exclusively to liver microsomal P-450. This problem could be overcome by synthesizing a spin label with a very high affinity for the cytochrome. Some progress has been made toward purifying mammalian P-450 systems. When they become available, spin labels of the metyrapone type should be extremely useful for monitoring the reconstitution of the enzyme.

Rosen and Rauckman (1977) examined the metabolism of compounds **X–XII** by rat liver microsomes. Compound **X** was not oxidized to **XII** in the

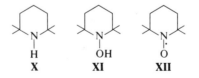

X   **XI**   **XII**

presence of NADPH, oxygen, and microsomes. However, incubation of **XI** with microsomes and NADPH under aerobic conditions led to the formation of **XII**, which was detected by ESR. The observed reaction rate was not enhanced by pretreatment of the rats with phenobarbital, a known inducer of *P*-450. Furthermore, the rate of **XII** formation was not affected by SKF 525A (*β*-diethylaminoethyl diphenylpropylacetate) and was not completely abolished by carbon monoxide exposure of dithionite-reduced microsomes. This led to the conclusion that some other system, possibly a mixed-function amine oxidase, was responsible for the oxidation of **XI** to **XII**. In contrast, piperidine nitroxide **XII** underwent NADPH-dependent reduction to **XI**, which was sensitive to SKF 525A, dithionite/carbon monoxide, and phenobarbital pretreatment, suggesting that it was mediated by the cytochrome *P*-450 reductase system.

The molecular organization of cytochrome *P*-450 reductase was probed by Stier and Sackmann (1973) with the aid of a phosphate ester of the piperidinol nitroxide **XIII** and 5-doxylstearic acid(**XIV**). Both labels were

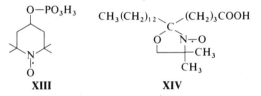

reduced by the reductase, but Arrhenius plots of reduction rate constants revealed a striking difference in the behavior of the water-soluble label **XIII** and the more lipid-soluble label **XIV**. The activation energy of fatty acid reduction decreased rapidly below 32°C, whereas no break was seen in the Arrhenius plot of **XIII**. Stier and Sackmann (1973) proposed a mosaiclike structure for the liver microsomal membranes in which the reductase, surrounded by cytochrome *P*-450, was enclosed in a rigid phospholipid halo, which was quasi-crystalline below 32°C but underwent a crystalline–liquid phase transition at 32°C.

B. Spin Trapping of Reactive Intermediates of Drug Metabolism

Spin trapping is a technique in which short-lived free radicals react with a diamagnetic molecule (spin trap) to yield a more stable free radical (Janzen, 1971). Although this technique originally was developed to investigate the structure of radicals generated during chemical reactions, it now is being used in biological systems (Harbour and Bolton, 1975). Most of the spin-trapping agents generate nitroxide radicals upon reaction with the unstable radical, and so the technique deserves some mention here even though

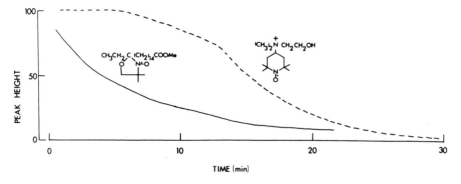

Fig. 1. The amplitude of the low-field line of two spin labels measured at 37°C in the presence of rat liver microsomes. The complete system contained microsomes (1 mg/ml), CCl₃Br (1 m*M*), NADPH, and an NADPH-regenerating system. (C. F. Chignell and G. Krishna, unpublished results, 1976.)

it is outside of the scope of this chapter. Since many drugs and other xenobiotics are thought to be converted to reactive free radicals during metabolism, it seems likely that the technique of spin trapping will be used in the future to determine the structure of these radicals.

Spin probes can themselves be used to detect free radicals generated during the metabolism of foreign chemicals by the liver microsomal *P*-450 system. One such application is shown in Fig. 1. In this experiment, the two spin probes were incubated with rat liver microsomes, CCl₃Br, NADPH, and an NADPH-regenerating system. The destruction of the 16-doxylstearic acid methyl ester label began almost immediately after mixing, whereas the ESR signal of the choline label remained unchanged for 6 min before declining. Neither label was destroyed if any one component was omitted from the incubation. It has been suggested that the first step in the metabolism of the halogenated methanes is homolytic bond fission according to the following reaction (Slater, 1972):

$$CCl_3Br \rightarrow \dot{C}Cl_3 + \dot{B}r$$

The destruction of the spin probes shown in Fig. 1 probably is due to reaction with free radicals generated during the metabolism of CCl₃Br. Since the choline spin probe is water soluble and cannot penetrate membranes, it probably reacts with secondary free radicals, possibly those resulting from lipid peroxidation (Slater, 1972), that reach the surface of the membrane. On the other hand, the stearate spin probe does penetrate the membrane and may react with the primary radicals formed from CCl₃Br. (Editor's note: for a more complete discussion of spin traps see Chapter 6, p. 279.)

IV. PHARMACOLOGICAL EFFECTS OF DRUGS

A. Effects Mediated via Receptors

For most drugs, combination with a tissue target-macromolecule or receptor usually is the first step in a series of events that culminates in an expression of pharmacological activity. The receptor can be an enzyme, nucleic acid, or perhaps a membrane-bound macromolecule.

1. DESIGN OF A SPIN-LABELED DRUG

In order to design a useful spin-labeled drug for monitoring drug–receptor interactions, it first must be recognized that the introduction of the label itself causes no small perturbation of the molecular structure. In addition to steric factors, the spin-label group, usually in the form of a five- or six-membered alicyclic ring, also alters the lipophilicity of the drug molecule. The effect that the introduction of a spin-label group into a drug molecule may have on the pharmacological activity of that drug can often be assessed from classic structure–activity relationships. As an example, let us consider the sulfonamide diuretics. These drugs are known to exert their pharmacological effect by inhibiting the enzyme carbonic anhydrase (Maren *et al.*, 1960; Maren, 1967). Structure–activity considerations indicate that only compounds with the general formula $ArSO_2NH_2$ (where Ar is heteroaromatic or homoaromatic) are good inhibitors of the enzyme. In the sulfanilamide series, it also is known that the introduction of alkyl groups into the molecule (**XV**, R = Me, Et, etc.) enhances biological activity (Maren, 1967).

NHR

SO_2NH_2

XV

From these elementary structure–activity relationships, it is obvious that the spin-label moiety should be introduced into the 4 position of the molecule. Such labels are in fact good probes for the active site of carbonic anhydrase (Erlich *et al.*, 1973).

A second consideration in the design of a spin-labeled drug is the availability of a suitable functional group on the parent drug molecule. Reactive groups, e.g., carboxyl, amino, hydroxyl, in a drug provide convenient points for attaching a spin label. In deciding which part of the drug molecule to modify, one should give some consideration again to the structure–activity relationships mentioned above. However, if the spin-labeled drug is to be used in a relatively crude biological system, it may be unwise to choose the ester linkage since it may be hydrolyzed by tissue esterases.

When the parent drug does not contain a suitable group for derivatization, it may be possible to obtain a suitable intermediate. For example, acetazolamide(**XVI**) does not have a suitable functional group for the

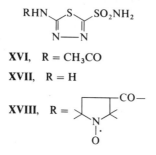

XVI, R = CH₃CO

XVII, R = H

XVIII, R =

attachment of a spin label. However, the intermediate **XVII** was available from the drug manufacturer and was used to prepare label **XVIII** (Chignell *et al.*, 1972a). If the drug does not contain a suitable group and if an intermediate is unavailable, it will be necessary to carry out a complete synthesis.

Enzymes are the target molecules for many drugs. Spin-label studies of such systems can be attempted either with analogues of the natural substrates or with spin-labeled inhibitors. Spin-labeled substrates usually are poor choices for such experiments if they can still be metabolized by the enzyme. Under these conditions, the concentration of enzyme–spin-probe complex at any one time will be quite low. In addition, the metabolic products bearing the spin label will be present free in solution, and their ESR signal often will overshadow that due to the Michaelis complex.

2. TESTING THE PHARMACOLOGICAL ACTIVITY OF A SPIN-LABELED DRUG

Once a spin-labeled drug has been synthesized, it becomes necessary to demonstrate that it has the same pharmacological properties as the parent molecule. In a situation in which the biological target is an enzyme and the spin-labeled drug has been designed as an inhibitor, the simplest way to test pharmacological activity is to determine whether the label inhibits enzymatic activity. For example, Wee *et al.* (1976a) employed a series of bis-quaternary ammonium analogues of hexamethonium(**XIX**) and decamethonium (**XX** and **XXI**) as probes for the active site of acetylcholinesterase. It was anticipated that these spin probes would be good inhibitors of acetylcholinesterase, since it was known that an increase in bulk around the quaternary nitrogens tends to enhance biological activity (Bergman and Wurzel, 1953; Wilson, 1952; Taylor and Jacobs, 1974). This hope was

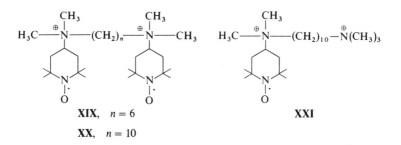

XIX, $n = 6$

XX, $n = 10$

XXI

fulfilled in that both of the decamethonium analogues were more active than the parent drug (Table I). In contrast, the phenyltrimethylammonium spin probe **XXII** is less active than the parent molecule, **XXIII** (Table I). Wilson

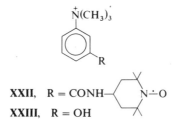

XXII, R = CONH—⟨ ⟩N—O

XXIII, R = OH

and Quan (1958) suggested that the activity of **XXIII** results from the formation of a hydrogen bond between the phenolic hydroxyl group and some other group within the active site of the enzyme. It therefore is easy to see why spin probe **XXII** is a poor inhibitor.

Spin labels **XXV** and **XXVII** are analogues of β-adrenergic antagonists dichlorisoproterenol(**XXIV**) and propranolol(**XXVI**), respectively. When their ability to inhibit isoproterenol stimulation of adenylyl cyclase was tested, both spin probes were found to be less active than the corresponding

TABLE I

INHIBITION OF TORPEDO CALIFORNICA
ACETYLCHOLINESTERASE BY QUATERNARY
AMMONIUM COMPOUNDS

Compound	$K_1 \times 10^6\ M$
XIX	3.3
XX	0.18
XXI	0.25
Decamethonium iodide	6.00
XXII	100
XXIII	0.31

parent molecules (Sinha and Chignell, 1975). Structure–activity studies by Grunfeld *et al.* (1975) revealed the importance of the *N*-isopropyl function for biological activity in these compounds.

When the target molecule is not an enzyme, other techniques must be employed to demonstrate that both drug and spin-labeled drug occupy the same binding site. One of the most useful methods is to demonstrate direct competition between the spin label and some other ligand known to bind to the site. In a spin-label study of the biotin binding sites on egg white avidin, two such techniques were employed (Chignell *et al.*, 1975). In the first, direct competition was demonstrated between biotin(**XXVIII**) and the biotin spin probe (**XXIX**). The second procedure involved the displacement by the spin

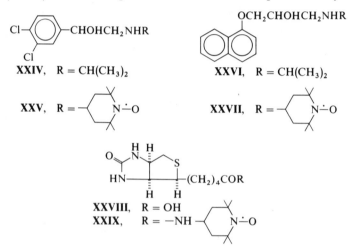

probe of a dye, 2-(4′-hydroxyazobenzene)benzoic acid, known to occupy the biotin binding site on avidin (Fig. 2). In such studies, both stoichiometry and competitive displacement are important for the demonstration of a common binding site.

Competition for a common binding site also was used by Weiland and co-workers (1976) to monitor the interaction of decamethonium spin probe **XXI** with cholinergic-receptor-enriched membranes from *Torpedo californica*. However, in these experiments the displacing agent was not another small molecule but a polypeptide, cobra α-toxin, from *Naja naja Siamensis*, which has been shown to bind specifically to the cholinergic receptor (Weiland *et al.*, 1976).

3. INFORMATION DERIVED FROM SPIN-LABELED DRUGS

1. *Binding-Site Topography.* The use of spin-labeled ligands to study the topography of combining sites was discussed at length by Morrisett (1975).

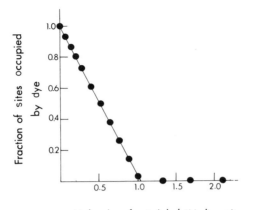

Molecules of spin label / binding site

Fig. 2. Titration of mixture containing avidin $(5 \times 10^{-6} \ M)$ and 2-(4′-hydroxyazobenzene)benzoic acid $(9 \times 10^{-5} \ M)$ in 50 mM sodium phosphate buffer with biotin spin label **XXIX**. [Reproduced from Chignell *et al.* (1975) with permission.]

This approach was first employed by Hsia and Piette (1969) to measure the depth of the combining site of immunoglobulins directed against the 2,4-dinitrophenyl hapten. These workers synthesized a series of 2,4-dinitrophenyl spin labels (**XXX**) in which the distance d was gradually increased.

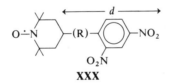

XXX

A dramatic increase in the mobility of the piperidine moiety of the antibody-bound probe was detected by ESR when d was 10 Å. Since the antibody recognizes only the dinitrophenyl portion of **XXX**, this finding sets an upper limit of 10 Å for the depth of the combining site.

This "molecular dipstick" approach has been used to examine the topography of the active sites of a number of mammalian erythrocyte carbonic anhydrases (Chignell, 1971; Chignell *et al.*, 1972b; Wee *et al.*, 1976b) with the aid of a series of spin-labeled sulfonamide inhibitors (Table II). The ESR spectra of some of these labels (**XXXI–XXXV**) bound to human carbonic anhydrase II(C) are shown in Fig. 3. The corresponding rotational correlation times (τ_c) for enzyme-bound spin probes **XXXII–XXXV** are given in Fig. 4 for human carbonic anhydrase II(C) as well as for three other mam-

TABLE II

STRUCTURES OF SULFONAMIDE SPIN LABELS

Spin label	Structure	d (Å)
	$\longleftarrow \quad d \quad \longrightarrow$	
XXXI	—CONH—⟨◯⟩—SO_2NH_2	8.16
XXXII	—$CONHCH_2$—⟨◯⟩—SO_2NH_2	8.88
XXXIII	—$CONHCH_2CONH$—⟨◯⟩—SO_2NH_2	11.28
XXXIV	—$NHCO(CH_3)_2CONH$—⟨◯⟩—SO_2NH_2	12.72
XXXV	—$NHCO(CH_2)_3CONH$—⟨◯⟩—SO_2NH_2	14.72
XXXVI	O—N ⟨ ⟩—$NHCO$—⟨◯⟩—SO_2NH_2	7.86

malian erythrocyte carbonic anhydrases. If the pyrrolidine ring bearing the nitroxide moiety is close to the aromatic ring (XXXI) it becomes highly immobilized when the inhibitor binds to the enzyme. As the distance d (Table II) is increased, there is a concomitant increase in the mobility of the nitroxide portion of the molecule up to XXXIV. Since it is known that the sulfonamide group of these probes coordinates with a zinc atom at the bottom of the active site, this sets an upper limit of about 14 Å for the depth of the site. This finding agrees well with X-ray crystallographic data for the enzyme (Nostrand et al., 1975). The rather gradual increase in mobility along the series suggests that the active site of human carbonic

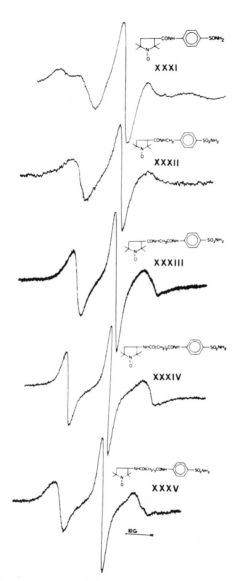

Fig. 3. The ESR spectra of spin labels **XXXI–XXXV** bound to human carbonic anhydrase II(C).

anhydrase II(C) is funnel-shaped rather than slitlike as is the case for the combining site of antibodies directed against the 2,4-dinitrophenyl hapten (Hsia and Piette, 1969).

It can be seen from Fig. 4 that other mammalian carbonic anhydrases also have active sites that are about 14 Å in depth. Bovine carbonic anhy-

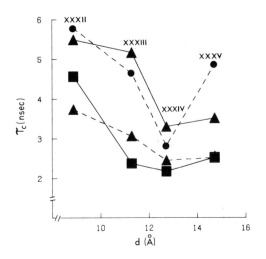

Fig. 4. Rotational correlation times τ_c (nanoseconds) of spin labels **XXXII–XXXV** bound to various mammalian carbonic anhydrases. Key: ▲–▲, human carbonic anhydrase I(B); – –▲-▲–, human carbonic hydrase II(C); – –●-●–, bovine carbonic anhydrase II(B); ■–■, carbonic anhydrase I. For the definition of d, see Table II. [Adapted from Wee *et al.* (1976b) with permission.]

drase II(B) is of interest because there is an apparent decrease in mobility on going from label **XXXIV** to label **XXXV** (Fig. 4). This decrease may be due to accessory binding sites on the surface of the enzyme, or perhaps the connecting chain between the aromatic ring and the pyrrolidine ring is flexible enough to permit the latter to reenter the active site. This finding emphasizes the importance of increasing the chain length by increments as small as structurally possible; otherwise, false estimates of active-site depth may be obtained.

Human erythrocytes contain two isozymes of carbonic anhydrase that have been termed I(B) and II(C). Although they are closely similar in structure (Nostrand *et al.*, 1975), they differ in that the II(C) isozyme has a much higher turnover number. Spin label **XXXI** cannot differentiate between the two isozymes since it becomes highly immobilized on binding to both of them (Erlich *et al.*, 1973). In contrast, spin probe **XXXVI** (Table II) becomes highly immobilized on binding to human carbonic anhydrase II(C) but exhibits much greater mobility at the active site of human carbonic anhydrase I(B) (Fig. 5). Although the molecular mechanism for this difference is not understood, it has provided a means of rapidly differentiating between high- and low-activity forms of carbonic anhydrase found in the erythrocytes of other mammalian species (Wee *et al.*, 1976b).

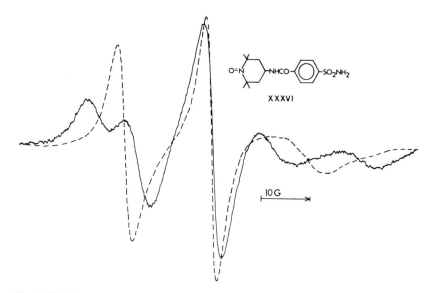

Fig. 5. The ESR spectra of spin label **XXXVI** bound to human carbonic anhydrase I(B) (---) and II(C) (——). [Reproduced from Wee *et al.* (1976b) with permission.]

2. Measurement of Molecular Distances. In the previous section, the use of spin-labeled drugs to determine the depth of their combining sites has been discussed. When multiple binding sites are present, it may be possible to determine the distance between adjacent sites. This approach is best illustrated by the biotin–avidin system. Avidin is a tetrameric protein that binds four molecules of biotin(**XXVIII**) (Green *et al.*, 1971). The biotin binding sites, one per subunit, are grouped in two pairs at opposite ends of the avidin molecule. The ESR spectrum of a 4 : 1 complex between biotin spin probe **XXIX** and avidin contains broad line components characteristic of a highly immobilized label (Fig. 6). Electron–electron dipole interactions between spin labels bound to adjacent sites split each of the major hyperfine lines into doublets with a separation of 13.8 G. The distance (r_{12}) between adjacent bound nitroxides was calculated to be 16 Å from the following equation,

$$r_{12} = \left(\frac{5.56 \times 10^4}{|2D|} \right)^{1/3}$$

where $|2D|$ is the maximal splitting of the peaks in Gauss (Jost and Griffith, 1972).

Molecular distances can also be calculated in a system containing a spin label and a paramagnetic metal. This approach has already been discussed

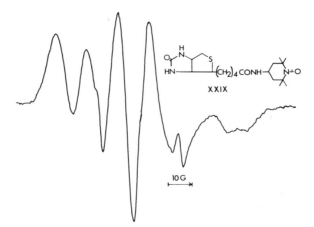

Fig. 6. The ESR spectrum of spin label **XXIX** bound to avidin. [Reproduced from Chignell *et al.* (1975) with permission.]

for the binding of metyrapone spin labels to cytochrome *P*-450 (Griffith *et al.*, 1974; Ruf and Nastainczyk, 1976). Other examples have been discussed by Morrisett (1975). Molecular distances can also be estimated by monitoring the effect of the spin label on the nuclear magnetic resonance (NMR) relaxation time of nuclei in nearby groups present in the binding site (Morrisett, 1975).

3. Kinetic Studies of Drug Binding to Receptors. Competition studies between various ligands and cobra α-toxin for cholinergic receptor binding sites suggest that there is a slow change of receptor state in which affinity for the ligand is increased. This phenomenon was demonstrated directly for decamethonium analogue **XXI** by monitoring the intensity of the low-field ESR line of the unbound label (Fig. 7).

B. Effects Not Mediated via Receptors

Goldstein *et al.* (1974b) identified three mechanisms of drug action that do not depend on a drug receptor. (1) The biological effect is a nonspecific consequence of the physical and/or chemical properties of the drug. (2) The biological effect is a consequence of direct chemical interaction between the drug and a small molecule or ion. (3) The biological effect is due to incorporation of a drug instead of a normal metabolite.

The most interesting drugs in the first category are the volatile anesthetics. Trudell and co-workers (1973a) examined the effect of two volatile anesthetics, halothane and methoxyfluorane, on egg lecithin–cholesterol vesicles containing spin-labeled phosphatidylcholine. They found that the

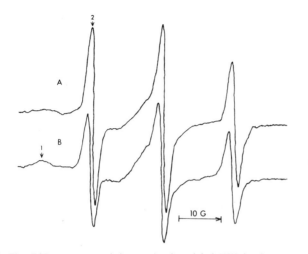

Fig. 7. The ESR spectrum of decamethonium label **XXI** in the presence of receptor-enriched membranes from *Torpedo californica*. The sample contained label **XXI** (8.87 × 10^{-6} M) and receptor membranes (10.3 × 10^{-6} M in α-toxin binding sites). Spectrum A was recorded 30 sec after the addition of the label, and spectrum B was recorded 30 min after label addition. Arrow 1 indicates the position of the low-field line of receptor-bound label, and arrow 2 shows the low-field line of the free label. [Reproduced from Weiland *et al.* (1976) with permission.]

molecular order of the vesicles decreased as the concentration of anesthetic increased. This finding suggests that the inhalation anesthetics cause general fluidization of the membrane rather than a disorder localized in a particular region of the membrane. In addition, measurements of the isotropic nitrogen hyperfine coupling constant indicated a decrease in the polarity of the environment of the spin label with increasing anesthetic concentrations. In a second series of experiments (Trudell *et al.*, 1973b), these workers found that the disorder induced by halothane could be reversed by the application of helium or hydrostatic pressure. This observation is in agreement with the known reversal of anesthesia induced by pressure which has been demonstrated *in vivo* for a number of biological systems.

Local anesthetics also are thought to exert their pharmacological effect by altering membrane structure. Several studies have employed spin-labeled membrane probes to monitor the effect of local anesthetics on various membrane systems. These have been reviewed by Smith and Butler (1975). An alternative approach is to employ a spin-labeled analogue of the local anesthetic. Giotta and co-workers (1973) synthesized three spin-labeled analogues of the local anesthetic parethoxycaine (**XXXVII**) and examined their interaction with the walking nerve of the lobster. They found that these labels partitioned into the membrane in the order ethoxy < butoxy <

hexyloxy. Since the duration of anesthesia was in the same order, it would appear that there is a relationship between duration of activity and ability to partition into the hydrocarbon region of the membrane.

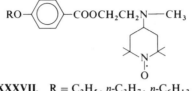

XXXVII, $R = C_2H_5$, $n\text{-}C_3H_7$, $n\text{-}C_6H_{13}$

It is also of interest that piperidine spin probe **XII** can itself induce reversible blockage of nerve conduction. Hsia and Boggs (1973) were able to show that at atmospheric pressure label **XII** dissolved in the apolar portion of the rat phrenic nerve membrane. However, at 150 atm helium, the label bound to a high-affinity polar site, which they suggested was probably a membrane protein. Hsia and Boggs (1973) claimed that high pressure increased anesthesia and suggested that this increase was due to the binding of the label to a new protein site.

The effect of both the local and general anesthetics on membrane and nerve conduction is reversible. However, there is also a group of nonspecific destructive agents that are used in disinfection and antisepsis as well as in certain methods of contraception. Detergents destroy the integrity of lipoprotein membranes and also cause dissociation of nucleoprotein complexes (e.g., ribososomes) whose integrity depends on ionic bonds. These phenomena also can be studied by the spin-label technique. For example, Kirkpatrick and Sandberg (1973) studied the effect of anionic surfactants (sodium dodecyl sulfate and sodium deoxycholate) and a nonionic surfactant (Triton X-100) on human erythrocytes covalently labeled with a maleimide spin label (**III**). Whereas the anionic detergents were found to induce conformational changes in membrane proteins, no such effects were observed for membrane proteins solubilized by the nonionic detergent.

Two examples of drug effects that result from chemical interaction with an ion are the use of the chelating agents calcium EDTA and pencillamine for lead and copper poisoning, respectively (Goldstein *et al.*, 1974b). To date, there are no examples in which spin-labeled chelating agents have been used to monitor this effect. However, when the metal is paramagnetic it may be possible to study the structure of such chelates by monitoring changes in the ESR spectra of the metal and/or the spin label.

Several drugs, e.g., 5-bromouracil, 5-fluorouracil, ethionine, and *p*-fluorophenylalanine, exert their pharmacological effect by a so-called counterfeit incorporation whereby the drug replaces a normal metabolite in the synthesis of an important cellular constituent (Goldstein *et al.*, 1974b). In this case, it is not the drug itself but the reaction product, e.g., an enzyme with modified biological activity, that produces the characteristic biological response. Theoretically, spin-labeled drug analogues that mimic the behavior of normal metabolites could be used to follow the biological fate of the drug. This approach has not yet been attempted.

V. FUTURE OF SPIN-LABELING STUDIES IN PHARMACOLOGY

From the foregoing discussion, it is obvious that spin labeling already has had a considerable impact on pharmacology. The question then arises as to whether the technique will continue to be a useful tool in the hands of the pharmacologist. In the area of drug absorption, distribution, and excretion, it seems unlikely that spin labeling will have much impact. The spin immunoassay already has been superseded by simpler procedures that require less expensive instrumentation. The use of spin-labeled drugs to monitor pharmacokinetic parameters has many limitations, and it seems unlikely that the procedure will gain wide acceptance. It is in the area of the pharmacological effect of drugs that spin labeling will continue to play an important role. Many new drug receptors have been isolated, and attempts to purify them are being made in many different laboratories. When these purification procedures become available, it seems logical to expect that studies with spin-labeled drug analogues will follow. This is particularly true of membrane-bound receptor systems, since many of the other spectroscopic techniques, e.g., fluorescence, are severely limited by the presence of particulate structures.

REFERENCES

Bergman, F., and M. Wurzel (1953). The active surface of pseudo-cholinesterase and the possible role of this enzyme in conduction. *Biochim. Biophys. Acta* 11, 440–441.
Chignell, C. F. (1971). The interaction of a spin-labeled sulfonamide with bovine carbonic anhydrase. *Life Sci.* 10, 699–706.
Chignell, C. F. (1974). Protein binding and drug action. *Annu. Rep. Med. Chem.* 9, 280–289.
Chignell, C. F. (1976). Ligand binding to plasma albumin. *In* "Handbook of Biochemistry and Molecular Biology" (G. D. Fasman, ed.), 3rd ed., Vol. II, pp. 554–582. CRC Press, Cleveland, Ohio.
Chignell, C. F. (1977). Protein binding. *In* "Drug Fate and Metabolism—Methods and Techniques" (E. R. Garrett and J. L. Hirtz, eds.), Vol. 1, pp. 187–228. Dekker, New York.
Chignell, C. F., D. K. Starkweather, and R. H. Erlich (1972a). The synthesis of some spin-labeled analogs of drug molecules. *J. Med. Chem.* 15, 876–878.
Chignell, C. F., D. K. Starkweather, and R. H. Erlich (1972b). The interaction of some spin-labeled sulfonamides with bovine erythrocyte carbonic anhydrase B. *Biochim. Biophys. Acta* 271, 6–15.

Chignell, C. F., D. K. Starkweather, and B. K. Sinha (1975). A spin label study of egg white avidin. *J. Biol. Chem.* **250**, 5622–5630.

Emanuel, N. M., N. P. Konovalova, and R. F. Djachkovskaya (1976). Toxicity, anti-tumor activity, and pharmacokinetics of spin-labeled ThioTEPA analogs. *Cancer Treat. Rep.* **60**, 1605–1609.

Erlich, R. H., D. K. Starkweather, and C. F. Chignell (1973). A spin label study of human erythrocyte carbonic anhydrases B and C. *Mol. Pharmacol.* **9**, 61–73.

Gillette, J. R. (1963). Metabolism of drugs and other foreign compounds by enzymatic mechanisms. *Prog. Drug Res.* **6**, 11–73.

Giotta, G. J., R. J. Gargiulo, and H. H. Wang (1973). Binding of spin-labeled local anesthetics to lobster nerves. *J. Membr. Biol.* **13**, 233–244.

Goldstein, A., L. Aronow, and S. M. Kalman (1974a). "Principles of Drug Action: The Basis of Pharmacology," 2nd ed., p. vii. Wiley, New York.

Goldstein, A., L. Aronow, and S. M. Kalman (1974b). "Principles of Drug Action: The Basis of Pharmacology," 2nd ed., pp. 111–117. Wiley, New York.

Green, N. M., L. Konieczny, E. J. Toms, and R. C. Valentine (1971). The use of bifunctional biotinyl components to determine the arrangement of subunits in avidin. *Biochem. J.* **125**, 781–791.

Griffin, B. W., S. M. Smith, and J. A. Peterson (1974). *Pseudomonas putida* cytochrome P-450. Binding of a spin-labeled analog of the inhibitor metyrapone. *Arch. Biochem. Biophys.* **160**, 323–332.

Grunfeld, C., A. P. Grollman, and O. M. Rosen (1975). Structure-activity relationships of adrenergic compounds on the adenylate cyclase of frog erythrocytes. *Mol. Pharmacol.* **10**, 605–614.

Harbour, J. R., and J. R. Bolton (1975). Superoxide formation in spinach chloroplasts: Electron spin resonance detection by spin trapping. *Biochem. Biophys. Res. Commun.* **64**, 803–807.

Hsia, J. C., and J. M. Boggs (1973). Pressure effect on the membrane action of nerve-blocking spin label. *Proc. Natl. Acad. Sci. U.S.A.* **70**, 3179–3183.

Hsia, J. C., and L. H. Piette (1969). Spin labeling as a general method in studying antibody active site. *Arch. Biochem. Biophys.* **129**, 296–307.

Janzen, E. G. (1971). Spin trapping. *Acc. Chem. Res.* **4**, 31–40.

Jost, P., and O. H. Griffith (1972). Electron spin resonance and the spin labeling method. *In Methods Pharmacol.* **2**, 223–276.

Kirkpatrick, F. H., and H. E. Sandberg (1973). Effect of anionic surfactants, nonionic surfactants and neutral salts on the conformation of spin-labeled erythrocyte membrane proteins. *Biochim. Biophys. Acta* **298**, 209–218.

Leigh, J. S. (1970). ESR rigid-lattice line shape in a system of two interacting spins. *J. Chem. Phys.* **52**, 2608–2612.

Leute, R. K., E. F. Ullman, A. Goldstein, and L. A. Herzenberg (1972). Spin immunoassay technique for determination of morphine. *Nature (London), New Biol.* **236**, 93–94.

Maren, T. H. (1967). Carbonic anhydrase: Chemistry, physiology, and inhibition. *Physiol. Rev.* **47**, 595–781.

Maren, T. H., A. L. Parcell, and M. N. Malik (1960). A kinetic analysis of carbonic anhydrase inhibition. *J. Pharmacol. Exp. Ther.* **130**, 389–400.

Montgomery, M. R., and J. L. Holtzman (1974). Determination of serum morphine by the spin-label technique. *Drug. Metab. Dispos.* **2**, 391–395.

Montgomery, M. R., J. L. Holtzman, R. K. Leute, J. S. Dewees, and A. Bolz (1975). Determination of diphenylhydantoin in human serum by spin immunoassay. *Clin. Chem.* **21**, 221–226.

Morrisett, J. D. (1975). The use of spin labels for studying the structure and function of enzymes. *In* "Spin Labeling: Theory and Applications" (L. J. Berliner, ed.), Vol. 1, pp. 274–338. Academic Press, New York.

Nostrand, B., I. Vaara, and K. K. Kannan (1975). Structural relationship of human erythrocyte carbonic anhydrase isozymes B and C. *In* "Isozymes" (C. L. Markert, ed.), Vol. 1, pp. 575–599. Academic Press, New York.

Ohnishi, S., and H. M. McConnell (1965). Interaction of the radical ion of chlorpromazine with deoxyribonucleic acid. *J. Am. Chem. Soc.* **87**, 2293.

Reichman, I. M., B. Annaev, and E. G. Rozantsev (1972). Studies of microsomal cytochrome P-450 with isocyanide spin label. *Biochim. Biophys. Acta* **263**, 41–51.

Rosen, G. M., and E. J. Rauckman (1977). Formation and reduction of a nitroxide radical by liver microsomes. *Biochem. Pharmacol.* **26**, 675–678.

Ruf, H. H., and W. Nastainczyk (1976). Binding of a metyrapone spin label to microsomal P-450. *Eur. J. Biochem.* **66**, 139–146.

Sherman, P. H., and W. C. Landgraf (1967). Spin-labeled albumin: A new non-radioactive biologic tracer tag. *Surg. Forum* **18**, 73–75.

Sinha, B. K., and C. F. Chignell (1975). Synthesis and biological activity of spin-labeled analogs of biotin, hexamethonium, decamethonium, dichlorisoproterenol and propranolol. *J. Med. Chem.* **18**, 669–673.

Slater, T. F. (1972). *In* "Free Radical Mechanisms in Tissue Injury," pp. 118–170. Pion Ltd., England.

Smith, I. C. P., and K. W. Butler (1975). Oriented lipid systems as model membranes. *In* "Spin Labeling: Methods and Applications" (L. J. Berliner, ed.), Vol. 1, pp. 441–453. Academic Press, New York.

Stier, A., and E. Sackmann (1973). Spin labels as enzyme substrates. Heterogeneous lipid distribution in liver microsomal membranes. *Biochim. Biophys. Acta* **311**, 400–408.

Taylor, P., and N. M. Jacobs (1974). Interaction between bisquaternary ammonium ligands and acetylcholinesterase: Complex formation studied by fluorescence quenching. *Mol. Pharmacol.* **10**, 93–107.

Trudell, J. R., W. L. Hubbell, and E. N. Cohen (1973a). The effect of two inhalation anesthetics on the order of spin-labeled phospholipid vesicles. *Biochim. Biophys. Acta* **291**, 321–327.

Trudell, J. R., W. L. Hubbell, and E. N. Cohen (1973b). Pressure reversal of inhalation anesthetic-induced disorder in spin-labeled phospholipid vesicles. *Biochim. Biophys. Acta* **291**, 328–334.

Wee, V. T., B. K. Sinha, P. W. Taylor, and C. F. Chignell (1976a). Interaction of spin-labeled bisquaternary ammonium ligands with acetylcholinesterase. *Mol. Pharmacol.* **12**, 667–677.

Wee, V. T., R. J. Feldman, R. J. Tanis, and C. F. Chignell (1976b). A comparative study of mammalian erythrocyte carbonic anhydrases employing spin-labeled analogs of inhibitory sulfonamides. *Mol. Pharmacol.* **12**, 832–843.

Weiland, G., B. Georgia, V. T. Wee, C. F. Chignell, and P. Taylor (1976). Ligand interactions with cholinergic receptor-enriched membranes from *Torpedo*: Influence of agonist exposure on receptor properties. *Mol. Pharmacol.* **12**, 1091–1105.

Wilson, I. B. (1952). Acetylcholinesterase XII. Further studies of binding forces. *J. Biol. Chem.* **197**, 215–225.

Wilson, I. B., and C. Quan (1958). Acetylcholinesterase studies on molecular complementariness. *Arch. Biochem. Biophys.* **73**, 133–143.

6

Spin Labeling in Biomedicine

LAWRENCE H. PIETTE

CANCER CENTER OF HAWAII
UNIVERSITY OF HAWAII
HONOLULU, HAWAII

and

J. CARLETON HSIA

DEPARTMENT OF PHARMACOLOGY
FACULTY OF MEDICINE
UNIVERSITY OF TORONTO
TORONTO, ONTARIO, CANADA

I.	Introduction	247
II.	Spin Assay	248
III.	Spin Immunoassay	254
IV.	Spin-Membrane Immunoassay	259
V.	Spin Labeling in Carcinogenesis	264
	A. Introduction	264
	B. Carcinogen Binding to Cellular Constituents	265
VI.	Spin Trapping	279
	A. Studies of Free-Radical Carcinogenesis	279
	B. Lipid Peroxidation	282
VII.	*In Vivo* Electron Spin Resonance	284
	References	287

I. INTRODUCTION

The spin-label method has become one of the most sensitive biophysical methods available for providing information on conformational properties of macromolecules. As an outgrowth of some of the fundamental spin-label work on membrane structure, protein active-site conformation, and anti-

247

body structure, an area of application of the spin-label technique in the more applied fields of clinical and biomedical research is emerging. Specifically, the spin-labeling technique is being applied to areas of pharmacological assay in drug treatment, detection of drug metabolites, antigenic response of certain disease states, and the mechanism of drug or toxin interactions with cellular constituents. To provide an inclusive review of the literature, we have chosen to illustrate by means of a few examples the direction that spin-label research in biomedicine is taking. Specifically, we will discuss the development of spin-assay and spin-membrane immunoassay techniques as applied to clinical studies and the utilization of spin labeling and spin trapping in studies of carcinogenesis.

In the first part of the chapter, we describe recent developments in the use of the spin-labeling technique in practical clinical assays. Both the potential and the limitations likely to be encountered in the use of spin labeling as a clinical assay technique are critically analyzed. The two distinct advantages in the use of spin labels are, first, the physical state of the biological specimen (for example, turbidity does not interfere with the measurements) and, second, the fact that no separation procedures are required. Like most other analytical techniques, however, specificity and sensitivity determine the eventual clinical usefulness of the assay.

When sensitivity is not a problem, say between milli- and micromolar range, such as in spin assays and spin immunoassays, specificity is the key. However, at micro- to picomolar concentrations both sensitivity and specificity requirements are stringent, and this is the subject of spin-membrane immunoassay.

II. SPIN ASSAY

Spin assay is distinguished by three main features: (1) A preexisting macromolecular binding site, such as occurs in enzyme receptors and in carrier proteins like serum albumin, must be present. (2) A spin label is then tailor-made to mimic the specific binding to these sites. (3) The spin assay is unique in that it can be used for the diagnosis of a pathological state and the improvement of patient care in a clinical setting.

There are a number of unique advantages to the electron spin resonance (ESR) method of determining binding parameters. The spin-label probe can be tailor-made for a particular application, requiring only that the compound contain a nitroxyl moiety so that the label is paramagnetic. The amount of label that is free or bound to a macromolecule can be determined without physically separating the free from the bound label. The spectra of the free and bound labels are distinct, and their concentrations can be determined easily through a calibration of the free peak height with standards (Hsia *et al.*, 1973; Stryer and Griffith, 1965). The ESR instrument itself is straightforward, requiring no day-to-day adjustments, 10- to 25-μl samples, a detectable concentration range down to 10^{-7}–10^{-6} M, and a "sample-in, read, sample-out" procedure. This ease of operation is the basis for the commercialization of the technique by SYNVAR with the development of FRAT. The problems, in general, do not rest with the instrumentation but with the biological system itself, e.g., interfering substances.

An example of its attempted use for diagnostic purposes is the development of a spin assay for reserve bilirubin loading capacity (Hsia *et al.*, 1978b).

Soltys and Hsia (1977) used a spin-labeled derivative of DNP-γ-aminobutyric acid (GABA-DNP-SL, **I**) as a probe and a model for the binding of

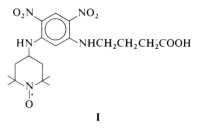

I

anionic ligands to human serum albumin (HSA). Although this anionic spin label binds to the bilirubin binding site, it also binds to other nonspecific sites. To reduce nonspecific binding, Wood and Hsia (1977) found that a dianionic piperidinesuccinic acid spin label (TOPS, **II**) can mimic the bind-

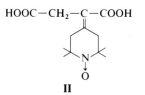

II

ing specificity of bilirubin to HSA. They proposed that this type of dianionic spin label could be developed into a spin assay for the diagnosis of bilirubin toxicity in jaundiced newborns, a dangerous condition that may result in severe neurological damage or even death (Amanullah, 1976).

The key to monitoring the dangerous levels of bilirubin is not simply the measurement of total serum bilirubin but rather the measurement of the reserve bilirubin binding capacity (RBBC) of HSA (Odell *et al.*, 1970). If the high-affinity bilirubin binding sites of HSA are fully occupied with bilirubin or any other substance(s) that compete with bilirubin, the excess bilirubin may diffuse into the central nervous system of the infant and the detrimental effects of icterus will result.

The binding of bilirubin to HSA is known to involve one high-affinity ($K_1 \simeq 10^8\ M^{-1}$) binding site and one or more secondary lower-affinity ($K_2 \simeq 10^5\ M^{-1}$) binding site(s) (Jacobsen and Wennberg, 1974). The capacity of the high-affinity binding site is taken as a measure of the capacity of HSA to bind and hold bilirubin in comparison to that of the body tissues and the secondary binding sites (Wosilait, 1974).

A number of methods to measure the excess bilirubin (Jirsova et al., 1967; Bratlid, 1972) and to evaluate an RBBC index have been proposed (Odell et al., 1969; Porter and Waters, 1966). Some of these procedures have been criticized (Jacobsen and Wennberg, 1974; Chan et al., 1971; Lucey et al., 1967; Levine, 1976), and many are experimentally complex and/or require large volumes of serum.

In addition, optical measurements usually are required to quantitate the amount of bilirubin that is present or to measure the concentration of any dyes used in simulated binding studies. These measurements are generally complex due to the less well defined spectral characteristics of the chromophores. Also, overlapping absorption and/or emission from other endogenous or exogenous ligands in the samples may lead to erroneous conclusions. It has been suggested that the spectral characteristics of bilirubin are a very complicated and poorly understood area (Lee and Gartner, 1976). Because of these difficulties, a method has been developed to measure the reserve high-affinity bilirubin binding capacity of HSA using ESR and the spin-labeling technique (Hsia et al., 1978b).

In addition to the piperdinesuccinic acid label (**II**) mentioned previously, another bifunctional anion DNP-aspartic acid spin label (TOPA, **III**) was

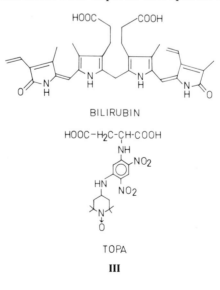

BILIRUBIN

TOPA

III

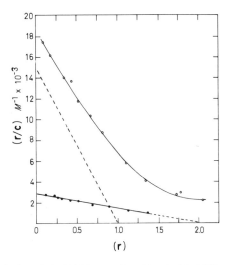

Fig. 1. Scatchard plot of the DNP-aspartic acid spin label **III** as a disodium salt with 5×10^{-4} M fraction V HSA in 100 mM phosphate buffer at pH 7.4 for solution with no added bilirubin (●) and 5×10^{-4} M added bilirubin (○). The dashed line is derived from the difference between the binding curves in the presence (○) and absence (●) of bilirubin by the graphic method described by Rosenthal and Hannah (1967) (Hsia *et al.*, 1978b).

prepared and selected because the dicarboxylic acid structure is similar to that of bilirubin. The presence of the aromatic residue increases the binding affinity and specificity of **III** to the bilirubin high-affinity binding site. The binding of **III** with HSA with 0 and 1 molar equivalent of bilirubin is shown in Fig. 1 in terms of a Scatchard plot (Scatchard, 1949). By means of the simple graphic analysis described by Rosenthal (1967), the affinity constant of **III** to the primary high-affinity bilirubin binding site is resolved into a straight line, which indicates the presence of one high-affinity **III** binding site in HSA with an association constant (K_1) of approximately 1.6×10^4 M^{-1}. In the presence of 1 molar equivalent of bilirubin, **III** appears to have two low-affinity binding sites with an association constant (K_2) of approximately 1.5×10^3 M^{-1}. If **III** shares the high-affinity bilirubin site, the quantitative blockade of the **III** binding site by bilirubin is to be expected since the K_1 of bilirubin to its high-affinity site is of the order of 10^8 M^{-1}.

It was shown that **III** binds specifically to the bilirubin binding sites in HSA (Fig. 2). A comparison of the shaded free peaks in the solid-line and dashed-line spectra shows that the addition of 1 molar equivalent bilirubin results in a marked difference in peak height (*a* to *b* in Fig. 2A). This difference indicates a definite decrease in the concentration of **III** that is bound to HSA. This conclusion is further supported by a decrease in the height of the

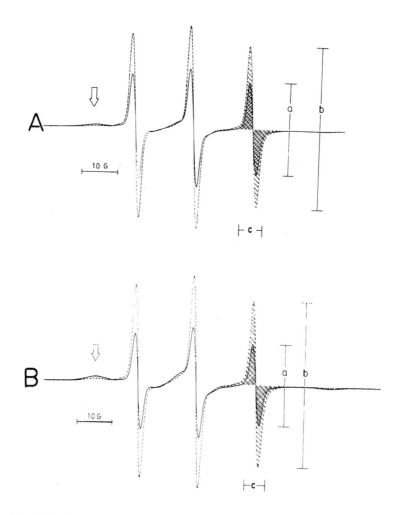

Fig. 2. (A) The ESR spectrum of 1.5×10^{-3} M **III** disodium salt with 5.0×10^{-4} M HSA in 100 mM phosphate buffer at pH 7.4 (solid line). Dashed line is the spectrum of the same solution but with the addition of 5.0×10^{-4} M bilirubin disodium salt. The sharp triplet is due to **III** that is free in solution, whereas the broad resonances near the baseline (the low-field peak indicated by an arrow) is the spectrum of **III** that is bound to HSA. The shaded high-field free peak (heights a and b) is directly related to the concentration of free **III** in each solution. The distance C is the minimal magnetic field sweep range required for this peak detection. (B) The same as above except when adult serum containing 5×10^{-4} M HSA is used in place of purified HSA. [From Hsia *et al.* (1978b).]

bound peak, indicated by the arrow in Fig. 2A. The added bilirubin has displaced **III** from binding to HSA. The displacement by bilirubin of **III** which is bound to HSA also can be seen in the ESR spectra of **III** in human serum without added bilirubin and with 1 equivalent of added bilirubin in Fig. 2B. It was shown that the relative displacement of **III** by adding 1 equivalent of bilirubin each to purified HSA and serum is similar. The similarity of this displacement in purified HSA and in serum suggests that **III** binds specifically to the bilirubin binding sites and that nonspecific binding to other serum components is minimal.

The bilirubin quantitatively displaces **III** from its primary high-affinity binding site in purified HSA and in serum alike (Figs. 3 and 4). The displacement of **III** by bilirubin yields the RBBC value for purified HSA and the reserve bilirubin loading capacity (RBLC) of serum in units of milligram percent (mg %). In order to demonstrate that bilirubin quantitatively displaces **III** from its primary binding site in HSA, these authors saturated the primary and secondary **III** binding site by adding 3 moles of **III** per mole of HSA (Fig. 3). Under these experimental conditions, the free **III** resonance peak intensity increases linearly up to 1 mole bilirubin per mole of HSA. However, displacement of **III** by bilirubin from its secondary binding site(s) is less specific, as evidenced by the fact that the displacement curve deviates from linearity after the addition of the first molar equivalent of bilirubin.

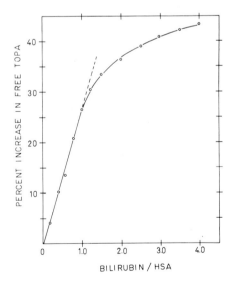

Fig. 3. Displacement curve of 1.5×10^{-3} M **III** by bilirubin in the presence of 5×10^{-4} M fraction V HSA in 100 mM phosphate buffer, pH 7.4. Similar results were obtained with defatted HSA monomers. [From Hsia *et al.* (1978b).]

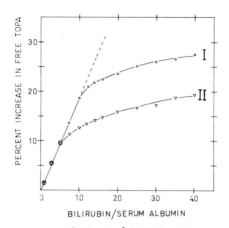

Fig. 4. Displacement curve of 1.5×10^{-3} *M* **III** in adult serum; albumin concentrations 5×10^{-4} *M* with no preexisting bilirubin (I) and with added 2.5×10^{-4} *M* (II) bilirubin disodium salt (bilirubin–HSA ratio 0.5 : 1). [From Hsia *et al.* (1978b).]

Therefore, the bilirubin to HSA ratio in the **III** displacement curve which deviates from linear displacement is a measure of RBBC of HSA, its maximal value being 1, i.e., one high-affinity bilirubin binding site per HSA. This hypothesis was verified in fresh adult sera with predetermined serum albumin concentration from three adult volunteers. A typical example is shown in Fig. 4, curve I. Similar to the displacement curve shown in Fig. 3, the RBBC of HSA in serum has a value of approximately 1. To show that the RBBC value in curve I was sensitive to preexisting bilirubin, the same serum sample was artificially jaundiced by the addition of 0.5 molar equivalent of bilirubin, and the RBBC decreased from 1 to 0.5.

Since a measurement of serum albumin concentration is not always available and other unknown ligands may be binding to the bilirubin binding sites, a simple index termed "reserve bilirubin loading capacity" was obtained by determining the point of deviation from the linear increase of **III** displacement by bilirubin measured in units of milligram percent.

III. SPIN IMMUNOASSAY

Spin immunoassay (SIA) is based on the fact that the resonance peaks of spin-labeled haptens that are antibody bound and free are well separated (Stryer and Griffith, 1965; Hsia, 1968; Hsia *et al.*, 1973).

At appropriate antibody to spin-labeled hapten concentrations the complex formation between the labeled haptens and antibody is exquisitely sensitive to the presence of free haptens, known standard or unknown, in the assay solution. Thus, displacement of the labeled hapten by the unknown can be determined by a standard curve. Spin immunoassay is very similar to radioimmunoassay techniques (RIA) in principle, with the significant difference that SIA does not require separation procedures.

Leute et al. (1972a) developed a rapid SIA for morphine based on this principle. The same authors also showed that SIA for morphine is more sensitive than thin-layer chromatography in a clinical study (Leute et al., 1972b). This and other SIA studies are reviewed in the preceding chapter. However, we shall present a rigorous model of SIA, on which we shall base our discussion of the advantages and disadvantages of SIA. This discussion, in turn, will be the basis for consideration of spin-membrane immunoassay (SMIA).

The simplicity and specificity of the SIA method have been tested against equilibrium dialysis in a model system using homogeneous mouse myeloma protein-315 and a spin-labeled dinitrophenyl ligand (IV) (Hsia et al., 1973).

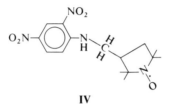

IV

The clear separation of major resonance peaks between antibody-bound and free spin labels (IV) is shown in Fig. 5. The M_{-1} peak intensity and the spin-label (IV) concentration (Fig. 6) in the presence and absence of nonimmune γG shows that the concentration of spin label IV can be determined accurately from its M_{-1} peak intensity, and there is no significant nonspecific binding by γG. However, in the presence of 1.1 mg/ml of protein-315, the M_{-1} peak intensity is significantly less than that of the standard due to the binding of spin label IV by protein-315. After most of the protein-315 binding sites were saturated, the net increase in M_{-1} signal intensity became parallel to that of the standard, indicating no further binding. Therefore, the difference in M_{-1} peak intensity between the standard and that in the presence of protein-315 is a measure of protein-bound spin label IV. This assumption was verified by displacing each titration point with excess DNP-ε-aminocaproate or by denaturing the protein with 1% trichloroacetic acid. In all cases, the free ligand concentration in the presence of protein-315 as measured by M_{-1} peak intensity was identical to that found in the buffer chamber of a microequilibrium dialysis cell (Voss and Eisen, 1968), suggesting that the spin-labeled hapten titration is equivalent to an instantaneous equilibrium dialysis measurement. The above-described experiments clearly demonstrate that the dynamic equilibrium or the ex-

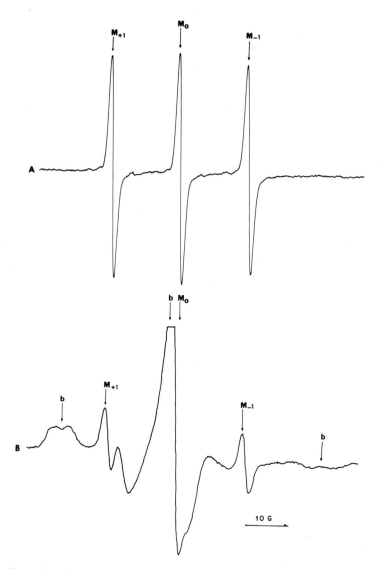

Fig. 5. The ESR spectrum of N-(1-oxyl-2,2,5,5-tetramethyl-3-methylaminopyrrolidinyl)-2,4-dinitrobenzene (1×10^{-5} M in buffered saline, 0.15 M NaCl–0.02 M phosphate, pH 7.4) at 22°C. (A) In the presence of 125 μg of nonimmune G; (B) same as (A) except 125 μg of protein-315 was used in place of G. Peaks arising from immunoglobulin-free signal are labeled as M_{+1}, M_0, M_{-1}, and those arising from immunoglobulin-bound signal are labeled b. [From Hsia *et al.* (1973).]

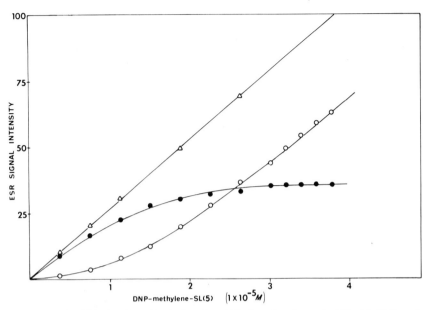

Fig. 6. The ESR titration of polynitrophenyl binding site of protein-315 with N-(1-oxyl-2,2,5,5,tetramethyl-3-methylaminopyrrolidinyl)-2,4-dinitrobenzene in buffered saline, 0.15 M NaCl–0.02 M phosphate, pH 7.4, at 22°C. Immunoglobulin-free signal intensity (M_{-1}, see legend of Fig. 5) in the presence of 55 μg of nonimmune γG (-△-△-) and 55 μg of protein-315 (-○-○-) and immunoglobulin-bound signal intensity (b) in the presence of 55 μg of protein-315 (-●-●-). [From Hsia *et al.* (1973).]

change processes between the protein-bound (AB–H) and free (H) hapten do not result in the averaging of the resonance spectra.

$$AB + H \rightleftharpoons AB-H$$

The Scatchard and Sips plots calculated from these binding studies were reported (Hsia *et al.*, 1973). The linear slope of the Scatchard plot indicates that the homogeneity and the number of binding sites ($n = 1.9$) of this protein are in excellent agreement with previous equilibrium dialysis determinations (Underdown *et al.*, 1971). The Sips plot also indicates that the binding site of protein-315 is homogeneous with respect to spin label **IV** ($a = 0.94$); however, the association constant for spin label **IV** ($K = 7.0 \times 10^5\ M^{-1}$) is considerably less than that for ε-DNP-L-lysine ($K = 1 \times 10^7\ M^{-1}$) (Eisen *et al.*, 1970). This decrease in association constant probably is due to the steric perturbation introduced by replacing the L-lysine with a bulky pyrrolidine nitroxide spin label (Hsia and Little, 1971). The lower affinity of spin-labeled hapten to the binding protein or antibody as compared to the homologous hapten is actually advantageous because this

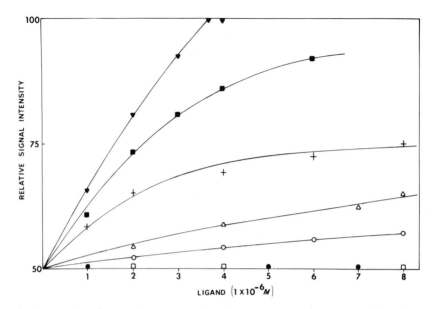

Fig. 7. Displacement of protein-315-bound DNP-methylene-SL by N-trinitrophenyl ε-aminocaproate (-▼-▼-), N-dinitriphenyl ε-aminocaproate (-■-■-), N-(isopropyl)-2,4-dinitroaniline (-+-+-), 2,4-dinitroaniline (-△-△-), p-nitrophenyl ε-aminocaproate (-○-○-), o-nitroaniline (-●-●-), and dansylglycine (-□-□-). Total DNP-methylene-SL concentration was 2×10^{-6} M; protein concentration was 0.225 mg/ml. [From Hsia et al. (1973).]

difference in affinity permits the homologous hapten to displace the spin-labeled hapten with greater ease.

The results from the displacement of protein-bound spin label **IV** by various structural analogues of 2,4-dinitrophenyl ligands and a structurally unrelated dansylglycine are shown in Fig. 7. The sensitivity and specificity of SIA are determined largely by the affinity of the binding protein or the specific antibody to the spin-labeled hapten and the ability of the antibody or binding protein to discriminate against structural analogues and un-related haptens. The concentrations required to displace 50% of the protein-bound spin label are as follows: TNP-ε-aminocaproate, 1.6×10^{-6} M; DNP-ε-aminocaproate, 2.3×10^{-6} M; N-(isopropyl)-2,4-dinitroaniline, 8×10^{-6} M. The concentrations for dinitroaniline, p-nitrophenyl ε-amino-caproate, o-nitroaniline, and dansylglycine were not presented since they are $\geqslant 8 \times 10^{-6}$ M. These differences in the concentrations demonstrate re-markable specificity and sensitivity of the SIA.

However, the sensitivity of SIA is in the range 10^{-6}–10^{-7} M, which is much less sensitive than RIA, although SIA is very suitable for use with water-soluble ligands. Unfortunately, most ligands of biological interest are lipophilic. To overcome the limitations of water solubility and low sensitivity of SIA, a third type of assay using the spin-labeling technique has been developed.

IV. SPIN-MEMBRANE IMMUNOASSAY

The recent development of membrane immunoassay (MIA) (Hsia and Tan, 1977), particularly spin-membrane immunoassay (SMIA), has provided a simple, sensitive immunoassay method based on liposome immunochemistry (Kinsky, 1972) and the ESR spin-labeling technique (Humphries and McConnell, 1974).

The MIA method is based on the fact that a lipid-linked antigen (sensitizer) may be solubilized in a lipid bilayer membrane matrix to form sensitized liposomes. The liposomes consist of concentric shells of lipid bilayer, the aqueous interspaces of which contain trapped markers. Once the liposomes are sensitized, they can react with specific antibodies. In the presence of complement, the trapped markers are released due to specific complement lysis. The sensitized liposomes with trapped markers act as an amplifier for the assay similar to radioactive hapten for RIA and enzyme activity for enzyme immunoassay.

The SMIA technique involves the use of the spin label as a marker (Humphries and McConnell, 1974; Wei et al., 1975; Hsia et al., 1976; Resenquist and Vistnes, 1977) and makes use of the ESR spectrometer for its detection. A detailed description of the principle and applications of MIA and SMIA is presented below.

The key to SMIA is the preparation of specific sensitizers. A sensitizer is any derivative of an amphiphilic molecule that is linked by covalent coupling of an antigen to the polar-head group of the amphiphile. The resulting molecule can serve as a sensitizer only if it can form an integral part of the membrane with its antigenic moiety protruding into the aqueous phase, available for binding by specific antibody. Various drugs and biological substances can be attached covalently, via a suitable coupling reagent, to an entire class of amphiphilic molecules (Fig. 8). This overcomes the water solubility problem encountered in SIA of spin-labeled haptens. One of the most convenient ways of incorporating the sensitizers into the membrane is through the use of liposomes. When the liposomes are immunologically complete, i.e., when they contain an appropriate sensitizer, they can react with specific antibody; when complements are present, specific complement lysis releases the trapped markers (Fig. 9). Various markers have been used,

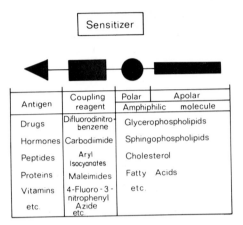

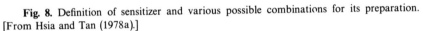

Fig. 8. Definition of sensitizer and various possible combinations for its preparation. [From Hsia and Tan (1978a).]

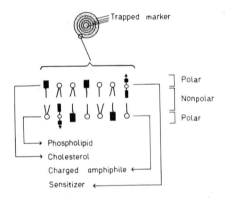

Fig. 9. Schematic representation of an immunologically responsive liposome. [Adapted from Kinsky (1972).]

including glucose (Kinsky, 1972), enzymes (Knudson *et al.*, 1971), and spin labels. Spin labels have been found to be among the most effective markers for the quantification of liposome immune lysis. The ESR spectrum of the spin label TEMPOcholine chloride (**V**) trapped inside the aqueous interspaces of liposomes showed broad and weak signals (resulting from spin exchange interactions between unpaired electrons on the nitroxide radicals)

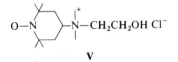

V

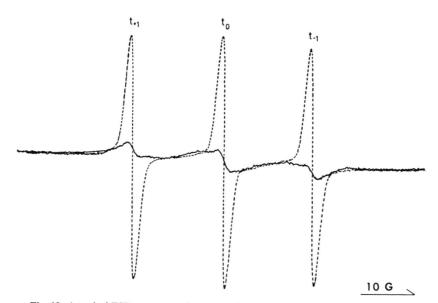

Fig. 10. A typical ESR spectrum of a preparation of liposomes loaded with TEMPOcholine chloride (**V**) (spin label, as marker): before lysis (low-amplitude line) and after lysis (high-amplitude line).

when the spin-label concentration was high (Fig. 10). These broad resonance peaks represent the background value of the assay. After lysis, the spin labels are released into the bulk phase, reducing their local concentrations and eliminating the spin exchange interaction so that a sharp, intense signal is detected. The magnitude of this peak intensity in the ESR spectrum is directly proportional to the number of spin-label molecules released from the liposomes. Thus, the degree of lysis can be determined accurately by measuring the change in this signal intensity. In the quantification of antigens, antiserum is mixed with an aqueous-phase antigen (a known standard or unknown sample), as depicted in step 1 of Fig. 11. To determine the degree of inhibition of antibody by the aqueous-phase antigen, one adds specifically

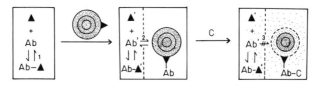

Fig. 11. Schematic description of SMIA method. Each small dot represents a spin label; ▲, antigen; Ab, antibody; C, complement. Steps 1 and 2 are reversible (equilibrium) processes. Step 3 is an irreversible (end-point) process (Hsia and Tan, 1978a).

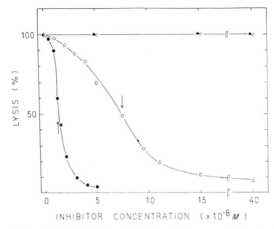

Fig. 12. Specificity of SMIA in the inhibition of immune lysis of ε-DNP-capro-PE-sensitized liposomes by ε-DNP-caprolate (-●-), α-DNP-glycine (-○-), o-nitroaniline (-▲-), and ε-dansyllysine (-▽-). Seventy to 80% maximal immune lysis for the ε-DNP-capro-PE-sensitized liposomes were normally observed. This maximal lysis was defined as 100% to represent the true end point of specific immune lysis, and the inhibition curves were all normalized. [From Chan *et al.* (1971).]

sensitized liposomes containing a high concentration of spin label in the aqueous interspaces (step 2) and then guinea pig complement (step 3) to initiate complement lysis of antibody-associated liposomes. The degree of lysis is inversely proportional to the amount of aqueous-phase antigen in the assay tube.

A preparation of ε-DNP-capro-PE-sensitized liposomes (Six *et al.*, 1974) was used as a model to demonstrate the specificity and sensitivity of SMIA. Two structurally related DNP derivatives (ε-DNP-caprolate and α-DNP-glycine) inhibited immune lysis of ε-DNP-capro-PE-sensitized liposomes by anti-DNP antiserum in the presence of guinea pig serum (Fig. 12). ε-DNP-Caprolate strongly inhibited immune lysis by 75 nl anti-DNP antiserum in the presence of optimal concentration (2.5 μl) of guinea pig serum. The amount of inhibitors required to inhibit 50% immune lysis (I_{50}) for ε-DNP-aminocaproic acid and α-DNP-glycine was approximately 1.5×10^{-8} and 7×10^{-8} M, respectively. However, ε-dansyllysine and o-nitroaniline, which are structurally unrelated to ε-DNP-lysine, were not inhibitory.

Spin-membrane immunoassay not only preserves the simplicity of SIA when separation is not required, but also has the added advantage of greatly increased sensitivity. Although the maximal sensitivity achieved in this model system is only 10^{-8} M (mainly because the lower limit of the affinity constant of anti-DNP antibody is approximately 10^8 M^{-1}), it is certainly not the limit of SMIA. In fact, Humphries and McConnell (1974) reported achieving a sensitivity of 10^{-11} M using another system. An augmentation phenomenon also was observed by these authors, but as yet it has not been confirmed by them in subsequent publications or by others.

An example of a specific application of SMIA is the thyroxine (T_4) assay (Hsia and Tan, 1978a). When T_4 sensitizer is incorporated into liposomes they become susceptible to immune lysis. Immune lysis of sensitized liposomes is dependent on the concentration of anti-T_4 antiserum and complement. Immune lysis is an all-or-none process, reaching a plateau in the presence of excess antisera and complement. Furthermore, heat inactivation of the complement completely abolishes the immune lysis of T_4-sensitized liposomes by anti-T_4 antisera. The degree of inhibition of immune lysis depends on the concentration of T_4 in the solution (Fig. 13). The sensitivity range for T_4 by SMIA (1–9 ng, with a 20-μl sample) is comparable to that by RIA. To appreciate the potential of SMIA more fully, a brief comparison with RIA is useful. The ideal clinical immunoassay technique for determining drugs, hormones, vitamins, peptides, or proteins in biological fluids should be simple, specific, sensitive, inexpensive, safe, and readily automated. It also should be applicable to a wide range of compounds and readily adaptable for multiple assays. Until recently, the method that most nearly satisfied these criteria was RIA (Skelley et al., 1973). One of its major limitations is the need to separate free and antibody-bound labels (Minden et al., 1969).

Separation procedures are time-consuming and relatively expensive and

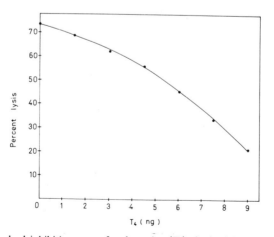

Fig. 13. Standard inhibition curve for thyroxine (T_4) obtained by SMIA. The inhibition curve was not normalized to 100% as were those shown in Fig. 12. [From Hsia and Tan (1978a).]

increase the possibility of error. The hazard, short shelf life (e.g., ^{125}I and ^{131}I), and cost of radioactive materials are further limiting factors. Unlike the RIA method, SMIA does not require separation procedures and involves no radiation hazard. It has the added advantages of speed, possibility of full automation, simplicity, and small sample volume (20 μl). Therefore, SMIA may cost less overall than RIA. Furthermore, the SMIA method also offers the great potential of performing multiple assays, because different markers (e.g., spin labels) and sensitizers can be used to detect different antigens simultaneously.

Of particular interest is the clinical application of SMIA for thyroid hormones since this is one of the most common applications of RIA. For example, recent studies have shown that approximately 1 of every 5000 infants in North America is born with congenital hypothyroidism, and mass screening for this disorder in neonates has been recommended (Fisher et al., 1976). Undoubtedly, the unique advantages of SMIA could well be applied in other new areas of investigation.

V. SPIN LABELING IN CARCINOGENESIS

A. Introduction

Carcinogenesis is an extremely complex process. The sequence of molecular steps leading to the manifestation of malignancy has yet to be elucidated (Weisburger, 1973). A great deal is known, however, about various aspects of the process. We know, for example, that most carcinogens must be activated before they are able to induce a malignancy. The site of activation most likely involves the hydroxylase system at the microsomal membrane level. However, we do not know all the details of this enzymatic activation, e.g., whether it is an ionic or free-radical mechanism or both. We also know that some of the early events in carcinogenesis probably involve interactions of the activated carcinogens with nuclear material, either with DNA directly or with the gene controlling nuclear proteins. Moreover, we know that some of the early events in cell transformation from normal to malignant occur at the cellular membrane level. Thus, it would seem reasonable that research that can provide new information or new techniques in the study of these early events would be a valuable contribution to the ultimate goal of a definitive mechanism for molecular carcinogenesis. For this reason, ESR spin-label methods are being exploited in an attempt to provide any additional information that may lead to a better understanding of this complex sequence of biological reactions.

B. Carcinogen Binding to Cellular Constituents

1. CELLULAR MEMBRANES

The ESR spin-label technique is an extremely powerful tool for studying small-molecule–macromolecule interactions (Hamilton and McConnell, 1968; Berliner, 1976; Jost *et al.*, 1972). The characteristics and sensitivity of the technique are such that binding orientation, conformation, and extent of binding of small spin-labeled molecules to macromolecules are readily deducible from a careful analysis of their ESR spectra. The method, in general, requires the synthesis of special labels that are components of the biological small molecules of importance, which in this case is either a chemical carcinogen or the macromolecule or array of macromolecules being probed such as the cell membrane or the enzymes involved in carcinogen activation. The spin-label technique takes advantage of the fact that most labels yield highly anisotropic ESR spectra as a result of either their structure or the environment in which they are attached, and it is the change in this anisotropy that is correlated with the interactions being probed. Not all spin-label techniques require the synthesis of special extrinsic probes. Instead, many intrinsic paramagnetic molecules or atoms within the macromolecular system can be used to probe the system or its interaction with small molecules. An example is an active one-electron redox intermediate involved in carcinogen activation. Such intermediates, although not spin labels as such, can be studied indirectly through spin-trapping techniques, which will be discussed later.

Both methods are being investigated actively in attempts to probe the events associated with molecular carcinogenesis (Hong and Piette, 1976; Saprin and Piette, 1977; Lai and Piette, 1977). Specifically, carcinogen spin labels that are derived from polyaromatic amines (Section V,B,2) and estradiol derivatives (**VI**) have been synthesized (Hong and Piette, 1976). These labels are being used to probe the binding characteristics of the carcinogens with microsomal membranes, DNA, and specific receptor proteins. There is probably no better single method for studying binding characteristics than

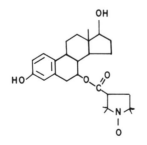

VI

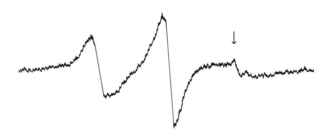

Fig. 14. The ESR spectrum of a suspension of smooth microsomal membranes from rat liver incubated with 5×10^{-5} M AFSL. The arrow ↓ indicates unbound free spin label.

spin labeling. Binding is monitored by taking advantage of the dramatic changes in ESR spectral anisotropy of the label as a result of a change in label correlation time in going from the unbound or freely tumbling state to the bound or rigidly confined state. Competition for binding by the spin-labeled carcinogen with the parent or another nonlabeled carcinogen molecule also is easily monitored by following the reverse change in label correlation time upon release of the label from the bound state. In effect, the technique offers a rapid method for assaying carcinogen binding. It has been shown that a spin-labeled derivative (AFSL) (**VII**) of the carcinogen 2-aminoacetylfluorene (AAF) binds quite strongly to rat liver microsomal membranes, either smooth or rough (Fig. 14). The binding can be blocked or the label displaced by either pre- or postincubation with unlabeled AAF. Estradiol is also competitive with **VII** for binding on the microsomal mem-

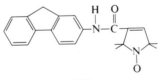

VII

brane as suggested and shown by Williams and Rabin (1971), using ³H-labeled estradiol and cold AAF for their studies. Williams and Rabin (1971) suggested identical binding sites for carcinogens and estradiol on smooth and rough microsomes. They further suggested that the carcinogen induces degranulation of rough microsomes and inhibits regranulation of smooth microsomes, whereas estradiol protects against degranulation by the carcinogen. The estradiol effect takes place only with male rat microsomes, and testosterone is equally effective with female microsomes. The spin-label method has verified the competitive binding characteristics of estradiol and

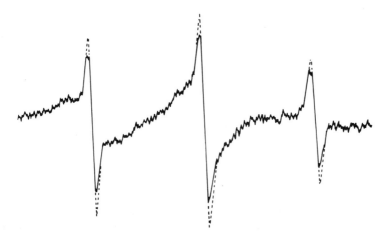

Fig. 15. The ESR spectra of a suspension of smooth microsomal membranes from rat liver incubated with 5×10^{-5} M AFSL (solid line) followed by reaction with 5×10^{-4} M estradiol (dashed line).

spin-labeled 2-aminoacetylfluorene (**VII**) (Fig. 15) and furthermore that **VII** is capable of degranulation of liver microsomes, but the estradiol inhibition is not sex specific. The labeling method has also shown from studies of partially purified preparations of cytochrome P-450 from rat liver that almost identical binding patterns for **VII** occur with this protein, suggesting that it may be a possible target for direct interaction with the carcinogen. Spin labeling, along with further fractionation of the membranes and probing, could in this way possibly yield more information on the specific receptor.

In many cases, in addition to straightforward binding data, the ESR spin-label method also can give structural information on the nature and geometry of the specific binding site. This has been demonstrated clearly in studies on carcinogen and ethidium bromide (EB) spin-label binding with calf thymus DNA (Hong and Piette, 1976).

2. NUCLEIC ACIDS

The early studies of Lerman (1964), in which an intercalation mechanism was postulated for the DNA–acridine dye complex to explain the strong binding of these dye molecules to DNA and their resultant frame-shift mutations, have stimulated considerable interest in applying similar mechanisms to the noncovalent and covalent interactions of carcinogens with DNA (Boyland and Green, 1962; Lesko *et al.*, 1969; Nagata *et al.*, 1966).

There appears to be convincing evidence that in some systems the initial step in chemical carcinogenesis is the noncovalent or covalent binding of the carcinogen to DNA. In the case of DNA the simplest and probably most attractive mechanism suggested is the somatic mutation hypothesis, wherein a base sequence is altered by the intercalation and subsequent covalent binding of a carcinogen into the DNA helix.

The recent studies of Ames *et al.* (1973) and McCann *et al.* (1975) have clearly shown that many proximate and suggested ultimate carcinogens are in fact mutagenic. These authors proposed that polycyclic hydrocarbons are carcinogenic because of the mutagenicity of epoxide intermediates formed during metabolism and that the mechanism of action most likely involves intercalation of the activated carcinogen followed by covalent binding. It therefore remains to demonstrate clearly that with these carcinogens intercalation is in fact a mode of binding to DNA.

The physical methods used to date to demonstrate intercalation mechanisms in DNA are varied and have been applied primarily to the acridine dyes and similar molecules with a strong affinity for DNA (Arcos and Argus, 1974; Waring, 1972). However, none of them appears to demonstrate unequivocally that intercalation has occurred. The problem of demonstrating intercalation of carcinogens becomes even more difficult because the extent of binding of most carcinogens at saturation is far lower (Magee, 1974) than that of the well-known intercalative dyes such as acridine (Lerman, 1964) and EB (Waring, 1972).

The use of spin labeling to detect small-molecule intercalation into DNA is not new. Using a somewhat related method, Ohnishi and McConnell (1965) showed that the stable free-radical ion of chlorpromazine intercalates into DNA. The spin-label technique was actually an outgrowth of these experiments.

The application of the method by Hong and Piette (1976) to demonstrate intercalation of carcinogens in DNA involved the synthesis of spin labels of known intercalative dye molecules such as EB with a paramagnetic nitroxide attached and a series of modified aromatic amine carcinogens with similar nitroxides attached. The anisotropy of the hyperfine interaction of these spin labels is such that their ESR spectra are highly dependent on the orientation of the label with respect to the molecule to which it binds. Intercalation of these molecules into DNA implies a fixed orientation relative to the DNA helical axis. Electron spin resonance spectra obtained as a result of orientation of the DNA–spin-label complex relative to the magnetic field will clearly show this spectral anistropy if intercalation has occurred.

Examples of the types of spin labels used in these studies are as follows. Ethidium bromide spin label (EBSL, VIII) [3-*N*-(3′-carbonyl-1′-oxyl-2′,2′,5′, 5′-tetramethylpyrroline)-3,8-diamino-5-ethyl-6-phenylphenanthridinium bromide] was used as a model ligand. For EB there already exists a considerable body of evidence for intercalative binding with DNA (Waring, 1965;

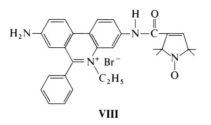

VIII

Fuller and Waring, 1964; LePecq and Paoletti, 1967), and thus the compound stands as a type-specific example of an intercalating drug. The aromatic amine carcinogen spin labels used in this study were aminofluorene spin label (AFSL, **VII**; Section V,B,1) [2-N-(3'-carbonyl-1'-oxyl'2',2',5',5'-tetramethylpyrroline)aminofluorene], aminoanthracene spin label (AASL, **IX**) [2-N-(3'-carbonyl-1'-oxyl-2',2',5',5'-tetramethylpyrroline)aminoanthracene] and aminochrysene spin label (ACSL, **X**) [6-N-(3'-carbonyl-1'-oxyl-2', 2',5',5'-tetramethylpyrroline)aminochrysene].

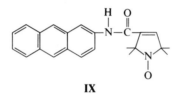

IX

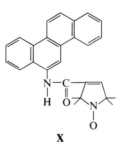

X

These spin labels were chosen because (1) the parent non-spin-labeled compounds are aromatic amines consistent with the model ligand EB, and their intercalative nature with DNA is most likely (Ames et al., 1973; McCann et al., 1975; Kriek, 1974); (2) they were known to be carcinogenic in human beings or animals (Ames et al., 1973; Lambelin et al., 1975; Miller and Miller, 1971); and (3) they were shown to be activated by human liver homogenates to form potent frame-shift mutagens (Ames et al., 1973).

The usefulness of the spin-label technique to obtain information about the presence or lack of an orientation in binding of small molecules such as chemical carcinogens to large macromolecules such as DNA derives from the dependence of the ESR spectra of the reporter group (a nitroxide attached to the small molecule) and its orientation in space. Thus, when a spin label such as a modified carcinogen is included in a highly ordered structure such as is found with DNA and is constrained to reflect the structural arrangement wherein it is bound, the splitting of the hyperfine lines of the ESR spectrum will depend on the angle between the helical axis of the DNA and the applied magnetic field (Ohnishi and McConnell, 1965). In order that the spin label reflect its orientation dependence, it must be rigidly held in the DNA and possess a well-defined geometric relationship between the nitroxide group and the rest of the molecule. It is the latter requirement that is difficult to establish in the labels used here. The linkage between the nitroxide group and the planar aromatic rings of the intercalating molecule is through an amide bond. This linkage would allow rotation of the nitroxide around this bond; however, from space-filling molecular models it would be a highly hindered rotation. Thus, it is assumed that, for all labels used except the aminochrysene derivative, the nitroxide is approximately coplanar and nonrotating in some cases or partially rotating in others with respect to the aromatic ring when binding occurs. The angular dependence of the hyperfine splittings of the spin label is related directly to the orientation and motion of the principal axes of the hyperfine and $\mathbf{g}$ tensors of the nitroxide moiety with respect to the plane of the DNA fiber through the spin Hamiltonian $\mathcal{H}$,

$$\mathcal{H} = |\beta| \mathbf{H} \cdot \mathbf{g} \cdot \hat{\mathbf{S}} + h\hat{\mathbf{S}} \cdot \mathbf{T} \cdot \hat{\mathbf{I}} \tag{1}$$

where $\mathbf{g}$ and $\mathbf{T}$ are tensors for the spectroscopic splitting factor and hyperfine interactions, respectively. The electron and nuclear spin operators are designated as $\hat{\mathbf{S}}$ and $\hat{\mathbf{I}}$, $|\beta|$ is the Bohr magneton, and $\mathbf{H}$ is the magnetic field vector. For the nitroxides used here the hyperfine tensor is axially symmetrical $T_{xx} = T_{yy}$, and one describes the hyperfine splitting in the plane perpendicular to the p π orbital containing the nitroxide unpaired electron by $T_\perp$ and that parallel by $T_\parallel$. The scheme in Fig. 16 shows a possible orientation of the DNA fibers relative to the magnetic field and the orientation of the p π orbital of the nitroxide if it is intercalated between base pairs and thus perpendicular to the helical axis. In this representation, T_{zz} is aligned parallel to the field, and T_{xx} and T_{yy} would be perpendicular if perfect orientation of the DNA were achieved. In single-crystal studies of nitroxides, $T_\parallel$ is 30.8 G and $T_\perp$ is 5.8 G with the isotropic coupling constant 14.1 G. These values vary slightly depending on the nitroxide and its environment.

The ESR anisotropy data for the carcinogens was obtained using a device that allowed the oriented spin-label–DNA complex to be rotated relative to the magnetic field. The device consists of a rectangular quartz

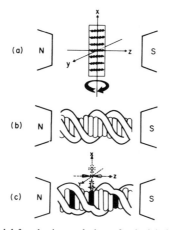

Fig. 16. Schematic model for the intercalation of spin labels into DNA and the relative geometric orientation of the spin label in the complex. (a) Sketches representing the complex fiber oriented on a quartz plate in the ESR resonance cavity and the direction to which this fiber can be rotated around the x axis with respect to the applied magnetic field. (b) The secondary structure of normal DNA the helical axis of which is oriented parallel to the magnetic field. (c) The spin-label–DNA complex fiber, in the same orientation, in which the nitroxide ring plane is coplanar with respect to its constraining ligand moiety (in black). The x axis is taken as parallel to the NO bond, with z parallel to the nitrogen $2p$ π orbital that holds the odd electron, and y is perpendicular to the xz plane. The helix is drawn as viewed from a remote point, so that the base pairs and the intercalated spin labels appear only in an edgewise projection, and the phosphate–deoxyribose backbone appears as a smooth coil. [From Hong and Piette (1976); reprinted by permission of the publisher.]

plate on both sides of which the DNA fiber complexed with spin label, which is a highly viscous solution, was stretch-oriented by drawing the solution with a brush across the plate in a direction perpendicular to the long axis of the plate, and the plate was then placed into a specially constructed goniometer attached to the ESR cavity. Rotations were made around the long axis of the plate (see Fig. 16a). After each stepwise rotation of 10°, from 0° to 180° to the applied magnetic field, the ESR spectrum was recorded. Orientation of the complexes also can be achieved by flowing solutions of complex through a capillary tube at a shear rate of approximately 3000 sec^{-1} in a direction either parallel or perpendicular to the magnetic field. This method of complex orientation requires a specially designed cavity with sample access ports perpendicular and parallel to the field modulation coils. The latter technique gave the same results as the former but was limited to only two field orientations because the magnet could not be rotated. In general, this method is not as convenient to use. Finally, a third method was used whereby condensed complexes were deposited as thin films onto quartz plates, and the plates were then placed into the goniometer. A clear demon-

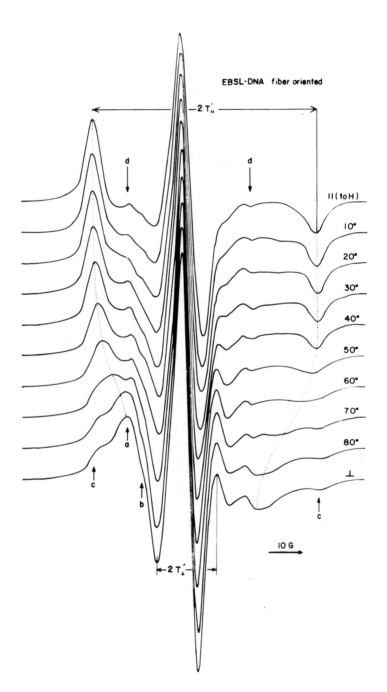

EBSL-DNA fiber oriented

$2 T_\parallel'$

d   d

$\parallel$(to H)

10°

20°

30°

40°

50°

60°

70°

80°

$\perp$

a

c   b

c

10 G

$2 T_\perp'$

stration of the technique is shown in Fig. 17 for the model intercalator EBSL derived from ethidium bromide.

The orientation dependence of EBSL can be determined as shown in Fig. 17, which illustrates the spectral changes observed as a function of orientation of the EBSL–DNA with the magnetic field. The first spectrum shows that the fibers are oriented parallel to the field, and in the subsequent spectra the fibers are rotated through 90° from this orientation in 10° intervals to the perpendicular orientation. In the parallel configuration, the spectrum clearly delineates $T_\parallel$ but is broad and resembles a powder spectrum, suggesting that motion of the label is restricted and that perfect orientation has not been achieved. With rotation of the sample about the x axis (Fig. 16a), the two outer extreme lines characteristic of the maximal anisotropic hyperfine splitting, where T_{zz} is parallel to the field, slowly move toward the center. Figure 18 illustrates the angular dependence of these outer lines. In the perpendicular configuration, $T_\perp$ is readily observed. However, the broad peak a in Fig. 17a about 18 G from the center line, which is at the minimum of the curve in Fig. 18, and the lack of a peak at b in Fig. 17 are suggestive of a distribution of fibers in a static condition not all oriented perpendicular to the field but distributed across a rather wide angle (Schreier-Muccillo and Smith, 1973). The majority of the fibers are reasonably oriented in the plane of the plate as suggested in Fig. 16a, and they give a $2T_\parallel'$ of 64.3 g and $2T_\perp'$ of 17.3 G.

This orientation dependence of the spectrum, assuming the nitroxide to be coplanar with the EB ring, only can be interpreted as evidence of a fixed orientation of the EB, namely, intercalation between base pairs in the DNA helix.

The anisotropy of the spectra clearly demonstrates that for most of the complexes the EBSL is oriented in the DNA such that the z axis of the nitroxide is parallel to the fiber axis. As well, the plane of the nitroxide is parallel to the fiber axis and the plane of the nitroxide ring is parallel to the base-pair layers of the DNA and perpendicular to the DNA helical axis. (Editor's note: further discussions of Figs. 17 and 18 may be found in Chapter 7, p. 317.)

Fig. 17. Angular dependence of the ESR spectrum of oriented EBSL–DNA (**VIII**) complex fibers with respect to the laboratory magnetic field. (a) Observed position of the low-field peak with a fiber orientation perpendicular to the magnetic field. (b) Expected position of the peak that would be observed when the fibers are ideally oriented. (c) Those complexes that are nonoriented and therefore field independent and probably the result of an edge effect in the sample cell. (d) Those labels unbound and in equilibrium with the bound spin. Dotted line, angular dependence of the outer extrema of the spectra. The separation between the low-field peak and center peak is taken as a measure of the angular dependence of the hyperfine splitting of the spectrum. [From Hong and Piette (1976); reprinted by permission of the publisher.]

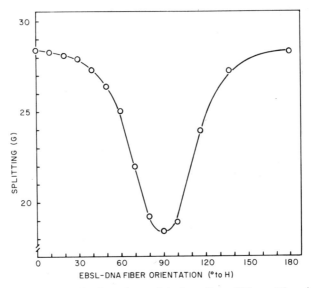

Fig. 18. Plot of the angular dependence of the hyperfine splittings of the oriented EBSL–DNA complex fiber with respect to the laboratory magnetic field. [From Hong and Piette (1976); reprinted by permission of the publisher.]

The results obtained when this technique was applied to the carcinogen spin labels AFSL, AASL, and ACSL are described in the following paragraphs.

In Fig. 19, we see the superpositioning of spectra of all three aromatic amine complexes in both the parallel and perpendicular orientation of the stretched fibers relative to the applied field. Upon inspection of the spectra in Fig. 19, it can be seen that both the strongly immobilized component *a* and the partially immobilized component *b* of AFSL and AASL are field orientation dependent, but in addition there is a third complex in low concentration similar to that found in the EBSL spectra that is strongly immobilized and field orientation independent. It is felt that the third species results from random orientation of the complexes and not from nonoriented binding of the amines. As with EBSL, the results for the carcinogens suggest that

Fig. 19. Angular dependence of ESR spectra of the oriented AFSL–DNA (a), AASL–DNA (b), and ACSL–DNA (c) complex fibers. Solid and dashed spectra obtained with the complex fiber orientation parallel and perpendicular to the applied magnetic field, respectively. Point *a*, the outer extrema of the strongly immobilized component; point *b*, partially immobilized component of the bound spin species; point *c*, trace of unbound spin component. $A_{\parallel}$ and $A_{\perp}$ measure the splitting between outer extrema of the corresponding resonance components when their fibers are oriented parallel and perpendicular to the magnetic field, respectively. [From Hong and Piette (1976); reprinted by permission of the publisher.]

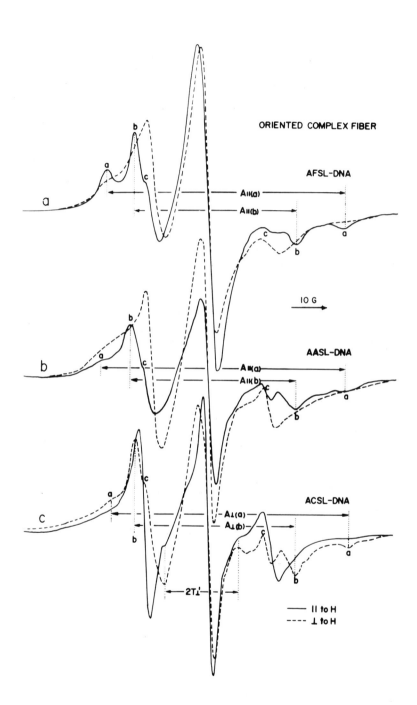

ORIENTED COMPLEX FIBER

AFSL-DNA

AASL-DNA

ACSL-DNA

10 G →

—— ‖ to H
---- ⊥ to H

the nitroxide moiety is approximately coplanar with the aromatic amine for VII and IX (Section V,B,2), with the nitroxide x axis aligned parallel to the fiber axis, and is immobilized upon insertion of the amines by intercalation between base pairs of the DNA. Orientation of the complexes is not as complete for the aromatic amines as it is for EBSL–DNA complexes. In the perpendicular configuration the minimal splitting $T_\perp$ is not observed. The peak at 17 G, similar to point a in Fig. 17, to the low-field side of the center line in the perpendicular configuration (Fig. 19a) suggests a wide angular distribution of fibers in a static condition relative to the perpendicular orientation. The fact that two species of complexes with different degrees of immobilization are observed that are field orientation dependent suggests two distinct binding modes for these amines, one in which the nitroxide is strongly immobilized and the other in which there is freedom of motion about the nitroxide long axis independent of the DNA, giving rise to a partially immobilized spectrum.

Aminochrysene label (X) showed a completely opposite field orientation dependence compared to the other aromatic amines. The chrysene derivative also was different in that the amine is in position 6 compared to position 2 of the others, and the conjugated aromatic ring system is larger. As can be seen in Fig. 19c, changes in the spectra in going from the parallel configuration to the perpendicular one are just the opposite of those observed for the other two amines. The characteristics of the spectral changes are similar in that there appear to be two field-dependent species, a strongly immobilized form (A_{max} 63.5 G) and a partially immobilized form (A_{max} 44.0 G), except that this maximal splitting or $2T_{\parallel}'$ as well as A_{min} or $2T_\perp'$ of 19.9 G are observed when the plane of the plate on which the fibers are stretched is perpendicular to the field. This result suggests that the aromatic rings are indeed oriented within the DNA structure and are most likely intercalated but that the nitroxide is rotated almost 90° to the plane of the rings and restricted in its motion.

Although the field orientation studies of these carcinogen spin labels appear to be convincing evidence of an intercalative mechanism for these molecules, a second spin-label study has been used to support these conclusions, namely, thermal denaturation. It was shown previously by means of optical methods that the melting characteristics of native double-stranded DNA are altered upon intercalation of specific dye molecules.

The melting profiles are cooperative for the complexes and shift to higher melting temperatures upon intercalation. The melting characteristics of the DNA generally are monitored by following the increase in hypochromicity of the DNA at 260 nm as a function of temperature.

In the case of the spin-labeled carcinogen complexes with DNA, similar melting profiles can be measured merely by following the release of the spin

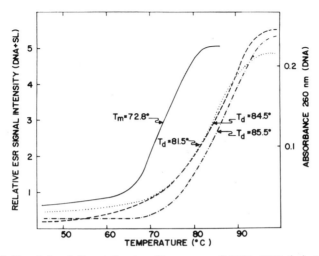

Fig. 20. Heat-induced dissociation transition curves of AFSL–DNA (···), AASL–DNA (– – –), and ACSL–DNA (–·–) complexes. All three broken lines were obtained by plotting I_{-1} versus temperature of the corresponding composite ESR spectrum for each spin label–DNA complex: AFSL–DNA with deoxyribose/phosphate ratio (D/P ratio) of approximately 1×10^{-2}; AASL–DNA with D/P ratio of approximately 3×10^{-3}; ACSL–DNA with D/P ratio of approximately 3×10^{-2}. Solid line, melting curve for 7.6×10^{-5} M DNA obtained by plotting hyperchromicity at $A_{260\,nm}$ versus temperature. [From Hong and Piette (1976); reprinted by permission of the publisher.]

label from the DNA as the helical structure unfolds during temperature denaturation or melting. This release is easily followed by monitoring the change in the spin-label anisotropy going from a characteristic bound spectrum or broad line to the sharp isotropic spectrum of a freely tumbling label in solution. In Fig. 20, we see examples of these cooperative melting curves for the three carcinogen spin labels discussed above. For each label there is an increase in the temperature for dissociation T_d, which is related to T_m, the actual melting temperature of the DNA. These increased melting curves are further evidence for intercalation.

There is no question that the ESR data for these carcinogen spin labels demonstrate a strong noncovalent binding to DNA, probably through intercalation. It remains, however, to demonstrate that the labels are in fact predictive of what occurs for the nonlabeled carcinogen. This can be partially demonstrated through competitive binding studies using the spin-labeled parent compound. In Fig. 21, we see such a competitive binding curve for AFSL and the parent AAF with native DNA. Similar competitive studies also could be performed with other carcinogens not structurally related to the spin label. Such studies perhaps could suggest what the structural requirements are for binding.

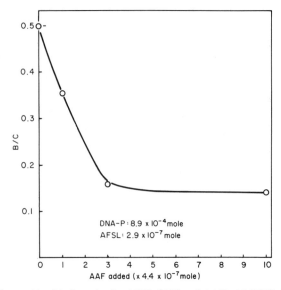

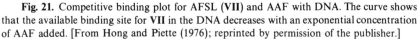

Fig. 21. Competitive binding plot for AFSL (**VII**) and AAF with DNA. The curve shows that the available binding site for **VII** in the DNA decreases with an exponential concentration of AAF added. [From Hong and Piette (1976); reprinted by permission of the publisher.]

In Fig. 19, it is observed that the carcinogen spin labels are in fact binding to more than one site as determined by the two different correlation times apparent in each of the spectra recorded. Assuming that the labels reflect the binding of the parent compound, a possible interpretation for the two different resonance components in the ESR spectra of **VII**–DNA, **IX**–DNA, and **X**–DNA complexes (Section V,B,2) would be that there are two distinct binding modes for these aromatic amines, one in which the nitroxide is strongly immobilized in a specific binding site, presumably a G–C base-pair region, and the other in which the nitroxide is partially immobilized in a base-pair region other than G–C. Further insight into this problem probably could be obtained by repeating these types of experiments on synthetic polynucleotides of known base composition and comparing the results with native DNA.

Although a large amount of data is being gathered on the interaction of carcinogens with various cell components, and in particular with the fundamental macromolecules, and although it is obvious that somewhere there must be the decisive step leading to the generation and appearance of tumors, the precise significance of these reactions for carcinogenesis is unclear. The problem of the interaction between the carcinogen and the cellular receptor has two facets. The first facet concerns the nature of the principal receptor: Is it a nucleic acid or a protein? The second facet concerns the mechanism of the interaction, particularly the predominance of the physical or chemical interaction (Pullman, 1974).

These studies on the interaction of carcinogen and DNA have strongly demonstrated the feasibility of the spin-label technique as a powerful tool for providing not only clear-cut evidence of physical binding, i.e., intercalation of the carcinogen with DNA, but also other information such as base-preferential binding and the postbinding structural changes of the host DNA molecule. Moreover, carcinogenic and mutagenic activity of most of the carcinogens used were found to persist even after attachment of the nitroxide reporter on the respective ligand molecule.

VI. SPIN TRAPPING

A. Studies of Free-Radical Carcinogenesis

Two of the most perplexing problems in carcinogenesis are (1) the exact mechanism of carcinogen activation and (2) the relationship between carcinogenesis and aging. These two problems may be interrelated in that carcinogen activation may be dependent on the same processes that lead to aging in cells.

Covalent binding of carcinogens such as the polynuclear aromatic hydrocarbons to cellular constituents, the initiating process in the series of steps necessary to induce malignant transformation, requires bioactivation by monooxygenase enzymes. These oxygenase enzymes containing iron (Fe^{3+}) in the form of cytochrome P-450 as the oxygen-activating component and substrate binding site are found in most microsomes and nuclei of mammalian cells and are responsible for metabolism of a variety of endogenous as well as foreign compounds including aromatic hydrocarbons. Although metabolic oxidation of these molecules before their excretion constitutes, in general, a regulatory or detoxifying pathway for the host, some intermediates of hydrocarbon oxidation are strongly implicated as causative agents in carcinogenesis. The mechanism of such critical metabolic activation, however, continues to be a subject of controversy and speculation.

The predominantly local tumorigenic effects evoked by hydrocarbons suggest that some major activation mechanism of the hydrocarbon itself is the promoter of biological activity. In view of the strong electron-donating properties of polycyclic hydrocarbons and the existence *in vivo* of many potential oxidants, the oxidation of the hydrocarbon to reactive radical intermediates capable of binding to cellular nucleophiles might constitute the critical first step in hydrocarbon carcinogenesis.

At present, there is very little direct evidence to suggest a one-electron or free-radical activation mechanism for carcinogens except perhaps for the benzo[a]pyrene and nitroquinoline derivatives. The primary reason for the

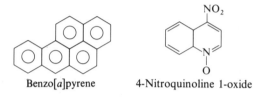

Benzo[a]pyrene   4-Nitroquinoline 1-oxide

lack of this evidence is probably that, except for the two cases mentioned, such one-electron oxidations are extremely fast and the intermediates short-lived, and thus specialized techniques are required to observe them. Until now such techniques were not readily available for biological systems. With the development by Janzen and Blackburn (1971) of the spin-trapping technique for studying labile free radicals and its application by Harbour and Bolton (1975) and Saprin and Piette (1977) to biological systems, it appears that a method is now available that can probe these one-electron reactions.

The spin-trapping technique in many respects can be viewed as a modified spin-label method. The technique requires the use of a nonparamagnetic molecule that can react rapidly with a labile free radical, leading in turn to a stable paramagnetic product or free radical. The family of molecules used as spin traps by Janzen and Blackburn (1971) in their early work were nitrone derivatives. One advantage of using nitrones is that the product formed upon reaction with a labile free radical is a stable nitroxide—thus the association with spin labeling. Another advantage is that the radical products are in most cases addition products, and the characteristics of the hyperfine splitting constant and g values of the resultant nitroxide are dependent on the structure of the free-radical adduct. The general reaction scheme is as follows:

$$R\cdot + R_1-\underset{\underset{H}{|}}{C}=\underset{\underset{O}{|}}{N}-R_2 \longrightarrow R_1-\underset{\underset{H}{|}}{\overset{\overset{R}{\diagdown}}{C}}-\underset{\underset{O}{|}}{N}-R_2 \qquad (2)$$

For the most part, only two nitrones have been used in most spin-trapping studies to date. They are phenyltertiary butylnitrone (PBN, **XI**) and

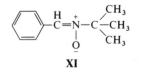

XI

5,5′-(dimethyl-1-pyrroline 1-oxide) (DMPO, **XII**). Both compounds upon

XII

addition of a free-radical adduct across the double bond yield stable nitrox-ides. The resultant hyperfine splitting is dependent on the radical adding. The β-hydrogen splitting in particular changes with the structure of the adduct reacting. The changes in β splitting with **XII** are slightly more sensi-tive than with **XI**. The major limitation of the technique, in general, is the low solubility of the spin traps in aqueous media.

Saprin and Piette (1977) attempted to probe the possible one-electron oxidation of the carcinogenic nitrosoamines using the PBN spin trap. They were able to show that the NADPH mixed-function oxidases of microsomes, which are known to activate most hydrocarbon carcinogens to ultimate carcinogens, can also oxidize the dialkylnitrosoamines in a one-electron oxidation to a reactive free radical as detected by the PBN spin trap. In Fig. 22, we see the stable nitroxide formed by the apparent addition of a reactive R· adduct across the double bond, the most probable reaction being

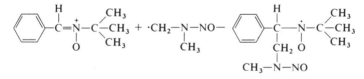

The large a_β^H splitting of 5.75 G is the largest observed so far with PBN. The second splitting in this spectrum is due to the solvent ethanol, which gives an a_β^H splitting of 3.35 G resulting from the structure

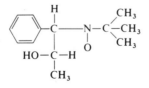

Until now, the accepted mechanism for nitrosoamine activation involved demethylation by a liver demethylase enzyme, resulting in the formation of a reactive carbonium ion that serves as the primary alkylating agent. The spin-trapping studies, however, suggest that an alternate mechanism may be the production of a labile radical that can also act as an alkylating agent.

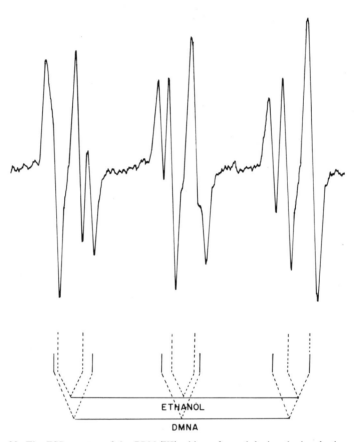

Fig. 22. The ESR spectra of the PBN (**XI**) adducts formed during the incubation of a 2%
dimethylnitrosoamine (DMNA) solution in ethanol for 20 min at 37°C with 0.5 ml of microsomes in 0.05 M PO_4^{2-} buffer, pH 7.4, 2 mg of protein per milliliter, and the lipid peroxidation
system [0.3 mM NADPH, 1.2×10^{-5} M Fe^{2+}, 2.0×10^{-4} M pyrophosphate, and 0.15 M
PBN (**XI**) final concentration]. [From Saprin and Piette (1977); reprinted by permission of
the publisher.]

B. Lipid Peroxidation

Considerable evidence is accumulating that suggests a relationship between lipid peroxidation aging and carcinogenesis (Harman, 1968; Frankfurt *et al.*, 1967). The best evidence has been obtained from inhibition studies
using antioxidants in feeding experiments with mice. Both prolongation of
life and inhibition of tumorigenesis have been reported (Harman, 1968;
Shamberger, 1972). Lipid peroxidation is a free-radical reaction, and antioxidants act as free-radical inhibitors, thus inhibiting lipid peroxidation. The
liver microsomal membrane system is an excellent model for studying lipid

peroxidation. Hochstein and Ernster (1963) demonstrated an induced NADPH-dependent enzymatic lipid peroxidation in liver microsomes. The extent of peroxidation can be altered if the animals from which the microsomes have been isolated are given large doses of antioxidants (Wattenberg, 1972; McCay et al., 1962).

The exact sequence of events leading to lipid peroxidation are unknown. It is suggested, however, that a labile ·OH or $HO_2·$ radical is the primary radical produced by a one-electron oxidation involving cytochrome P-450 reductase (McCay et al., 1976). Saprin and Piette, in their studies of the NADPH-dependent oxidation of substrates using the spin trap XI, were not able to detect ·OH or $HO_2·$ as an addition product in the reaction. Lai and Piette (1977), however, in a follow-up to the Saprin study, showed that ·OH is produced in this system and can be trapped if one uses XII as the trap coupled with a rapid mixing and detection system. In Fig. 23, the stable

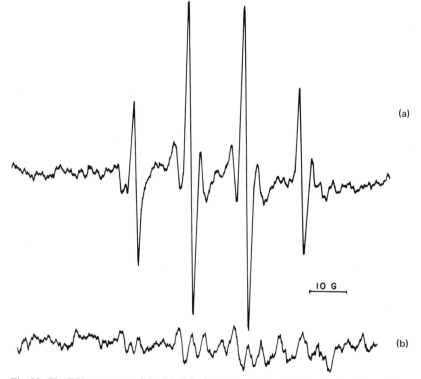

Fig. 23. The ESR spectrum of the DMPO–OH radical adduct obtained when the reaction mixture containing 1.8 mg/ml microsomes, 2.2×10^{-5} M Fe^{2+}, 4.4×10^{-5} M EDTA, 7.0 mM DMPO (XII) at pH 7.4 in 0.15 M KCl solution was mixed (a) with 0.74 mM NADPH, (b) without NADPH, $a^N = a_\beta^H = 15.0$ G and $g = 2.0062$. [From Lai and Piette (1977).]

nitroxide from **XII** is shown as a result of addition of an OH· radical across the double bond. The most probable reaction is the following:

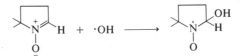

The OH· radical was identified as the adduct by the characteristic a_β^H hyperfine splitting of 15.0 G and by comparison with spectra obtained by trapping the OH· radical with **XII** in UV-photodissociated H_2O_2 (Harbour *et al.*, 1974).

Lai and Piette suggest a reaction scheme for lipid peroxidation as follows:

(a) Oxidized *P*-450 reductase + NADPH $\longrightarrow$ reduced *P*-450 reductase + NADP$^+$

(b)   Reduced *P*-450 reductase + O_2 $\longrightarrow$ oxidized *P*-450 reductase + $\cdot O_2^-$

(c) $2 \cdot O_2^- + 2H^+ \longrightarrow H_2O_2 + O_2$ (3)   (3)

(d) $\cdot O_2^- + H_2O_2 \longrightarrow O_2 + OH^- + OH\cdot$

(e) $Fe^{2+} + H_2O_2 \longrightarrow Fe^{3+} + OH^- + OH\cdot$

The OH· radical so generated then reacts by abstracting a hydrogen from the methylene carbon atom of the unsaturated fatty acid of phosphatidylcholine followed by oxygen addition and lipd peroxidation.

The significance of these spin-trapping experiments is that free-radical events can be probed directly in the biological system of interest without any special techniques being required. With the advent of more water-soluble derivatives, the sensitivity of the system for biological reactions will be improved. Modifications of the traps should also allow for the trapping of radical ions as well as neutral radicals.

VII. *In Vivo* ELECTRON SPIN RESONANCE

One of the most frequent questions raised in assessing the utility of ESR measurements of biological systems is whether the *in vitro* measurement in fact reflects the actual situation *in vivo*. Probably the most controversial area in this regard is the vast number of ESR measurements that have been made on excised tissue or cell suspensions in which one tries to correlate the presence or absence of some paramagnetic intermediate with a metabolic condition or disease state, e.g., cancer. It has been thought that the answer to this question would be provided if one could conduct the ESR measurements *in vivo*. Such experiments, of course, would require a radical modification of existing equipment with some sacrifice in sensitivity.

Feldman *et al.* (1975) reported *in vivo* ESR studies in the liver of rats. The technique uses a traveling wave structure or helix in place of a standard microwave cavity. Such structures are broad-band devices with a very low Q; thus, the sensitivity inherent in the system is reduced by this loss in Q. The devices are characterized by a parameter S, called the slowing factor, given by

$$S = \frac{2\pi a}{p} = \cot \psi$$

where a is the mean radius of the coil, p is the pitch, and ψ is the pitch angle. If a and p are chosen so that $S \gg 1$, it can be shown (Volino and Csakvary, 1968) that the axial velocity v_z is given by

$$v_z \pm c/S \ll c$$

In addition, $\lambda_g \approx \lambda_0/S \ll \lambda_0$, where λ_g is the wavelength along the helix and λ_0 is the free-space wavelength. Thus, the microwave radiation is slowed until the energy density inside the helix is large enough to permit its use for ESR. If $2\pi a < \lambda_0/10$, the helix acts as a straight wire and no slowing occurs. If $2\pi a > \lambda_0/2$, the helix acts as a radiative antenna. Therefore, the helices are operated in the range $\lambda_0/10 \leq 2\pi a < \lambda_0/2$ and are wound so that $S \gg 1$.

When a helix is perfectly matched to a microwave system, all of the incident power is reflected back toward the klystron from the coil. This is in direct contrast to a perfectly matched resonant cavity, in which all of the incident power is absorbed. The coils are never perfectly matched; consequently their Q is adversely affected. An advantage of the device is that the helical coil used can take almost any dimension and thus can be implanted in and around the tissue being studied. To ensure complete filling of the helix, Feldman *et al.* inserted the coil into a regenerating lobe of rat liver, allowing the regeneration process to envelope the entire coil within a few weeks (Fig. 24). Figure 25 shows the completed assembly with the coil inserted and the animal placed in the magnetic field.

Electron spin resonance signals were detected in the coils after injection of a standard TEMPOL spin-label solution intraperitoneally (Fig. 26). Signals also were detected for a spin-labeled drug derived from chlorpromazine. Estimates as to the sensitivity of the method in detecting signals in the liver were about 8×10^{-9} mole/gm body weight, which translates into a sensitivity of about 5×10^{-4} M spin or about 5000 times less sensitive than a standard *in vitro* experiment. The authors suggest that some of this sensitivity loss can be retrieved by using time-averaging methods since the *in vivo* experiment can be run for as long as 1–2 hr without any ill effects on the animals. Until such time as the overall sensitivity is improved, the practical application of *in vivo* ESR seems to be limited.

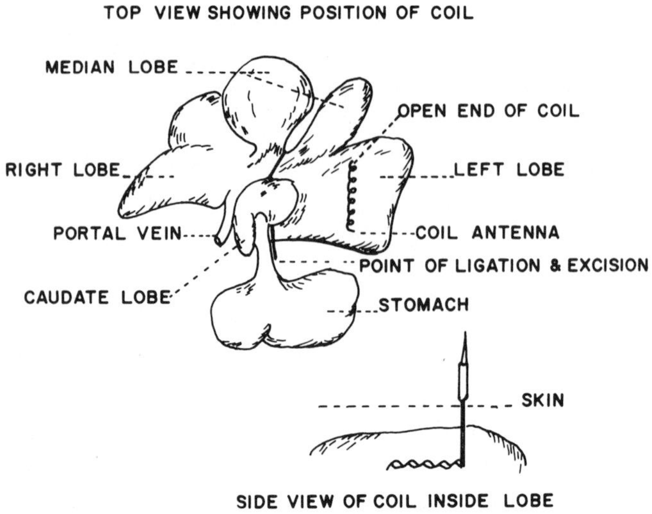

TOP VIEW SHOWING POSITION OF COIL

MEDIAN LOBE

OPEN END OF COIL

RIGHT LOBE

LEFT LOBE

PORTAL VEIN

COIL ANTENNA

POINT OF LIGATION & EXCISION

CAUDATE LOBE

STOMACH

SKIN

SIDE VIEW OF COIL INSIDE LOBE

Fig. 24. Liver showing location of ligation and excision for partial hepatectomy and position of coil and antenna. [From Feldman *et al.* (1975); reprinted from *Phys. Med. & Biol.*, Copyright 1975 by The Institute of Physics.]

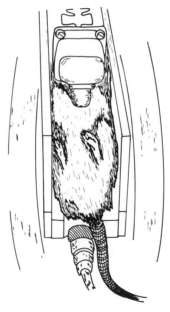

Fig. 25. Anesthetized rat in holder in magnetic field showing coax coupling to implanted coil. [From Feldman *et al.* (1975); reprinted from *Phys. Med. & Biol.*, Copyright 1975 by The Institute of Physics.]

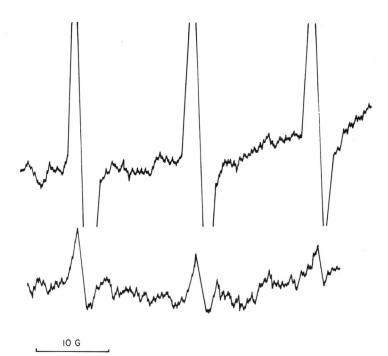

10 G

Fig. 26. Top: Initial spectrum as detected in implanted coil in the liver after intraperitoneal injection of 1.5×10^{-7} mole/gm body weight of TEMPOL into an anesthetized Wistar rat. Bottom: Spectrum obtained a short time after that shown in top. [From Feldman *et al.* (1975); reprinted from *Phys. Med. & Biol.*, Copyright 1975 by The Institute of Physics.]

REFERENCES

Amanullah, A. (1976). Neonatal jaundice. *Am. J. Dis. Child.* **130**, 1274–1280.

Ames, B. N., W. E. Durston, E. Yamasaki, and F. D. Lee (1973). Carcinogens are mutagens: A simple test combining liver homogenates for activation and bacteria for detection. *Proc. Natl. Acad. Sci. U.S.A.* **70**, 2281–2285.

Arcos, J. C., and M. F. Argus (1974). *In vitro* interaction of hydrocarbons with nucleic acids. *In* "Chemical Induction of Cancer" (J. C. Arcos, M. F. Argus, and G. Wolf, eds.), Vol. 2A; pp. 257–268. Academic Press, New York.

Berliner, L. J. (1976). "Spin Labeling: Theory and Applications." Academic Press, New York.

Boyland, E., and B. Green (1962). The interaction of polycycline hydrocarbons and purines. *Br. J. Cancer* **16**, 507–517.

Bratlid, D. (1972). Determination of conjugated and unconjugated bilirubin by methods based on direct spectrophotometry and chloroform extraction, a reappraisal. *Scand. J. Clin. Lab. Invest.* **29**, 91–97.

Chan, G., D. Schiff, and L. Stern (1971). Competitive binding of face fatty acids and bilirubin to albumin: Differences in HBABA dye vs. sephadex G-251 interpretation of results. *Clin. Biochem.* **4**, 208–214.

Eisen, H. N., M. C. Michaelides, B. J. Underdown, E. P. Schulenburg, and E. S. Simms (1970). Experimental approaches to homogeneous antibody populations. Myeloma proteins with antihapten antibody activity. *Fed. Proc., Fed. Am. Soc. Exp. Biol.* **29**, 78–84.

Feldman, A., E. Wildman, G. Bartolini, and L. H. Piette (1975). *In vivo* electron spin resonance in rats. *Phys. Med. & Biol.* **20**, 602–612.

Fisher, D. A., G. N. Burraw, J. H. Dussault, D. R. Hollingsworth, P. R. Larsen, E. B. Man, and P. G. Walfish (1976). Recommendations for screening progress for congenital hyperthyroidism. Report of a committee of the American Thyroid Association. *Am. J. Med.* **61**, 932–934.

Frankfurt, O. S., L. P. Lypchina, T. V. Bunto, and N. M. Emanuel (1967). Inhibition of carcinogenesis with antioxidants BHT. *Bull. Exp. Biol. Med. (Engl. Transl.)* **8**, 163.

Fuller, W., and M. J. Waring (1964). A molecular model for the interaction of ethidium bromide with DNA. *Ber. Bunsenges. Phys. Chem.* **68**, 805–808.

Hamilton, C., and H. McConnell (1968). Spin Labels. *In* "Structural Chemistry and Molecular Biology" (A. Rich and N. Davidson, eds.). Freeman, San Francisco, California.

Harbour, J. R., and J. R. Bolton (1975). Superoxide formation in spinach chloroplasts: Electron spin resonance detection by spin trapping. *Biochem. Biophys. Res. Commun.* **64**, 803–807.

Harbour, J. R., V. Chow, and J. R. Bolton (1974). An electron spin resonance study of the spin adducts of OH and HO_2 radicals with nitrones in the ultraviolet photolysis of aqueous hydrogen peroxide solutions. *Can. J. Chem.* **52**, 3549–3554.

Harman, D. (1968). Free radical theory of aging: Effect of free radical reaction inhibitors on the mortaility rate of small LAF mice. *J. Gerontol.* **23**, 476–482.

Hochstein, P., and L. Ernster (1963). ADP-activated lipid peroxidation coupled to the TPNH oxidase system of microsomes. *Biochem. Biophys. Res. Commun.* **12**, 388–394.

Hong, S. J., and L. H. Piette (1976). Electron spin resonance spin label studies of intercalation of ethidium bromide and aromatic amine carcinogens in DNA. *Cancer Res.* **36**, 1159–1171.

Hsia, J. C. (1968). Development of single and double spin labelling methods and their application to the study of antibody active sites. Ph.D. Thesis, University of Hawaii, Honolulu.

Hsia, J. C., and J. R. Little (1971). Alterations of antibody binding properties and active site dimensions in the primary and secondary immune response. *Biochemistry* **10**, 3742–3748.

Hsia, J. C., and C. T. Tan (1978a). Membrane immunoassay: Principle and application of spin membrane immunoassay. *Ann. N.Y. Acad. Sci.* **308**, 139–148.

Hsia, J. C., L. T. L. Wong, and W. Kalow (1973). Homogeneous murine myeloma protein 315 and spin labeled DNP as a model system for spin labelled hapten titration Technical and spin immunoassay. *J. Immunol. Meth.* **3**, 17–24.

Hsia, J. C., S. W. Chan, and C. T. Tan (1976). Development of a spin immunoassay dipolar spin immunoassay. *Clin. Chem.* **9**, 2 (abstr.).

Hsia, J. C., J. D. Wood, N. H. Kwon, S. S. Er, and G. W. Chance (1978b). Development of a spin assay for RBLC of human serum. *Proc. Natl. Acad. Sci. U.S.* **75**, 1542–1545.

Humphries, G. K., and H. M. McConnell (1974). Immune lysis of liposomes and erythrocyte ghosts loaded with spin label. *Proc. Natl. Acad. Sci. U.S.A.* **71**, 1691–1694.

Jacobsen, J., and R. P. Wennberg (1974). Determination of unbound bilirubin in the serum of newborns. *Clin. Chem.* **20**, 783–789.

Janzen, E. G., and B. J. Blackburn (1971). Detection and identification of short-lived free radicals by electron spin resonance trapping Technical (spin trapping) photolysis of organo lead, tin and mercury compounds. *Acc. Chem. Res.* **4**, 31–40.

Jirsova, V., M. Jersa, A. Heringova, O. Koldovsky, and J. Weichova (1967). Significance of sephadex gel filtration of serum from icteric newborn. *Biol. Neonat.* **11**, 204–208.

Jost, P., A. S. Waggoner, and O. H. Griffith (1972). Electron spin resonance and the spin labeling method. *Methods Pharmacol.* **2**, 223.

Kinsky, S. C. (1972). Antibody complement interaction with lipid model membranes. *Biochim. Biophys. Acta* **265**, 1–23.

Kriek, E. (1974). Carcinogenesis by aromatic amines. *Biochim. Biophys. Acta* **355**, 177–203.

Knudson, K. C., L. H. Bing, and L. Kater (1971). Quantitative measurement of guinea pig complement with liposomes. *J. Immunol.* **106**, 258.

Lai, C. S., and L. H. Piette (1977). Hydroxyl radical production involved in lipid peroxidation of rat liver microsomes. *Biochem. Biophys. Res. Commun.* **78**, 51–59.

Lambelin, G., J. Roba, R. Roncucci, and R. Parmentier (1975). Carcinogenicity of 6-aminochrysene in mice. *Eur. J. Cancer* **11**, 327–334.

Lee, K. S., and L. M. Gartner (1976). Spectrophotometric characteristics of bilirubin. *Pediatr. Res.* **10**, 782–788.

LePecq, J. B., and C. Paoletti (1967). A fluorescent complex between ethidium bromide and nucleic acids. Physical chemical characterization. *J. Mol. Biol.* **27**, 87–106.

Lerman, L. S. (1964). Acridine mutagens and DNA structure. *J. Cell. Comp. Physiol.* **64**, Suppl. 1, 1–18.

Lesko, S. A., Jr., P. O. P. Ts'o, and R. S. Umans (1969). Interaction of nucleic acids via chemical linkage of 3-4 benzypyrene to deoxyribonucleic acid in aqueous solution. *Biochemistry* **8**, 2291–2298.

Leute, R. K., E. F. Ullman, A. Goldstein, and L. A. Herzenberg (1972a). Spin immuno assay techniques for determination of morphine. *Nature (London)* **236**, 93–94.

Leute, R. K., E. F. Ullman, and A. Goldstein (1972b). Spin immunoassay of opiate narcotics in urine and saliva. *J. Am. Med. Assoc.* **221**, 1231–1234.

Levine, R. L. (1976). "Albumin-titratable bilirubin" as an index to bilirubin binding. *Clin. Chem.* **22**, 560–561.

Lucey, J. F., T. Valaes, and S. A. Doxiadis (1967). Serum albumin reserve PSP dye binding capacity in infants with kernicterus. *Pediatrics* **39**, 876–883.

McCann, J., N. E. Spingarn, J. Kobori, and B. N. Ames (1975). Detection of carcinogens as mutagens: Bacterial tester strains with R factor plasmids. *Proc. Natl. Acad. Sci. U.S.A.* **72**, 979–983.

McCay, P. B., P. M. Pfeifer, and W. H. Stipe (1962). Vitamin E protection of membrane lipids during electron transport functions. *Ann. N.Y. Acad. Sci.* **203**, 62–73.

McCay, P. B., R. A. Floyd, E. K. Lai, J. L. Poyer, and K. L. Fong (1976). Evidence for superoxide-dependent reduction of $Fe3+$ and its enzyme generated hydroxyl radical formation. *Chem.-Biol. Interact.* **15**, 77–89.

Magee, P. N. (1974). Molecular mechanisms in chemical carcinogenesis. *Recent Results Cancer Res.* **44**, 2–8.

Miller, E. C., and J. A. Miller (1971). The mutagenicity of chemical carcinogens: Correlations, problems, and interpretations. *Chem. Mutagens: Princ. Methods Their Detect.* **1**, 83–119.

Minden, P., F. A. Bascom, and R. S. Farr (1969). A comparison of seven procedures to detect the primary binding of antigen by antibody. *J. Immunol.* **102**, 832–841.

Nagata, C., M. Kotama, Y. Tagashima, and A. Imamura (1966). Interaction of polynuclear aromatic hydrocarbons, 4-nitroquinedine-1-oxides, and various dyes with DNA. *Biopolymers* **4**, 409–427.

Odell, G. B., S. N. Cohen, and P. C. Kelly (1969). Studies in kernicterus. II. The determination of the saturation of serum albumin with bilirubin. *J. Pediatr.* **74**, 214–230.

Odell, G. B., G. N. B. Storey, and L. A. Rosenberg (1970). Studies in kernicterus. III. The saturation of serum proteins with bilirubin during neonatal life and its relationship to brain damage at five years. *J. Pediatr.* **76**, 12–21.

Ohnishi, S., and H. M. McConnell (1965). Interaction of the radical ion of chlorpromazine with deoxyribonucleic acid. *J. Am. Chem. Soc.* **87**, 2293.

Porter, E. G., and W. J. Waters (1966). A rapid micro method for measuring the reserve albumin binding capacity in serum from newborn infants with hyperbilirubinemia. *J. Lab. Clin. Med.* **67**, 660–668.

Pullman, B. (1974). Summary of the chemical aspects of carcinogenesis. In "Chemical Carcinogenesis" (P. O. P. Ts'o and J. A. DiPaolo, eds.), pp. 713–727. Dekker, New York.

Resenquist, E., and A. I. Vistnes (1977). Immune lysis of spin label loaded liposomes incorporating cardiolipin; a new sensitive method for detecting anticardiolipin antibodies in syphilis serology. J. Immunol. Methods. 15, 147.

Rosenthal, H. E. (1967). A graphic method for the determination and presentation of binding parameters in a complex system. Anal. Biochem. 20, 525–532.

Saprin, A. N., and L. H. Piette (1977). Spin trapping and its application in the study of lipid peroxidation and free radical production with liver microsomes. Arch. Biochem. Biophys. 180, 480–492.

Scatchard, G. (1949). The attractions of proteins for small molecules and ions. Ann. N.Y. Acad. Sci. 51, 660–692.

Schreier-Muccillo, S., and J. C. P. Smith (1973). Spin labels as probes of the organization of biological and model membranes. Prog. Surf. Membr. Sci. 9, 136–148.

Shamberger, R. (1972). Increase of peroxidation in carcinogenesis. J. Natl. Cancer Inst. 48, 1491–1499.

Six, H. R., W. W. Young, Jr., K. Uemure, and S. C. Kinsky (1974). Effect of antibody compliment on multiple vs. single compartment liposomes. Application of a fluorometric assay for following changes in liposomal permeability. Biochemistry, 13, 4050–4058.

Skelley, D. S., L. P. Brown, and P. K. Besch (1973). Radioimmunoassay. Clin. Chem. 19, 146–186.

Soltys, B. J., and J. C. Hsia (1977). Fatty acid enhancement of human serum albumin binding Properties; a spin label study. J. Biol. Chem. 252, 4043–4048.

Stryer, L., and O. H. Griffith (1965). A spin labeled hapten. Proc. Natl. Acad. Sci. U.S.A. 54, 1785–1791.

Underdown, B. J., E. S. Simms, and H. N. Eisen (1971). Subunit number and structure of combining sites of the immunoglobulin, a myeloma protein product by mouse plasmacytoma. Biochemistry, 10, 4359–4368.

Volino, F., and F. S. G. Csakvary (1968). Resonant helices and their application to magnetic resonance. Rev. Sci. Instrum. 39, 1660–1665.

Voss, E. W., Jr., and H. N. Eisen (1968). Valence and affinity of IgM antibodies to the 2,4-dinitro-phenyl (DNP) Group. Fed. Proc., Fed. Am. Soc. Exp. Biol. 27, 684.

Waring, M. J. (1965). Complex formation between ethidium bromide and nucleic acids. J. Mol. Biol. 13, 269–282.

Waring M. J. (1972). Inhibition of nucleic acid synthesis. In "The Molecular Basis of Antibiotic Action" (E. F. Gale et al., eds.), pp. 173–227. Wiley, New York.

Wattenberg, L. W. (1972). Inhibition of carcinogenic and toxic effects of polycyclic hydrocarbons by phenolic antioxidants and ethoxyquin. J. Natl. Cancer Inst. 48, 1425–1430.

Wei, R., C. R. Alving, R. L. Richards, and E. S. Copeland (1975). Liposome spin immunoassay: A new sensitive method for detecting lipid substances in aqueous media. J. Immunol. Methods. 9, 165–170.

Weisburger, J. H. (1973). Carcinogenesis. In "Cancer Medicine" (J. Holland and E. Free, III, eds.), p. 45. Lea & Febiger, Philadelphia, Pennsylvania.

Williams, D. J., and B. R. Rabin (1971). Disruption by carcinogens of the hormone dependent association of membranes with polysomes. Nature (London) 232, 102–105.

Wolilait, W. D. (1974). A theoretical analysis of the binding of bilirubin by human serum albumin: The contribution of the two binding sites. Life Sci. 14, 2189–2198.

Wood, D. J., and J. C. Hsia (1977). Spin assay as a general method for studying plasma protein binding: Bilirubin-albumin binding. Biochem. Biophys. Res. Commun. 76, 863–868.

7

Applications of Spin Labeling to Nucleic Acids

ALBERT M. BOBST

DEPARTMENT OF CHEMISTRY
UNIVERSITY OF CINCINNATI
CINCINNATI, OHIO

I. Introduction	291
II. Spin Labeling–Spin Probing of Nucleic Acids	293
A. Spin Labeling by Chemical Modification	293
B. Enzymatic Synthesis of Spin-Labeled Nucleic Acids	302
C. Spin Probing of Nucleic Acids	304
III. Electron Spin Resonance Spectra of Nucleic Acids Containing Spin Labels or Spin Probes	308
A. Measurements	308
B. Theoretical Analysis of Nitroxide Spectra	311
C. Analysis of Nitroxide–Manganese(II) Ion Interaction Spectra	317
IV. Monitoring Conformational Changes in Nucleic Acids by Electron Spin Resonance	319
A. RNA Systems	320
B. DNA Systems	325
V. Molecular Association Studies with Spin-Labeled Nucleic Acids	329
A. Interaction with Small Molecules	329
B. Interaction with Polypeptides and Proteins	331
C. Interaction with Mammalian Cell Monolayers	336
D. Bioassay for Detecting $(A)_n$ Tracts in RNA's	339
VI. Conclusion	341
References	342

I. INTRODUCTION

The properties of nucleic acids have been studied using a great number of techniques, and much is understood about the physicochemical characteristics of nucleic acids in simple solutions. However, little is known, particularly in quantitative terms, about these biopolymers in more complex environments, which can be as complex as a living cell. Even at the level of studies of proteins and nucleic acid interactions, which are not nearly as complex as interactions occurring in living cells, investigations have been limited because the nucleic acid signals monitored by such techniques as ultraviolet (UV), fluorescence, circular dichroism (CD), infrared, or nuclear magnetic resonance (NMR) spectroscopy are too complicated to be analyzed properly due to interference from the surrounding molecules.

Although the properties of nucleic acids have been carefully studied by UV, CD, and NMR spectroscopy, the advent of spin labeling signaled a need to systematically investigate the behavior and characteristics of spin-labeled nucleic acids to determine whether this approach could contribute significantly to our understanding of the structure–function relationships of nucleic acids in complex biological systems. Initially, the spin-label method was complementary to UV and CD spectroscopy in determining whether a relationship exists in the structural information gained by these various techniques about nucleic acids in a simple environment. Of course, one must be aware that UV and CD data in particular reflect information about the averages of the total nucleic acid matrix, whereas nitroxide radicals provide information on events of the local environment only. Subsequently, the studies were extended to include the properties of spin-labeled nucleic acids in more complex situations. It was clearly established that with spin-labeled nucleic acids valuable information can be gained about nucleic acid–nucleic acid, nucleic acid–protein, and nucleic acid–living cell interactions. Recently, it was found that nitroxide radicals can be incorporated into a nucleic acid matrix by enzymatic copolymerization. The specificity gained by this method should further strengthen the electron spin resonance (ESR) approach and make it possible to gain deeper insight into nucleic acid characteristics.

Spin-labeled mononucleotides have been reviewed (Gaffney, 1975) and are therefore not discussed here. A recent review of spin-labeled nucleic acids (Dugas, 1977) gives exhaustive attention to spin-labeled tRNA's but does not treat in detail other aspects of spin-labeled nucleic acids. For that reason, spin-labeled tRNA's are not discussed here in depth. The chapter is divided into four main topics. First, the various ESR-signal-generating reagents used for chemical modification or probing of nucleic acids and the biosynthesis of nitroxide-containing nucleic acid systems are discussed. Second, ESR data obtained on nucleic acid systems are examined from an experimental and theoretical viewpoint. Third, examples of monitoring conformational changes of nucleic acids in simple environments by ESR are discussed. Finally, results on spin-labeled nucleic acids in more complex systems are presented. An attempt is made to present the material in each section in chronological order.

II. SPIN LABELING–SPIN PROBING OF NUCLEIC ACIDS

A. Spin Labeling by Chemical Modification

Studies of the specificities of current and potential types of spin-label reagents are of great interest. Such studies are rather difficult in view of the complexity of nucleic acid chemistry. Namely, the nucleic acid building blocks to be modified chemically consist of bases, sugars, and phosphates, all of which are potential target sites for spin labels. Figure 1 shows a random segment of a polynucleotide chain containing a pyrimidine and a purine base with the numbering of the bases according to the IUPAC system. The

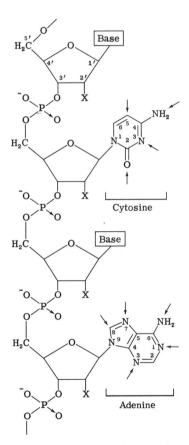

Fig. 1. A random segment of a polynucleotide chain with a cytosine and an adenine base in the RNA molecule (X = OH) and DNA molecule (X = H). Arrows indicate the positions at which chemical modifications by site-nonspecific labeling via electrophilic attack can occur in the bases.

numbering of the sugar moiety is also given, whereby position 2' is sub-stituted with X to include RNA and DNA systems as well as polynucleotide analogues. For instance, poly(2'-deoxy-2'-fluoro)uridylic acid, $(dUfl)_n$, con-sists of the base uridine and X = F. The arrows next to the bases indicate the positions where chemical modification by electrophilic attack may occur. Although it generally is assumed that, with most of the spin-label reagents listed in Fig. 2, the base is attacked, the sugar–phosphate backbone also can be the target site of some chemical modification. The reader also should realize when extrapolating from building block chemistry to polymer chemistry that not only the effect of the primary structure of a nucleic acid (Fig. 1) on the chemical reactivity of the base and sugar–phosphate back-bone toward spin-label reagents must be considered. In addition, one must be aware that the secondary and tertiary structures of nucleic acids have a pronounced effect on the chemical reactivity of the building blocks.

Furthermore, it should be noted that there still exist many uncertainties about the basic chemistry of various nucleic acid building blocks, and hasty analogies can lead to erroneous conclusions. For instance, until recently it generally was believed that a nucleoside lacking amino groups, such as uridine or thymidine, would not react as a monomer or when present in a polymer with alkylating reagents except in a highly alkaline solution. This belief was shattered after a careful study of simple model systems of nucleo-sides and nucleotides with common alkylating reagents clearly demonstrated that uridine and thymidine bases can be extensively alkylated, even near pH neutrality (Singer, 1975).

Therefore, it is not surprising that site-specific labeling of a nucleic acid chain has proved to be very difficult to achieve. The degree of difficulty is somewhat reduced if the spin labeling is done on a polynucleotide chain of known sequence containing one or more minor bases with chemical proper-ties that make them a specific target site, as in the case of 4-thiouracil. Such modifications are discussed in Sections II,A and B.

Figure 2 shows the chemical reagents containing one nitroxide radical that have been used so far for the chemical modification of nucleic acids. Chemical modification is believed to consist of site-nonspecific and/or site-specific labeling, which results in a covalent linkage of variable stability between nitroxide label and nucleic acid matrix. An attempt will be made to discuss site-nonspecific and site-specific labeling separately, although the distinction between both possibilities is not always straightforward.

1. SITE-NONSPECIFIC LABELING

Smith and Yamane (1967, 1968) were the first to show that spin-label reagents of the type **I, II,** and **III** can covalently bind to highly polymerized yeast RNA as well as to unfractionated tRNA from yeast at pH 5.5 in 0.1 *M*

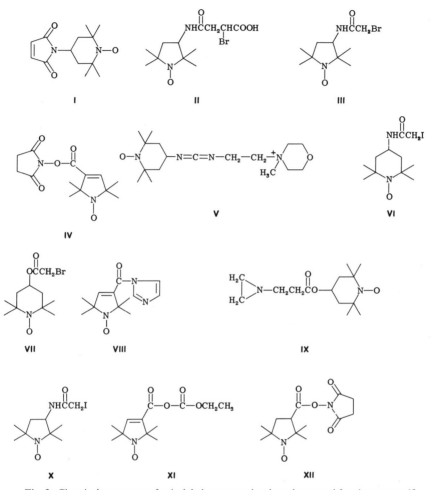

Fig. 2. Chemical structures of spin-label reagents that have been used for site-nonspecific or site-specific labeling of nucleic acids: **I**, piperidine maleimide; **II**, pyrrolidine bromo acid; **III**, pyrrolidine bromoacetamide; **IV**, pyrroline hydroxysuccinimide; **V**, piperidine carbodiimide; **VI**, piperidine iodoacetamide; **VII**, piperidine bromoacetoxy; **VIII**, pyrroline imidazole; **IX**, piperidine ethyleneimine; **X**, pyrrolidine iodoacetamide; **XI**, pyrroline anhydride; **XII**, pyrrolidine hydroxysuccinimide.

acetate buffer at 25°C. Exhaustive dialysis to remove noncovalently bound nitroxide radicals gave the largest percent labeling, about 5%, with the pyrrolidine bromoacetamide **III**. Considerably less covalent binding of reagents **I**, **II**, and **III** to calf thymus DNA than to RNA's was noticed under the same reaction conditions. Reagent **III** was reacted with the bases adenine, guanine, cytosine, uracil, and thymine at pH 5.5, and the extent of labeling was as follows: G, 20%; A, 3%; C, U, and T less than 1%. The preferential alkylation of the purines was attributed to reactive nitrogen 7.

The pyrroline hydroxysuccinimide **IV** was first used to spin label the α-amino group of Val-tRNA from *Escherichia coli* (Hoffman *et al.*, 1969). Subsequently, it was used to spin label the α-amino group of Phe-tRNA from *E. coli* as well as the amino groups and/or heterocyclic nitrogen atoms of some synthetic polyribonucleotides (Schofield *et al.*, 1970). Polymers $(C)_n$, $(A)_n$, $(U)_n$, and $(U-G)_n$ (3 : 1) were incubated with the pyrroline hydroxysuccinimide **IV** in a 0.1 *M* phosphate buffer (pH 6.8) for 20 hr at 37°C. The nucleic acids were then separated from excess spin label by precipitation with ethanol and chromatography on Sephadex G-25. The extent of modification (in mole of nitroxide per mole of nucleotide) was as follows: $(C)_n$, about 10^{-3}; $(U-G)_n$ (3 : 1), about 3×10^{-4}; $(A)_n$ and $(U)_n$, less than 10^{-4}.

Girshovich *et al.* (1971) studied the kinetics of the reaction between the piperidine carbodiimide **V**, a spin-labeled analogue of the water-soluble *N*-cyclohexyl-*N'*-β-methylmorpholinoethylcarbodiimide (Kumarev and Knorre, 1970) and the nucleosides uridine, guanosine, and inosine in *N*-methylmorpholinium buffer (pH 7.5) at 25°C. Rate constants of 260, 170, and 220 M^{-1} min^{-1} were measured for uridine, guanosine, and inosine, respectively. Considerably lower values were obtained for the two polymers $(U)_n$ and tRNA under the same conditions. Polymer $(U)_n$ can be completely modified with reagent **V** (1 mole of label per mole of uridine), although the nitroxide radicals are completely released at pH 10. Reagent **V** was also used to completely modify the dinucleotide UpU (Leventahl *et al.*, 1972). The reaction product was isolated and purified by thin-layer chromatography (TLC) on cellulose and by Dowex-1 (Cl$^-$) column chromatography. The product, which was homogeneous on paper chromatography, was assumed to have been modified at the N-3 position of uridine.

The piperidine iodoacetamide **VI** proved to be very valuable for irreversibly labeling nucleic acids (Bobst, 1971, 1972). Polymers $(A)_n$, $(G)_n$, $(C)_n$, and $(U)_n$ were incubated with spin label **VI** (at a mole label/mole of residue ratio of 2) in 0.15 *M* phosphate buffer (pH 7.5) for 20 hr at 37°C with stirring. The reaction products were carefully separated from noncovalently bound label by Sephadex G-200 chromatography. In Fig. 3 a typical elution profile of such a purification is shown. Rechromatography of the high molecular weight peaks gave the same elution profile except that no absorption and no ESR signal indicating the presence of free spin labels were detected between 100 and 140 ml. At that time, this was considered to be evidence that all the labels were attached covalently to the nucleic acids. Subsequently, a more rigid test was developed to check for the presence or absence of free spins in spin-labeled nucleic acid preparations (see below, this section).

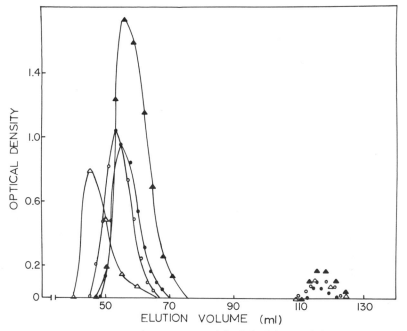

Fig. 3. Sephadex G-200 chromatography of $(A)_n$, $(G)_n$, $(C)_n$, and $(U)_n$ after they were spin labeled with piperidine iodoacetamide **VI**. Conditions: All separations were performed at 25°C in a column (2.5 × 75 cm) equilibrated with 0.001 M Tris–HCl buffer (pH 7.0). The same buffer was used for the elution. Key: △–△, 1.5 μmoles labeled $(G)_n$ (252 nm); ○–○, 1.5 μmoles labeled $(A)_n$ (260 nm); ●–●, 1.5 μmoles labeled $(C)_n$ (268 nm); ▲–▲, 3 μmoles labeled $(U)_n$ (260 nm). [From Bobst (1972); reprinted by permission of publisher.]

Under these spin-labeling conditions, the degree of labeling was about one nitroxide radical for every 600 residues for $(U)_n$ and $(A)_n$. Less labeling was obtained with $(G)_n$, most likely due to $(G)_n$ aggregation masking the reactive groups. It is interesting that the extent of labeling of $(C)_n$ under these reaction conditions was less than the sensitivity of the ESR system used (a Varian E-4 interfaced with a Fabri-Tek 1074). The reaction conditions have since been slightly modified to increase the extent of labeling of some polynucleotides that were routinely needed for spin bioassays to detect polyadenylate tracts in RNA's (Bobst et al., 1975) or to evaluate the nucleic acid affinity of nucleic acid-binding proteins (Bobst and Pan, 1975). The phosphate buffer contained 16.6% ethanol under the new labeling conditions, the reaction temperature was varied between 40° and 48°C, and the reaction was carried out for 48–96 hr. For the purine-containing polynucleotides, the lower temperature and time limits usually were used, whereas for the pyrimidine-containing polynucleotides the harsher conditions normally were used. The ratio of moles of reagent **VI** to moles of nucleic acid building block was kept between 1.5 and 2.5. These conditions were used routinely to spin label $(U)_n$, $(dUfl)_n$, $(A)_n$, and $(dA)_n$ with freshly prepared reagent **VI** to an extent of one label for every 75–100 residues. A new sensitive method was developed to detect the presence of unbound labels in the spin-labeled polynucleotides (Pan, 1975). It is based on the ability of polynucleotides to form complexes with poly-L-lysine. We observed that complex formation of poly-L-lysine with a spin-labeled nucleic acid strongly broadens the ESR line shapes of covalently bound nitroxide radicals (Fig. 4), whereas the mobility of unbound nitroxide radicals is not affected by the presence of poly-L-lysine. This method proved to be of considerable value in detecting a small free nitroxide radical contamination in spin-labeled polynucleotides. The reason is readily apparent from Fig. 4. Namely, in the case of a small free spin contamination the broad ESR spectrum of the complex contains three sharp ESR signals indicating the presence of some freely rotating nitroxide radicals. For instance, a small contamination of spin labeled $(A)_n$ is difficult to detect without this test, because the line shape characteristics of slightly contaminated spin-labeled $(A)_n$ are very similar to those of spin labeled $(A)_n$ that gives a negative free spin test with poly-L-lysine. In our hands, extensive dialysis of a spin-labeled nucleic acid always gave a positive free spin test

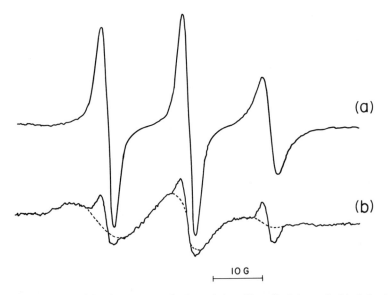

Fig. 4. ESR control for the presence of unbound nitroxide radicals in a spin-labeled polynucleotide. (a) Spin-labeled $(A)_n$ contaminated with a small amount of free nitroxide radicals; (b) after complex formation with poly-L-lysine of a free spin contaminated spin-labeled $(A)_n$ (solid line) and of an extensively purified spin-labeled $(A)_n$ (dashed line).

with poly-L-lysine, strongly suggesting that many studies done on spin-labeled nucleic acids purified by dialysis reflect not only the motion of covalently bound nitroxide radicals, but also that of free spin label. One purification over a long Sephadex G-200 column usually gave a negative free spin test; repeating the chromatographic step always resulted in a negative test.

Chemical modification of T2 phage and calf thymus DNA was attempted with the piperidine bromoacetoxy **VII** (Artyuk *et al.*, 1972). It was found to react slowly at pH 6.5 and 37°C in the presence of 0.1 M NaCl. However, the rate of modification increased considerably at low NaCl concentration, presumably because of destabilization of the DNA duplex. The extent of modification of phage T2 DNA in 10^{-4} M NaCl was about 10^{-2} spin label per nucleotide.

The pyrroline imidazole **VIII** was found to attack the free hydroxyl groups of the glucose moieties occurring in T2 phage DNA because of glucosylation of the hydroxymethylcytosine groups (Petrov *et al.*, 1972; Kamzolova *et al.*, 1973). Subsequently, this label was used to acylate the 2'(3')-OH groups of RNA building blocks (Petrov and Sukhorukov, 1975).

An ethyleneimine spin-label derivative, compound **IX**, was utilized to label DNA preparations in 0.15 M acetate buffer (pH 5.6) for 40 hr at 22°C, whereupon the unincorporated label was removed by exhaustive dialysis (Mil' *et al.*, 1973). The DNA was found to contain 1–2% label. Zavriev *et al.* (1973) spin labeled 16 S RNA isolated from *Bacillus subtilis* with the piperidine ethyleneimine **IX** in 0.1 M acetate buffer (pH 5.5) at room temperature for 3 days. After dialysis a spin-labeled RNA containing no less than one label per 50–70 residues was obtained. Compound **IX** also was used to spin label some synthetic polynucleotides (Zavriev *et al.*, 1974a).

The chemical reactivity of the pyrrolidine iodoacetamide **X** with homopolyribonucleotides and *E. coli* DNA has been investigated more recently (Caspary *et al.*, 1976). When the alkylation was allowed to take place at room temperature for 7 days in 40% ethanol and 60% phosphate buffer (0.0025 M Na$_2$HPO$_4$, 0.005 M NaH$_2$PO$_4$, adjusted with 1 N NaOH to pH 6.7) in the presence of a fivefold excess of compound **X**, the following extent of labeling was determined (number of labels per 10,000 bases): $(A)_n$, 67; $(U)_n$, 3.6; $(A)_n \cdot (U)_n$, 2.6; $(I)_n$, 7.7; $(G)_n$, 4.4; $(C)_n$, 24; $(G)_n \cdot (C)_n$, 6.8; denatured DNA, 25; and native DNA, 2.8. When the reaction was carried out for 10 days at room temperature in 60% ethanol and 40% buffer in the presence of a tenfold excess of compound **X**, a definite increase in the extent of labeling was observed. Under these conditions the following ratios were determined: $(A)_n$, 237; $(U)_n$, 14.9; $(I)_n$, 8.8; $(C)_n$, 152; denatured DNA, 100; native DNA, 4.2. The spin-labeled nucleic acids were recovered by ethanol precipitation and separated from free label by exhaustive dialysis.

2. SITE-SPECIFIC LABELING

In view of the difficulty of introducing by chemical means a nitroxide radical that is site specific for a major base in a polynucleotide, specific labeling studies usually are carried out on some well-selected tRNA's containing minor bases in positions known from sequencing. Thus far, three strategies have been reported for achieving site-specific labeling of tRNA systems. First, an attempt was made to selectively attack a functional group (usually the relatively basic α-amino group) of an amino acid acylated at the 3' end of a tRNA. Second, a minor base or bases were selectively modified (note that a minor base can be incorporated into a tRNA via a nucleotidyl-transferase; see below for an example). Third, the 3'-CCA end of unfractionated tRNA was spin labeled by chemically modifying the vicinal hydroxyl groups of the ribose ring of adenosine.

The first attempt to selectively spin label tRNA was reported by Hoffman et al. (1969) and Schofield et al. (1970). The pyrroline hydroxysuccinimide **IV** was used to acylate the α-amino groups of E. coli Val- or Phe-tRNA. Nonspecific labeling of the tRNA's was noticed, but fractionating the tRNA on hydroxyapatite after the spin-labeling reaction decreased the proportion of nonselectively base-labeled tRNA.

The pyrroline iodoacetamide **X** was used to modify the SH group of cysteine in yeast Cys-tRNA^{Cys} (Kabat et al., 1970). It was noticed that uncharged tRNA^{Cys} reacted slowly with the spin label, whereas the reaction proceeded rapidly with tRNA^{Cys} charged with cysteine (Cys-tRNA^{Cys}). Incubation of spin-labeled Cys-tRNA^{Cys} in mildly alkaline solution (pH 8.0) resulted in the loss of approximately 50–70% of the spins, suggesting that this population of labile spins was specifically bound to the amino acid.

Hara et al. (1970) found that 4-thiouridine can be spin labeled under mild conditions with the pyrrolidine bromoacetamide **III**. Ultraviolet spectroscopy studies indicated that treating 4-thiouridine 2'(3')-phosphate with a fivefold excess of reagent **III** in a buffer containing 0.005 M Na$_2$CO$_3$ and 0.09 M NaHCO$_3$ (pH 8.9) at 20°C caused the 4-thiouridylic acid characteristic UV band at 330 nm to disappear completely within 30 min. These authors showed that under similar reaction conditions the 4-thiouridine occurring in some E. coli tRNA's can be selectively labeled. Selective labeling of tRNA was proved by digesting the modified tRNA and comparing the UV absorption and ESR spectra of spin-labeled 4-thiouridine 2'(3')-phosphate with the spectra of the isolated spin-labeled nucleotide. The selective spin labeling with reagent **III** of s^{4}U at position 8 of E. coli tRNA$_{f1}^{Met}$ and tRNA$_{f3}^{Met}$ was also accomplished, and these spin-labeled tRNA's were used for extensive NMR studies (Daniel and Cohn, 1975, 1976). The extent of spin labeling of either tRNA was followed by the decrease in absorbance at 335 nm. Double integration of the nitroxide ESR signal of the spin-labeled tRNA's yielded in either case 1 ± 0.05 mole of unpaired spin per mole of tRNA.

McIntosh *et al.* (1973) presented strong evidence for selective modification of a minor base of *E. coli* tRNA^{Glu} with the pyrroline anhydride **XI**. This mixed anhydride label is believed to specifically acylate the rare base 2-thio-5-(N-methylaminomethyl)uridine in the wobble position of the anticodon of *E. coli* tRNA^{Glu} within 2 hr at 25°C. The reaction was carried out under conditions in which the tRNA is presumed to assume its native conformation. Although only a moderate reaction yield was determined, the elution profile of a complete RNase digest of labeled tRNA^{Glu} revealed that almost 90% of the total ESR signal was contained in one peak only. On the basis of the position of the peak, which was consistent with that obtained with other acylating agents, it was assumed to contain the spin-labeled rare base.

The piperidine iodoacetamide **VI** was introduced specifically into a yeast tRNA^{Phe} modified at the 3' terminus (Sprinzl *et al.*, 1974). The 3' terminus was modified by incorporating a 2-thiocytidine with tRNA nucleotidyltransferase, which gave a tRNA^{Phe}-C-s^{2}C-A. This tRNA was incubated (with label **VI** dissolved in 10% aqueous tetrahydrofuran) in 0.2 M sodium cacodylate buffer, pH 6, 10 mM MgCl$_2$, for 26 hr at 37°C. The reaction product was isolated and purified by chromatographing twice on Sephadex G-25 and precipitation with ethanol. The 2-thiocytidine of tRNA^{Phe}-C-s^{2}C-A reacted to the extent of about 50% with reagent **VI**. Evidence for site-specific modification of tRNA^{Phe}-C-s^{2}C-A was obtained through a control experiment in which tRNA^{Phe}-C-C-A was reacted with reagent **VI** under identical reaction conditions. The control revealed no covalent binding of reagent **VI** to the tRNA^{Phe}-C-C-A.

Various positions of yeast tRNA^{Val} and yeast Val-tRNA^{Val} were labeled with reagents **IV**, **V**, and **VI** (Vocel *et al.*, 1975) using the approaches mentioned above. The four pseudouridine residues were modified by the piperidine carbodiimide **V** according to Girshovich *et al.* (1971), and it was observed that the tRNA^{Val} preparation containing four labels per tRNA^{Val} retained 30% acceptor activity as compared with unmodified tRNA^{Val}. The pyrroline hydroxysuccinimide **IV** was used to modify the α-amino group of the aminoacyl-tRNA^{Val} according to Schofield *et al.* (1970). Modification of the thiouridylic acid residue with the piperidine iodoacetamide **VI** according to the procedure of Hara *et al.* (1970) resulted in a modified tRNA preparation that retained its acceptor activity almost completely. Vocel *et al.* (1975) also spin labeled unfractionated tRNA and unfractionated tRNA aminoacylated with phenylalanine from *E. coli* with reagents **VI** and **IV**, respectively.

More recently, the detailed preparation and characterization of several spin-labeled tRNA's with reagents **III**, **V**, **XI**, and **XII** have been reported (Caron and Dugas, 1976a). The tRNA's were recovered by precipitation with cold ethanol, resuspended in buffer, and stored at $-20°C$. Extensive enzymatic digests of the spin-labeled material, subsequently chromatographed on DEAE-Sephadex A-25, were used to identify labeling sites on the tRNA's. An interesting procedure was also described for labeling the 3'-CCA end of unfractionated tRNA by the same authors. Namely, $NaIO_4$, 4-amino-2,2,6,6-tetramethylpiperidine-1-oxyl, and $NaBH_4$ were used to chemically modify the vicinal hydroxyl groups of the adenosine ribose ring, which is believed to form a morpholine spin-label derivative.

B. Enzymatic Synthesis of Spin-Labeled Nucleic Acids

A great impediment encountered in the chemical approach to site-specific spin labeling of a nucleic acid is the absolute necessity for the presence in the polynucleotide chain of one or more rare bases containing functional groups that have been proved to undergo specific reactions under mild conditions. Even if this requirement is met, one often faces the problem of the occurrence of background labeling of the functional groups of the major bases by the usually rather reactive spin label reagents **I–XII**.

For this reason the need arose to design a spin-labeled nucleic acid building block that would be accepted by a nucleic acid polymerase as substrate and hence would be incorporated into a nucleic acid chain. Such a label was synthesized recently, and it was shown unambiguously that the label was enzymatically incorporated into the polynucleotide (Bobst and Torrence, 1978). The structure of this spin label, ppRUGT (**XIII**), and the reaction scheme for its enzymatic incorporation are given in Fig. 5. The synthesis of ppRUGT has been published (Bobst and Torrence, 1978) and consists of the following reaction steps. 5-Hydroxyuridine 5'-diphosphate is alkylated with 4-(α-chloroacetamido)-2,2,6,6-tetramethylpiperidino-1-oxy in the presence of 1 equivalent of NaOH. (Recently, we noticed that first converting the chloro label to the iodo label with NaI in acetone according to the Finkelstein reaction increases the reaction yield.) The ppRUGT is purified on Whatman 3MM paper with the solvent system ethanol-1 M ammonium acetate, pH 7.5 (7:3). Digestion of ppRUGT by bacterial alkaline phosphatase gives as the only UV-absorbing product N-(1-oxyl-2,2,6,6-tetramethyl-4-piperidinyl)-O-[1-β-D-ribofuranosyluracil-5-yl)glycolamide (RU-glycolamidoTEMPO = RUGT); R_f of RUGT (silica gel TLC with 16% methanol–chloroform) 0.47; mp (uncorrected) 131°–133°C. *Analysis:* calculated for $C_{20}H_{31}N_4O_9$: C, 50.95; H, 6.62; N, 11.88. Found: C, 50.97; H, 7.25; N, 11.71.

Polynucleotide phosphorylase (PNPase) from *Micrococcus luteus* or *E.*

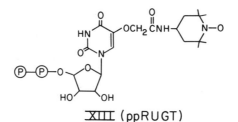

$$\underline{XIII} \ (ppRUGT)$$

$$RpN_1pN_2 + 2 \ P_i \ \underset{\longleftarrow}{\overset{PNPase}{\longrightarrow}} \ R + ppN_1 + ppN_2$$

$RpN_1pN_2 = (RUGT,U)_n$

$ppN_1 = ppU$

$ppN_2 = ppRUGT$

Fig. 5. Reaction scheme illustrating the copolymerization of ppRUGT (**XIII**) and uridine 5'-diphosphate (ppU) with polynucleotide phosphorylase (PNPase) from *Micrococcus luteus* or *E. coli* B.

coli can then be used to copolymerize ppRUGT with uridine 5'-diphosphate (ppU) (Fig. 5). This enzyme, which actually can be regarded as a depolymerase of RNA, does not require and cannot transcribe a template. PNPase has proved to be very useful for the synthesis of a great variety of high molecular weight polynucleotides containing natural bases as well as base analogues. It should be emphasized that, although this enzyme allows nitroxide labeling specificity with respect to the nucleic acid building block, no sequence-specific labeling of the polynucleotide is achieved. For that purpose, a template-dependent polymerase that can accept a nitroxide-labeled nucleoside triphosphate analogue as substrate and can incorporate it into a defined polynucleotide sequence would be useful.

The copolymerization of ppRUGT with ppU gave the synthetic polynucleotide $(RUGT,U)_n$ with weight-average degrees of polymerization that depended on the reaction conditions (Bobst and Torrence, 1978). The polymerization products were purified by Sephadex G-100 chromatography and analyzed for their nitroxide-label content by ESR. Samples of 33,000- and 66,000-dalton $(RUGT,U)_n$ were found to have a RUGT/U ratio of 1/50 to 1/100. Thus, these polynucleotides contain a random distribution of one to four RUGT molecules per chain. In Fig. 6, an Ealing CPK atomic model of $(RUGT,U)_n$ is shown. From this figure, it should become apparent that the steric perturbation caused by a small extrinsic nitroxide radical at every fiftieth to hundredth residue is likely to have a minimal effect on most physicochemical characteristics of the biopolymer. It is expected that the

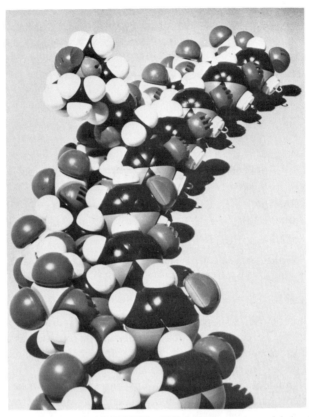

Fig. 6. Molecular structure of $(RUGT,U)_n$. Ealing CPK atomic model showing a small stretch of the $(RUGT,U)_n$ chain (one RUGT and eight U residues); the nitroxide radical is visible in the upper left of the chain. [From Bobst and Torrence (1978); reprinted by permission of publisher, IPC Business Press Ltd. ©.]

enzymatic approach described here will lead to the synthesis of a great variety of RUGT-containing nucleic acids by copolymerizing ppRUGT not only with uridylic 5'-diphosphate but also with other nucleoside 5'-diphosphates.

C. Spin Probing of Nucleic Acids

1. MANGANESE(II) IONS

Paramagnetic metal ions have been used in a variety of biological systems as probes to obtain structural information. In several studies, Mn(II) has been used to probe nucleic acid characteristics by ESR. So far, most studies have been done on the binding of Mn(II) to DNA (Reuben and perpendicular orientation of the aromatic plane of the drug to the axis.

mplexes between DNA and the 5-methylphenazinium cation radical re studied by Ishizu et al. (1969). Compound **XV**, like probe **XIV**, is le in aqueous solution at biological pH values and therefore is unac- le as a spin probe for nucleic acids.

e disadvantages of using an unstable radical as a DNA-intercalating finally were overcome when a series of spin-labeled analogues of intercalating agents was synthesized. A study on intercalating spin- d ethidium bromide into DNA was reported (Piette, 1974), and sub- tly its structure (**XVI**) was published (Hong and Piette, 1976). Sinha hignell (1975) described the synthesis of the two 9-aminoacridine spin s **XVII** and **XVIII**. These probes were synthesized from 4-amino- -tetramethylpiperidine and the corresponding 9-chloroacridine. A ed procedure was published a year later (Sinha et al., 1976) for the eses of **XVII** and **XVIII**, the additional spin-labeled 9-aminoacridines **XX**, and **XXI**, and the spin-labeled azaacridine **XXII**. In addition to the ure of the spin-labeled ethidium bromide **XVI**, the structures of var- spin-labeled aromatic amino carcinogens (**XXIII**, **XXIV**, and **XXV**) given, as were their ESR characteristics in the absence and presence of (Hong and Piette, 1976). Electron spin resonance evidence was pre- d which suggested that the spin probes **XVI**, **XVII**, **XVIII**, **XXIII**, , and **XXV** bind to DNA by intercalation (see Section III,B,2).

ELECTRON SPIN RESONANCE SPECTRA OF NUCLEIC ACIDS CONTAINING SPIN LABELS OR SPIN PROBES

Measurements

Instrumental considerations and operating procedures to measure oxide motion were discussed in detail by Jost and Griffith (1975). Elec- spin resonance measurements of nucleic acids containing spin labels or probes are done in aqueous media, and thus all the precautions when ling with aqueous samples should be taken. Sensitivity problems often encountered when spin-labeled nucleic acids are measured because the ent of labeling is usually low, and it is desirable to work at a low nucleic d residue concentration. In addition, the sensitivity of the instrument is uced because of the high dielectric constant of water, lowering the Q of cavity. Therefore, the measurements usually are done in commercial flat artz cells for optimal signal-to-noise ratio. Recently, Varian manufactured new E-238 ESR TM_{110} cavity, which can accommodate larger flat cell

Gabbay, 1975, and references therein; Jouve et al., 1975). Only one study has been reported about the interaction of Mn(II) with spin-labeled nucleic acids (Vocel et al., 1975).

The analysis of the ESR data is relatively straightforward when Mn(II) ions are the only paramagnetic species in solution. Electron spin resonance was found to be a convenient tool for monitoring the formation of Mn(II)–nucleic acid complexes, because a change in the symmetry of the manganese aquo ion considerably broadens the ESR lines. An Mn(II)–DNA complex displays a strongly broadened ESR spectrum of low intensity, whereas the Mn(II) ion, with its $3d^5$ electron configuration ($S = 5/2$) and an ^{55}Mn nucleus of a nuclear spin of 5/2, gives a sharp, six-line spectrum. For this reason, the intensity of the ESR spectra can be used as a direct measure of the concentration of free manganese aquo ions in solution. This strategy also was used to determine the binding of Mn(II) ions to tRNA in the concentration range of 1–28 bound Mn(II) per tRNA at 23°C (Danchin and Guéron, 1970). Knowing the total Mn(II) concentration and determining the free Mn(II) by ESR, it was concluded from a Scatchard plot that the first four Mn(II) ions bind cooperatively to unfractionated tRNA from E. coli at a free Mn(II) concentration of 3 μM.

Vocel et al. (1975) studied by ESR the coupling of Mn(II) with spin-labeled tRNA^{Val}, valyl-tRNA^{Val} modified with a spin label at the valyl residue, and spin-labeled charged and uncharged unfractionated tRNA from E. coli. Since the tRNA was site specifically spin labeled, the dipole–dipole coupling between spin label and Mn(II) was used to determine the sites of ion coordination on the tRNA. For the analysis of the data the authors assumed that a dipole coupling with short correlation time caused only broadening of the observed spectra. With this assumption it was concluded that, for some valyl-tRNA^{Val} molecules, the second Mn(II) ion coordinates on the α-amino group of the valyl residue at a distance of 15–25 Å from the spin label. Possibly a different conclusion would have been reached through the formalism developed by Leigh (1970), which shows that, if a spin label and a paramagnetic ion are present in the same solution, the ion will not necessarily always cause broadening of the nitroxide signal (see section III,C). (Editor's note: see also Chapter 2, p. 71 for a detailed discussion of the theory and applications of this technique.)

2. 2,2,6,6-TETRAMETHYLPIPERIDINE-1-OXYL-4-OL (TEMPOL)

Sukhorukov et al. (1967), reported an interesting attempt to study the state of the surface of nucleic acids with the spin probe TEMPOL. The aim of their work was to investigate the hydration of native and denatured calf thymus DNA and yeast tRNA under various relative humidities. TEMPOL was incorporated noncovalently into the polynucleotide matrix by combining an aqueous solution of nitroxide radicals with a solution containing one

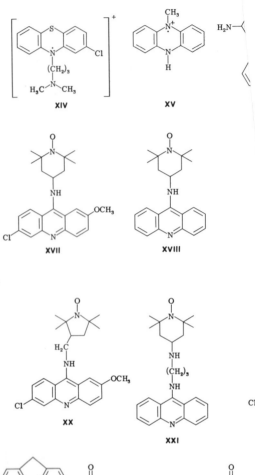
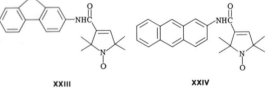

of the three kinds of nucleic acids. The concentrations were such that there was one probe for every nucleotide. After lyophilization of the solutions, the amount of electrolyte salts, mainly NaCl, in the dry lyophilized DNA and RNA preparations was 10%. All dry samples of the various nucleic acids gave essentially the same ESR spectra, which were characteristic of strongly immobilized spin probes. However, changing the relative humidity of the ampules containing the dried nucleic acids had a pronounced sample-specific effect on the ESR spectra. For the native DNA, the ESR spectrum indicative of an immobilized probe persisted up to 30% relative humidity, whereas for denatured DNA and tRNA relative humidities as high as 50 and 60%, respectively, were required to observe noticeable changes in the ESR line shapes. In a subsequent study (Sukhorukov and Kozlova, 1970) TEMPOL was used to investigate DNA preparations of varying water content over a wide temperature range. Discontinuities in the Arrhenius plots were noticed that are believed to indicate the existence of DNA in different as yet unknown conformations. The breakpoints arising from the several linear segments in the Arrhenius plot showed a small dependence on the water content of the DNA preparations and occurred at about 65° and 20°C. The high-temperature breakpoint most certainly reflects base-pair melting, which is commonly observed in UV melting studies. The low-temperature transition, however, indicates the presence of structural rearrangements observed so far only by ESR.

3. Nucleic Acid Intercalating–Binding Spin Probes

The chemical structures of spin probes that are known to have a strong affinity for nucleic acids are presented in Fig. 7. Most of these probes are believed to intercalate into the nucleic acid matrix.

The first demonstration of intercalation of aromatic molecules into native DNA by ESR came from Ohnishi and McConnell (1965). This was accomplished with the moderately stable, positive ion radical of chloropromazine (cpz$^+$), **XIV**. Its hyperfine interaction and g-factor anisotropy were sufficiently large that its binding to DNA had a pronounced effect on the ESR spectrum and yielded orientation-dependent information. However, cpz$^+$ showed many drawbacks as a general-purpose spin probe because it was only moderately stable, and its complex hyperfine structure was not suitable for a quantitative line shape analysis. Recently, spin probe **XIV** was used in a series of measurements to gain insight into DNA–chlorpromazine complexes (Porumb and Slade, 1976). Instead of DNA orientational effects being produced by passing the solutions through capillary tubes (Ohnishi and McConnell, 1965), measurements were done on DNA–cpz$^+$ fibers for a range of orientations of the fiber axis relative to the magnetic field. The ESR data were explained in terms of a strongly immobilized label and the prefer-

Fig. 7. Chemical structures of spin-probe reagents that have been acids: **XIV**, chlorpromazine cation; **XV**, methylphenazinium cation; **XVII–XXI**, aminoacridines; **XXII**, aminoazaacridine; **XXIII**, amino oanthracene; **XXV**, aminochrysene.

(active volume according to Varian is 0.039 ml) than those used for the multipurpose E-231 cavity (active volume 0.023 ml). Relative to the multipurpose E-231 cavity, the author was able to obtain a threefold increase in the ESR sensitivity with the E-238 cavity without affecting the signal-to-noise ratio. The E-238 cavity allows routine measurements of nitroxide radicals tumbling in the motional narrowing region at a concentration of 10^{-7} M on an E-4. Time averaging allows measurements even below that limit.

Optimal selection of microwave power, modulation amplitude, scan time, filter time constant, and gain is of particular importance when one is working with low concentrations of nitroxide radical to obtain correct line shapes necessary for correlation time calculations. Particular attention should be given to the modulation amplitude and the microwave power settings. Although there are many rules of thumb for setting these controls, it is safe to start out with a small modulation amplitude and low microwave power and to determine at which value changes in relative peak heights and/or linewidth occur. For spin-labeled nucleic acids with nitroxide radicals tumbling in the motional narrowing region, it was observed that microwave power saturation occurs at different power settings and is dependent on the type of glassware being used, i.e., large flat cell, regular flat cell, micro flat cell for variable temperature studies, or small capillary tube. With respect to the modulation amplitude setting, it was noticed that for values greater than 1.25 G some line broadening takes place, which significantly affects the calculated correlation time values. Normally, a value of 1 G was found to be safe, whereas a setting of 2 G started to affect the tumbling time values.

Often, ESR measurements as a function of temperature are reported for nucleic acids containing spin labels or spin probes. Usually, such studies are done either with a flat quartz microcell that fits directly into the Dewar flask or in a small sealed capillary placed into a standard ESR tube filled with toluene or xylene. Through experience with both approaches, the author became aware of the advantages and disadvantages of each (Pan and Bobst, 1973, 1974). The use of a microcell offers good sensitivity, but the cell is in direct contact with the temperature-controlled nitrogen flow and any small temperature fluctuation at the temperature control is immediately transmitted to the thin fluid layer between the two flat quartz plates [manufacturer's specifications are for temperature control of the gas flow: $\pm 1°C$ (Varian E-257 variable-temperature accessory); $\pm 0.5°C$ (Jeol Jes-VT-3A)]. Although we usually observed a fluctuation of less than 1°C with a Varian E-257 temperature control, the existence of such fluctuations is noteworthy, especially since it is difficult to directly monitor the temperature of the biological fluid without the risk of contamination. Exposing a dilute nucleic acid solution to a piece of metal or some other material increases the likelihood that observations will be artifactual. Caron and Dugas (1976b) inserted a small copper–constantan thermocouple into the liquid to a point just above the flat part of the cell and external to the microwave cavity to continuously monitor the temperature. We used an alternative approach with the microcell that consisted of calibrating the temperature dial of the control unit before and after the melting experiments in the appropriate temperature range by setting a thermistor probe at the bottom, center, and top of the cavity. This approach allows one to determine the temperature gradient. It is advisable to take this precaution because the temperature gradient in the Dewar can amount to several degrees at higher temperatures. The temperature in the microwave region is calculated by extrapolation of the measured temperature values. The alternative for recording spin melting data, namely, the use of capillary tubes inserted into a quartz tube filled with a fluid, is considerably less sensitive. However, the sample is now heated via a buffer (toluene or xylene) whose temperature can be continuously monitored by placing a thermistor or thermocouple in the heating fluid above the cavity. After the temperature gradient of the heating fluid is determined at various temperatures, the temperature in the microwave area can be calculated.

It should be emphasized that the temperature gradient for spin melting experiments is considerably larger than that present in jacketed cells used for circular dichroism or optical density melting studies. Since the temperature of spin melting experiments currently cannot be determined with the same degree of accuracy as the temperature existing during spectrophotometric melting studies, it is inadvisable to draw conclusions about the perturbation effect of spin labels on the local melting of a nucleic acid matrix unless a substantial difference exists between T_m^{sp} and T_m^{OD} of spin-labeled nucleic acids. In the case of spin-labeled $(A)_n \cdot (U)_n$ complexes, T_m^{sp} data obtained by the use of the microcell and by the use of small capillaries were very similar (Pan and Bobst, 1974), and those values agreed within the experimental error with the T_m values obtained from absorption melting curves.

Interfacing a small computer with an ESR spectrometer is particularly useful when one is working with relatively weak signals, as in the case of spin-labeled nucleic acids, or when one is analyzing complex spectra. In the first case, repeated scanning will help to reduce the signal-to-noise ratio of a weak signal recorded at relatively high gain setting. The improvement is proportional to $n^{1/2}$, where n is the number of scans. In the latter case, the analysis of the ESR spectrum is simplified, since the data are now available in a digitized form. This is almost a necessity when one is analyzing titration experiments with spin-labeled nucleic acids and proteins that give rise to composite spectra. For an excellent survey of the use of computers to process spin-labeling data, the reader is referred to Jost and Griffith (1976). Our relatively inexpensive approach is to transfer data from a Fabri-Tek 1074 18-bit signal-averaging computer, which stores the data from a Varian E-4 spectrometer, to a regular audio cassette magnetic tape via a KIM-1 microprocessor (Sinha, 1977). This makes possible the transfer of digitized ESR data to the Amdahl 470 for further data handling (for example, linear baseline correction, integration of first-derivative ESR spectra, resolution of a spectrum into spectral components, and comparison of experimental data with computer-simulated spectra).

B. Theoretical Analysis of Nitroxide Spectra

1. MOTIONAL NARROWING REGION SPECTRA

When a nitroxide radical is allowed to tumble fairly rapidly ($\tau_R \approx 10^{-10}$ sec), the spin label interacts very little with the nucleic acid matrix and thus does not provide detailed structural information about its environment. However, in this motional range an ESR spectrum of a weakly immobilized radical is obtained which can still provide valuable information about segmental motion of the nucleic acid (Bobst, 1977). When the spin-labeled nucleic acid project was initiated in the author's laboratory several years ago, it was hypothesized that it is not essential to achieve site-specific labeling of nucleic acids to analyze their characteristic properties as a function of pH or temperature or to follow their interaction with various biological systems. However, a stable covalent linkage attaching the nitroxide radical to the nucleic acid matrix would be essential. Appropriate motional range of the radical is achieved by attaching the N—O moiety to the macromolecule through a number of bonds ("legs") about which internal rotation can occur (Wallach, 1967). If such a "leg" is present, the rotational motion of the N—O moiety depends on both the internal rotation of the "leg" and the segmental mobility of the nucleic acid. It was found that the piperidine iodoacetamide **VI**, when attached to a single-stranded polynucleotide, allows the N—O moiety to tumble in the desired tumbling range of the motional narrowing region.

Although no detailed structural information is readily obtainable with fairly rapidly tumbling spin labels, it is desirable to stay in this motional range for many biological applications, because only in this tumbling range is it possible, at least with current ESR equipment, to design ESR experiments allowing one to work with label concentrations of 10^{-7} M or less. Such sensitivity enables many complex biological systems to be studied with little material, which offers, of course, many advantages.

An important requirement for properly analyzing an ESR spectrum of a nitroxide radical is to have accurate values for the magnetic tensors A (A_{xx},

A_{yy}, A_{zz}) and g (g_{xx}, g_{yy}, g_{zz}). Since the magnetic parameters of nitroxide radicals are rather solvent dependent, it is ideal to determine these parameters from viscous solutions in the same solvent as that in which the nitroxide spectrum is recorded (Freed, 1976). Unfortunately, such a system is not readily available for spin-labeled nucleic acids. Principal values of the hyperfine tensors (A_{zz}, $A_{xx} = A_{yy}$) for spin label VI were determined experimentally. The A_{zz} value was obtained from measurements in a 85% glycerol/15% water mixture at $-35°C$, and $A_{xx} = A_{yy}$ was calculated by using the determined A_{zz} value together with the measured hyperfine coupling constant a at low viscosity (Pan, 1975). Using Freed's program (Freed, 1976) for computer simulations of ESR spectra of nitroxide radicals, it is necessary to set $A_{xx} \neq A_{yy}$. Reasonable A_{xx} and A_{yy} values were calculated from the A_{xx}/A_{yy} ratio of similar nitroxide radicals with known hyperfine tensor values from single-crystal studies published by Snipes et al. (1974). The principal values of the g tensor of spin label VI also were assumed to be similar to those of related nitroxide radicals, whose g tensor values had been determined from single-crystal studies. This led to the following set of parameters for calculating tumbling times as outlined below: g_{xx}, 2.0088; g_{yy}, 2.0059, g_{zz}, 2.0026; A_{xx}, 7.15; A_{yy}, 7.35; A_{zz}, 35.6 G [for $A_{xx} = A_{yy}$ 7.2 G was adopted to calculate the constant in Eq. (2)]. The x axis is along the N—O bond; the z axis is along the 2p π orbital of the nitrogen; and the y axis is perpendicular to the other two.

Tumbling times can be calculated either by an approach developed by Stone et al. (1965) for rapidly isotropically tumbling nitroxide radicals or by Freed's more general formalism (Freed et al., 1971; Goldman et al., 1972) when the motion is slow and/or anisotropic. A correlation time τ can be determined by solving Eq. (1) (Stone et al., 1965) from the quadratic and linear terms:

$$T_2(m)^{-1} = \tau\{[3I(I+1) + 5m^2]b^2/40$$
$$+ (4/45)(\Delta\gamma\, H_0)^2 - (4/15)b\, \Delta\gamma\, H_0\, m\} + X \quad (1)$$

where H_0 is the applied field strength in gauss, $I = 1$ is the nuclear spin quantum number for ^{14}N, $b = (4\pi/3)(A_{zz} - A_{yy})$, $\Delta\gamma = (-|\beta|/h) \times [g_{zz} - \frac{1}{2}(g_{xx} + g_{yy})]$, and X represents contributions from other broadening mechanisms that do not depend on m. One should remember that, in order for the calculation to be valid, the following conditions must be met. (a) There must be axial symmetry in the anisotropic hyperfine interaction, i.e., $A_{xx} = A_{yy}$; (b) the tumbling motion must be isotropic; (c) the tumbling motion must be sufficiently slow so that $\omega^2\tau^2 \gg 1$, where $\omega = g|\beta|H_0 h^{-1}$; $(\pi a)^2\tau^2 \ll 1$, where a is the hyperfine splitting; $b^2\tau^2 \ll 1$.

Equation (2) is obtained by solving Eq. (1), considering the correlation time τ from the equation quadratic in m only, and using the parameters listed above. The τ values determined from the quadratic term are less sensitive to the applied microwave power (Hoffman et al., 1969), and in most of the literature these τ values are reported when Eq. (1) is used for correlation time calculations.

$$\tau = 5.47 \times 10^{-10} \, \Delta H_0[(h_0/h_{-1})^{1/2} + (h_0/h_{+1})^{1/2} - 2] \, \text{sec} \qquad (2)$$

Equation (2) is valid for correlation times shorter than approximately 5×10^{-9} sec; h_{+1}, h_0, h_{-1} are the peak heights in arbitrary units, and ΔH_0 is the peak-to-peak separation of the center peak in Gauss. A convenient way for calculating correlation times consists of programming Eq. (2) into an electronic calculator so that h_0, h_{+1}, h_{-1}, and ΔH_0 can be entered in any order and the calculations then performed by pressing a single key. With the Texas Instrument SR 52 this can be achieved with 70 key strokes. This approach is particularly useful when tumbling times are calculated at various temperatures for determining Arrhenius plots.

A more powerful route for calculating tumbling times consists of computer simulating the ESR spectra by using the formalism of Freed et al. (1971) appropriately generalized by Goldman et al. (1972) to include the completely asymmetric g and A tensors of the nitroxide radical. Such simulations are shown in Fig. 8 (Bobst et al., 1975), where it is apparent that the simulated spectra are in good agreement with the experimental data. The tumbling times are based on an axially symmetric rotational diffusion model with x', y', z' as the principal axes of the diffusion tensor R and with the z' axis as the symmetry axis of fast reorientation. The assumption is made that the x', y', z' axes are either the same as the molecular axes x, y, z or a cyclic permutation of them. Here $R_3 \, (= R_\parallel)$ is the rotational diffusion component about z', whereas $R_1 \, (= R_\perp)$ gives the components about x' and y'; N is the ratio of the rotational diffusion components and is equal to R_3/R_1. It follows that $N = 1$ for isotropic motion, whereas $N \neq 1$ for anisotropic motion. Essentially the same program, published by Freed (1976) $(R_1 = R, R_\parallel = R_z)$, was used to simulate the spectra shown in Fig. 8. For the simulations the symmetry axis of reorientation z' is assigned to either of the molecular axes x, y, z. When the Brownian diffusion model is used, the input parameters $R_\parallel$ and $R_\perp$ can be related to the correlation times $\tau_{R_\parallel}$ and $\tau_{R_\perp}$, respectively, by the following expressions: $\tau_{R_\parallel} = \frac{1}{6}R_\parallel$ and $\tau_{R_\perp} = \frac{1}{6}R_\perp$. Expressing N in terms of $\tau_{R_\parallel}$ and $\tau_{R_\perp}$ gives $N = \tau_{R_\perp}/\tau_{R_\parallel}$. Spectrum A can be simulated by taking $R_\parallel = R_\perp$ and choosing an isotropic reorientation value of $\tau_R = 3.5 \times 10^{-10}$ sec. Incidentally, with Eq. (1) one calculates a τ value that is very close to that obtained by computer simulation. However, upon hybridization of spin-labeled $(dUfl)_n$ with $(A)_n$ an axially symmetric Brownian rotational diffusion with $z' = y$ had to be assumed for good simulation.

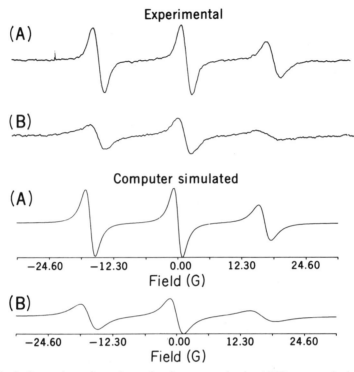

Fig. 8. Comparison of experimental and computer-simulated ESR spectra of spin-labeled poly(dUfl) (spectra A) and spin-labeled poly(dUfl) · poly(A) (spectra B) (at X band with center of spectra at 3383 G). [From Bobst *et al.* (1975); copyright © 1977 by the American Association for the Advancement of Science.]

The $\tau_{R_\parallel}$ and $\tau_{R_\perp}$ values chosen for simulating spectrum B are 6.3 × 10^{-10} and 22.1 × 10^{-10} sec, respectively. The resulting $N = 3.5$ suggests that in the duplex the nitroxide radical tumbles faster along its y axis than along the x and z axes. It should be stressed that the exact values of the g and A tensors are not known for nitroxide radicals bound to a nucleic acid. Therefore, the tumbling time calculations give approximate values only. However, small differences in g and A tensor values are not believed to strongly affect the experimental parameters of the spectra (h_{+1}, h_0, h_{-1}, ΔH_0). Instead, it is the molecular motion that strongly affects these parameters and allows one to draw definite conclusions about local mobility changes occurring in a nucleic acid chain upon complex formation.

In order to have a better idea of the effect of anisotropic motion of spin label **VI** in the motional narrowing region on the peak heights (h_{+1}, h_0, h_{-1}), N was varied from 1 to 4 for various $\bar{\tau}$ [$\bar{\tau} = \frac{1}{6}\bar{R}$; $\bar{R} = (R_\parallel R_\perp)^{1/2}$] and the x', y', z' axes were either the same as the molecular axes x, y, z or a cyclic

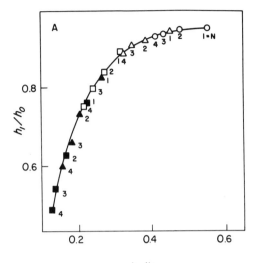

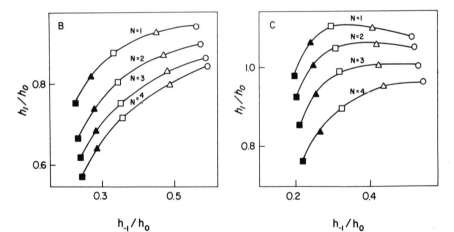

Fig. 9. Plots of h_1/h_0 versus h_{-1}/h_0 determined from computer-simulated ESR spectra for the following correlation times $\bar{\tau}$ (in seconds): 3×10^{-10} ($\bigcirc$), 5×10^{-10} ($\triangle$), 10×10^{-10} ($\square$), 15×10^{-10} ($\blacktriangle$), 20×10^{-10} ($\blacksquare$); N values vary from 1 to 4. The symmetry axis of reorientation z' is assigned to the molecular axis z in spectrum A, y in spectrum B, and x in spectrum C.

permutation of them (Pan, 1975). The result of such computer simulations are graphically plotted in Fig. 9. It is apparent that if the z molecular axis is taken as the symmetry axis of fast reorientation, various N values will generate the same curve (Fig. 9A) when h_1/h_0 is plotted versus h_{-1}/h_0. This is definitely not the case for $z' = x$ or $z' = y$ (Fig. 9B,C). Such plots clearly establish that anisotropic reorientation in this motional range can be detected only if the symmetry axis of fast reorientation (z') is along the x or y molecular axis. The plots were valuable to start with a set of fairly reasonable input parameters ($R_{\parallel}$, $R_{\perp}$, N) for simulating experimental ESR spectra of nucleic acid containing nitroxide radicals tumbling in the motional narrowing region.

Vasserman et al. (1971) similarly define the anisotropy in the rotation of nitroxide radicals. They use the model of an axially symmetric tensor for rotational diffusion ($D_1 = D_2 = D_{\perp}$; $D_3 = D_{\parallel}$) and define the anisotropy in the rotation as $d = D_{\parallel}/D_{\perp}$. The anisotropic character of the motion is expressed with the term ε [Eq. (3)], which is independent of the correlation time and the absolute linewidth.

$$\varepsilon = \frac{T_{2(+1)}^{-1} - T_{2(0)}^{-1}}{T_{2(-1)}^{-1} - T_{2(0)}^{-1}} = \frac{(h_0/h_{+1})^{1/2} - 1}{(h_0/h_{-1})^{1/2} - 1} \tag{3}$$

They show that ε is essentially dependent on two factors: the g and A tensor values and the anisotropic character of the motion. For average values of the A and g tensors, plotting ε versus d gives curves with an intersection at $\varepsilon = 0$, where the motion is considered isotropic. The anisotropic character of some spin-labeled nucleic acids has been analyzed in this manner by Sprinzl et al. (1974) and Caspary et al. (1976). Both groups noticed a small degree of anisotropy for spin-labeled nucleic acids and found that ε is temperature dependent.

2. RIGID-LIMIT SPECTRA

Some work has been reported on the use of the spin-label technique to obtain information about the orientation of spin probes upon binding to oriented DNA (Piette, 1974; Hong and Piette, 1976). For this purpose, the spin probe must be rigidly held in the DNA matrix, and the geometric relationship between the nitroxide group and the remainder of the molecule must be known. The latter requirement is difficult to establish because the linkage between the nitroxide group and the planar aromatic rings in the spin probes XVI, XXIII, XXIV, and XXV is not rigid. There is some degree of mobility between the nitroxide group and the remaining planar part of the molecule, which supposedly intercalates into the DNA. Without residual nitroxide motion the spectrum of a strongly immobilized nitroxide radical (rigid-limit spectrum) is directly dependent on the angle of orientation between the A and g tensors of the nitroxide moiety and the DNA fiber geometry according to the spin Hamiltonian, $\mathscr{H}$:

$$\mathscr{H} = \beta \mathbf{H} \cdot \mathbf{g} \cdot \mathbf{S} + \mathbf{S} \cdot \mathbf{T} \cdot \mathbf{I} \tag{4}$$

Here, g and T are the tensors of the nitroxide radical defined earlier (an axially symmetric hyperfine tensor is assumed, $T_{xx} = T_{yy} = T_\perp$, and $T_{zz} = T_\parallel$; note T is used here for A); S and I are the electron and nuclear spin operators, respectively; and β is the Bohr magneton. Figure 10 [spectrum (a)] shows the characteristic ESR spectrum of the uncomplexed nitroxide radical XVI where motional narrowing occurs. In spectra (b), (c), and (d) of Fig. 10 the nitroxide radical XVI is recorded in the presence of DNA under different conditions. Spectrum (b) is obtained after mixing probe XVI with a DNA solution; spectrum (c) is obtained after most of the unbound radical is filtered away; and spectrum (d) shows the dried DNA–probe XI precipitate. Spectrum (e) is typical of a powder spectrum obtained by dissolving probe XVI in 98 % glycerol (saturated) and recording the spectrum at $-30°C$. Comparing spectrum (b) with (d) reveals a slight increase in immobilization as measured by the increase in separation of the high- and low-field peaks. Orientation of the DNA–spin probe XVI complex relative to the applied magnetic field either by "flow" or by stretch-orienting the fibers on a thin plate allows one to measure the angular dependence of the ESR spectrum. This is illustrated in Fig. 17, p. 272, and Fig. 18, p. 274, of Chapter 6, Section V,B,2. In Fig. 17, p. 272, the first spectrum shows the DNA fibers oriented parallel to the field. In subsequent spectra the fibers are rotated in 10° intervals until the DNA fibers are perpendicular to the field. It is apparent from this figure that the fibers are reasonably well oriented, although perfect orientation has not been achieved. Rotation of the sample causes the two outer extreme lines, characteristic of the maximal anisotropic hyperfine splitting when T_{zz} is parallel to the field, to move slowly toward the center.

Figure 18, p. 274, shows the angular dependence of these outer lines. Since it seems that the majority of the fibers are reasonably well oriented in the plane of the plate, values of 64.3 and 17.3 G for $2T_\parallel'$ and $2T_\perp'$, respectively, were determined. Furthermore, because the nitroxide radical is assumed to be coplanar with the ethidium bromide ring, the orientation dependence of the ESR spectrum is evidence for the intercalation of the ethidium bromide ring between the base pairs in the DNA duplex. The anisotropy of the spectra indicates that for most of the complexes spin probe XVI is oriented in the DNA so that the z axis of the nitroxide is parallel to the DNA fiber axis.

C. Analysis of Nitroxide–Manganese(II) Ion Interaction Spectra

This topic is covered in great detail in Chapter 2 by Hyde and co-workers. Leigh (1970) established that if a nitroxide radical and a paramagnetic ion are present in the same solution, the ion will not necessarily cause broadening of the nitroxide signal. For two paramagnetic groups that are relatively rigid on a macromolecule and have a long reorientation time

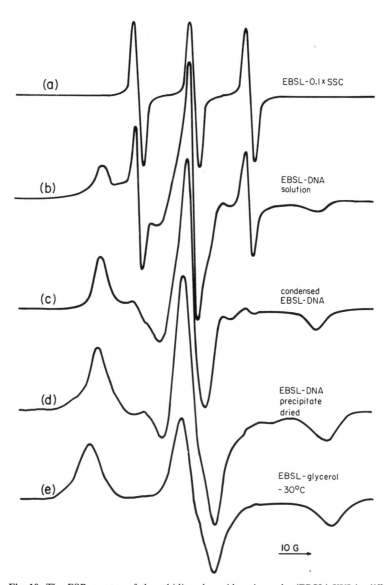

(a)   EBSL-0.1×SSC

(b)   EBSL-DNA
                                                       solution

(c)   condensed
                                                       EBSL-DNA

(d)   EBSL-DNA
                                                       precipitate
                                                       dried

(e)   EBSL-glycerol
                                                       -30°C

                                          IO G

Fig. 10. The ESR spectra of the ethidium bromide spin probe (EBSL) **XVI** in different environments. (a) Spectrum of 1×10^{-4} M spin probe **XVI** in $0.1 \times$ SSC at room temperature; (b) spectrum of the solution obtained after mixing 5×10^{-2} ml of 1×10^{-3} M spin probe **XVI** in acetonitrile with 5 ml of 8.7×10^{-3} M DNA in $0.1 \times$ SSC; (c) the same mixture as (b) after condensing the complex by ultrafiltering most of the unbound spin away; (d) the precipitate obtained by adding 2 volumes of ethanol plus 0.1 volume of 2 M sodium acetate to the condensed mixture of (c) followed by collection and drying; (e) the spin probe **XVI** alone dissolved in 98% glycerol (saturated) and the spectrum recorded at $-30°C$. [From Hong and Piette (1976); reprinted by permission of publisher.]

compared to the electron spin, the linewidth of the ESR signal can be derived as follows:

$$\delta H(\theta, \phi) = C(1 - 3 \cos^2 \theta_R{}')^2 + \delta H_0 \qquad (5)$$

$$C \equiv g\beta\mu^2\tau/r^6\hbar$$

The angles θ and ϕ are the azimuthal and polar angles of the applied field H in the molecule fixed coordinate system; g, β, and $\hbar$ have their usual meanings; τ is the electron spin-relaxation time of the manganese(II) ion [assumed to equal the correlation time for the dipolar interaction between the manganese(II) ion and water protons]; μ is the magnetic moment of the Mn(II) ion; r is the Mn(II)–nitroxide distance; $\theta_R{}'$ is the azimuthal angle between the dipole position and the magnetic field direction; and δH_0 is the residual linewidth of the nitroxide without the presence of the Mn(II) ion as the relaxing spin. It is obvious that Eq. (5) is sensitive to orientation changes. Either the first term on the right side of the equation is small, yielding a signal governed by δH_0 (an unperturbed nitroxide spectrum), or the first term is large so that the other fraction of the signal is too broad to be observed. Thus, in any spin-labeled nucleic acid system containing manganese(II) ions that meets the prerequisites for a valid Eq. (5), the quenching of the ESR signal depends on the variables r and τ. The value of τ can be determined from proton-relaxation enhancement, and the Mn(II)–nitroxide distance can then be determined according to Eq. (5). So far, there has been only one reported study on the interaction of Mn(II) with spin-labeled nucleic acids (Vocel et al., 1975), and the data were not analyzed according to Leigh but by assuming that a dipole coupling with short correlation time causes only broadening of the nitroxide signal (Section II,C,1).

IV. MONITORING CONFORMATIONAL CHANGES IN NUCLEIC ACIDS BY ELECTRON SPIN RESONANCE

Several studies have been concerned with monitoring conformational properties of nucleic acids with spin labels or spin probes. So far, most of the investigations have been done on RNA systems, although some work on DNA systems, published to a large extent by Russian workers, has also been reported. A recent review by Dugas (1977) gives much attention to spin-labeled tRNA's (Dugas, 1976), and therefore the reader is referred to that article for in-depth treatment of results obtained with spin-labeled tRNA's by ESR. Since the ESR data frequently are presented here in the form of Arrhenius plots, a few general comments about such plots should be made. They have been used widely for studying structural changes occurring in nucleic acids as well as other systems. The reader should, however, be aware that the numerical values of the used correlation times are only approximate, in particular if the motion displays some anisotropic character as is certainly the case in double-stranded nucleic acid systems (Section III,B,1). When the transition is accompanied by a discrete, measurable change in the motion of the spin label, the state before and after the transition can be characterized by a set of closely related spin-label motions, each state containing nitroxide radicals with an average activation energy Δ. In such cases, T_m^{sp} corresponds reasonably well to T_m^{OD} (Section III,A and IV,A). If the transition is not accompanied by a discrete, measurable change in spin-label mobility, as in the case of spin-labeled tRNA's, only a breakpoint is observed, which is often referred to as T_{crit}. Breakpoint T_{crit} occurs at temperatures considerably lower than T_m^{OD}. It seems likely that the ESR spectrum reflects, in such instances, a mixture of quite distinct spin-label motions, in particular in the temperature ranges in which changes in the nucleic acid structure are believed to occur. This section deals with data obtained on RNA and DNA systems by ESR via nitroxide radicals covalently bound to the nucleic acid matrix as well as via nitroxide radicals acting as spin probes through their covalent linkage to small nucleic acid intercalating–binding molecules.

A. RNA Systems

The potential of order–disorder studies of spin-labeled $(A)_n$ and $(G)_n$ by ESR was reported by Smith and Yamane (1967, 1968). A detailed study on spin-labeled samples of $(A)_n$ was subsequently published (Bobst, 1971, 1972). Conformational properties of spin-labeled $(A)_n$ in aqueous solution were investigated by ESR, circular dichroism, and absorption spectroscopy (Bobst, 1971, 1972). The objective of the study was to examine conformational transitions of this polynucleotide as a function of proton uptake by adenine bases. The ESR data provided strong evidence that the transition from single strand to double strand takes place in three steps. In addition to the already known two forms of double-stranded $(A)_n$ in acidic solution, called A (fully protonated state) and B (partly protonated state), it was suggested that a third phase, possibly consisting of large aggregates, is involved in the transition of the less protonated double strands to those of complete protonation. Subsequently, the ESR data were confirmed by evidence from polarography which indicated the existence of three different $(A)_n$ forms at acidic pH values.

Leventahl et al. (1972) studied the temperature dependence of the ESR spectrum of spin-labeled UpU and $(U)_n$. The dinucleotide as well as the polynucleotide were considered to be completely modified with spin label V. In the case of spin-labeled UpU, which is a biradical system believed to have a spin label in the N-3 position of each of the two U residues, the ESR spectrum at 20°C is similar to that of spin label **V** except for some line broadening. However, at 85°C it is a typical spectrum of a biradical system with $\omega_e > a$ (ω_e is the frequency of spin exchange, and a is the nitrogen hyperfine coupling constant). This is a direct demonstration of the possibility of spin exchange interaction between spin labels residing on adjacent nucleoside residues. In the case of completely spin labeled $(U)_n$, i.e., modified to the extent of 92% as determined by spectrophotometric and ESR intensity studies, a triplet with strongly broadened components was observed. Above 50°C the spectrum changed into a singlet, the linewidth of which decreased upon further increase in temperature. At 85°C the frequency of spin exchange was calculated to be 7×10^8 sec^{-1}, a value close to that found for the completely modified UpU. This was interpreted as indicating that spin exchange in the polymer occurs between nearest neighbors. Finally, it should be noted that the authors encountered some experimental difficulties with this spin-labeled system due to the fact that hydrolysis of the spin label occurs already at pH 5.0 at high temperatures.

Temperature-dependent conformational transitions of various polynucleotides spin labeled with **VI** were systematically investigated (Pan and Bobst, 1973). In Fig. 11 the relationship between $-\log \tau$ and the reciprocal absolute temperature for spin-labeled $(A)_n$ and the spin-labeled $(A)_n \cdot (U)_n$ duplex is presented. A linear relationship that was independent of the adenylic acid concentration used was determined for spin-labeled $(A)_n$. This is

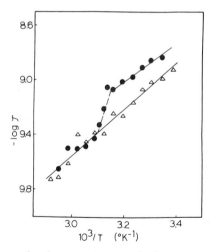

Fig. 11. Dependence of $-\log \tau$ on inverse absolute temperature for spin-labeled $(A)_n \cdot (U)_n$ (●-●) and spin-labeled $(A)_n$ (△-△) at a total polymer concentration of 8.9×10^{-4} and 3.58×10^{-4} M, respectively. [From Pan and Bobst (1973); reprinted by permission of publisher.]

good evidence for one independent mode of motion of the nitroxide radical with one temperature dependency. With the relationship $\tau = \tau^\circ e^{\Delta/RT}$ an activation energy Δ of 8.3 kcal/mole was calculated from the slope. On the other hand, the spin melting of the duplex revealed quite a different pattern. It is apparent from Fig. 11 that, besides an increase in the τ value by about 40% at 26°C, a sharp, almost stepwise change in the $-\log \tau$ values occurs in a certain temperature range that is flanked on both sides by linear segments. The temperature at the midpoint of the discontinuity of both segments is called the spin denaturation temperature (T_m^{sp}). From the two segments two activation energies that correspond to the single- and double-stranded systems can conceivably be calculated. Two comments are appropriate here. First, the activation energy Δ as well as the absolute tumbling times after T_m^{sp} is reached correspond to those of spin-labeled $(A)_n$. This is exactly what one would expect if the T_m^{sp} reflects the temperature-dependent transition from double to single strands. Second, the T_m^{sp} is very close to T_m^{OD} obtained from absorption melting curves measured under the same solvent conditions. (The reader is referred to Section III,A, in which the experimental uncertainty about exact T_m^{sp} values is analyzed.)

Results obtained with spin-labeled $(U)_n$ and the duplex consisting of spin-labeled $(U)_n \cdot (A)_n$ are reported in Fig. 12. As with spin-labeled $(A)_n$,

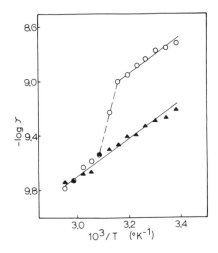

Fig. 12. Dependence of $-\log \tau$ on inverse absolute temperature for spin-labeled $(U)_n \cdot (A)_n$ (○-○) and spin-labeled $(U)_n$ (▲-▲) at a total polymer concentration of 11.6×10^{-4} and 5.8×10^{-4} M, respectively. [From Pan and Bobst (1973); reprinted by permission of publisher.]

one straight-line segment is obtained for spin-labeled $(U)_n$. It is interesting that the τ value for nitroxide radicals attached to $(U)_n$ is less than that for the same label bound to $(A)_n$. This observation corroborates models for $(A)_n$ and $(U)_n$ derived from other physicochemical techniques. Namely, $(U)_n$ is believed to have little or no base stacking at temperatures above 15°C, whereas $(A)_n$ shows strong stacking interactions at ambient temperature under the given salt conditions. As mentioned earlier the duplex formation slows down the mobility of the label considerably, but after T_m^{sp} is reached the τ values are identical for spin-labeled $(U)_n$ in the presence or absence of $(A)_n$. It is also noteworthy that the same T_m^{sp} values were found experimentally for both duplexes, spin-labeled $(U)_n \cdot (A)_n$ and spin-labeled $(A)_n \cdot (U)_n$. Caspary et al. (1976) essentially repeated the experiments given in Fig. 11 with an $(A)_n$ spin labeled with spin label **X** instead of **VI**. They observed a very similar dependence of $-\log \tau$ versus inverse absolute temperature as shown in Fig. 11. In the hands of both groups, the activation energy for the duplex system turned out to be lower than that for the single-stranded $(A)_n$. It is interesting that both groups obtained an Arrhenius plot of the spin melting of $(A)_n \cdot (U)_n$ that is characterized by a stepwise change in the τ values over a narrow temperature range (Fig. 11). In case such a stepwise change occurs during spin melting, it was found to be appropriate to define a new term, T_m^{sp} (Pan and Bobst, 1973), to characterize the transition and not to refer to a critical temperature, T_{crit}. The term T_{crit} was used by Zavriev et al. (1974a) to describe the spin melting of spin-labeled RNA systems that are characterized by Arrhenius plots with a single discontinuity formed by two linear segments with different activation energies, and T_{crit} was found to be usually substantially smaller than T_m^{OD} (see next paragraph). Unfortunately, Caspary et al. (1976) discuss their spin-melting data by referring to critical temperatures, although their spin-melting data for the duplexes display a stepwise change in the τ values over a narrow temperature range. It should be noted that stepwise changes in τ values occurring during spin melting of nitroxide-labeled nucleic acids have not often been observed. Incidentally, Caspary et al. (1976) found it surprising that $(U)_n$ could be spin labeled to a significant extent [nitroxide/base = 1/600 (Pan and Bobst, 1973)]. In view of the careful study by Singer (1975) (see Section II,A), which undermined traditional thinking about the reactivity of uridylic and thymidylic acids toward alkylating agents, successful alkylation of $(U)_n$ under the reaction conditions of Bobst (1972) should no longer be surprising. Actually, the conditions given in Section II,A,1 allowed the author in subsequent years to increase the ratio of 1/600 to 1/75–1/100 for $(U)_n$ as well as for some other polynucleotides.

Zavriev et al. (1973) examined spin melting of 16 S RNA spin labeled with **IX** at various ionic strengths. An Arrhenius plot gave two segments intersecting at a point referred to as T_{crit}. A linear relationship between T_{crit} and the logarithm of the ionic strength of the solution was observed. The authors suggested that T_{crit} reflects a critical state temperature which precedes melting of the 16 S RNA structure since T_{crit} is considerably smaller (by 14°–23°C depending on ionic strength) than T_m^{OD}. Zavriev et al. (1974a) also reported spin-melting data obtained on synthetic polyribonucleotides spin labeled with **IX**. An Arrhenius plot of such a system gave only two linear portions with a breakpoint referred to as T_{crit}, whereas monitoring the melting of such duplexes with the spin label **VI** (Pan and Bobst, 1973) or **X** (Caspary et al., 1976) gave a sharp, almost stepwise change in the tumbling time values, and the midpoint of the discontinuity of both segments was called T_m^{sp} (Pan and Bobst, 1973). The value of T_m^{sp} was found to be very close to that of T_m^{OD} obtained from absorption melting curves (Pan and Bobst, 1973). It is interesting that the spin-labeled $(A)_n \cdot (U)_n$ of Zavriev et al. (1974a) has a T_m^{OD} of 52.5°C and a T_{crit} of 42°C. One likely explanation for this different spin-melting behavior could be the difference in " leg " size between label **IX** and the two labels **VI** and **X**. It is apparent from the chemical structure of these reagents that the nitroxide moiety is much closer to the duplex with **VI** and **X** than with **IX**.

Temperature-dependent conformational transitions of spin-labeled $(U)_n$ (nitroxide/base $= 1/600$) are shown in Fig. 13, where an interesting feature is noticeable (Pan and Bobst, 1974). Plotting $-\log \tau$ versus the inverse absolute temperature reveals two segments with different slopes and a discontinuity caused by an increase in τ over a narrow temperature range. The measured

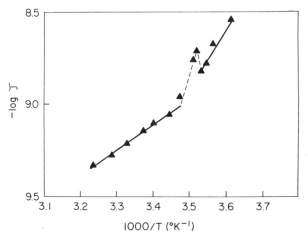

Fig. 13. Dependence of $-\log \tau$ on inverse absolute temperature for spin-labeled $(U)_n$ in 0.42 M CsCl, 8×10^{-5} M EDTA, 8×10^{-3} M sodium cacodylate, pH 6.6. [From Pan and Bobst (1974); reprinted by permission of publisher.]

increase in τ at $T_{\text{intermed}}^{\text{sp}}$ reflects the value for the lower limit of the correlation time of the intermediate form if the kinetics of the interconversions allow only a small steady-state concentration of the intermediate form. The activation energies are about 14 kcal/mole for the ordered form and 6.5 kcal/mole for the random coil, the latter value being the same as that obtained from Fig. 12 for the single-stranded $(U)_n$. As in the case of the spin-labeled duplexes in Figs. 11 and 12, it is believed that the segments correspond to two different modes of motion of the spin label, each macromolecular structure allowing the nitroxide radical to tumble with a particular activation energy Δ. The increase in τ at $T_{\text{intermed}}^{\text{sp}}$ suggests that the order–disorder transition in $(U)_n$ is not a simple process involving only two states as was concluded from UV melting studies. Some evidence for the formation of an intermediate state during the melting of ordered $(U)_n$ has also come from temperature-jump kinetic and viscosity experiments. It has been suggested that the intermediate state could consist of a multihairpin system with a compact, branched structure. It is conceivable that such a proposed structure would display an increased microviscosity as evidenced by the decreased local mobility of the nitroxide radical.

In the first spin melting studies of tRNA's (Hoffman *et al.*, 1969; Schofield *et al.*, 1970; Kabat *et al.*, 1970) the discontinuity usually observed for the two segments was not characterized by a substantial change in the tumbling times at the breakpoint. This observation turned out to be valid for all the tRNA's subsequently studied (McIntosh *et al.*, 1973; Sprinzl *et al.*, 1974; Caron and Dugas, 1976b). A typical set of ESR spectra of spin-labeled tRNA's is illustrated in Fig. 14. These are the kind of ESR spectra usually observed at ambient temperature with tRNA systems chemically modified with a spin label as shown in Fig. 2. For spin-labeled tRNA^{Glu}, $\text{tRNA}_f^{\text{Met}}$, and tRNA^{Phe} the dependence of the spin label correlation time on absolute temperature is reported in Fig. 15. It is apparent that over the temperature range $5°–80°C$ discontinuities in the slope as well as differences in slope values for the segments occur. Such Arrhenius plots revealed an interesting feature (Dugas, 1977). The tRNA's having a spin label at the s^4U_8 position show two thermal transitions, one at $32°C$ and the other at $53°C$. In contrast, spin labels in the anticodon of tRNA's or in regions other than the s^4U_8 position show only one transition around $52°–54°C$.

B. DNA Systems

Only a few detailed studies have been reported on spin-labeled DNA systems. Arthyuk *et al.* (1972) studied the properties of calf thymus DNA and phage T2 DNA spin labeled with the piperidine bromoacetoxy VII. Both spin-labeled samples gave similar ESR spectra with a slightly immobilized nitroxide radical ($\tau \simeq 1 \times 10^{-10}$ sec). A pH titration study on spin-

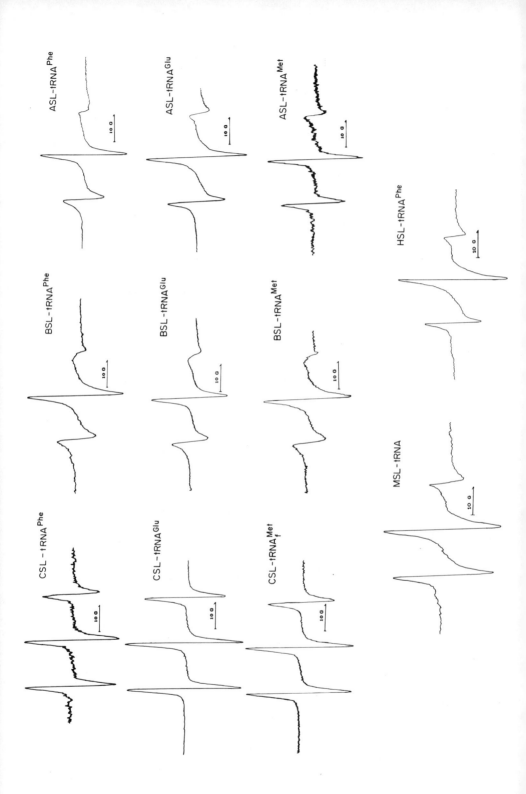

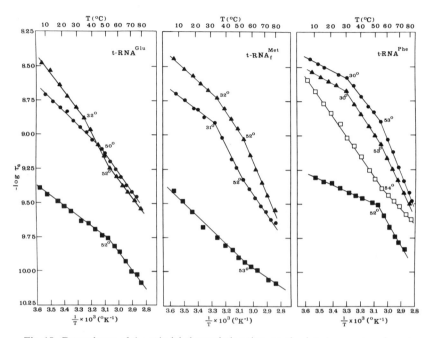

Fig. 15. Dependence of the spin-label correlation time on absolute temperature for spin-labeled tRNA^{Glu}, tRNA$_f^{Met}$, and tRNA^{Phe} in 0.02 M Tris–HCl buffer (pH 7.5), 10 mM MgCl$_2$. Key: ●–●, spin labeled with pyrroline anhydride **XI**; ▲–▲, spin labeled with pyrrolidine bromoacetamide **III**; ■–■, spin labeled with piperidine carbodiimide **V**; □–□, spin labeled with pyrrolidine hydroxysuccinimide **XII**. [From Caron and Dugas (1976b); reprinted by permission of publisher.]

labeled T2 DNA revealed that lowering the pH results in a sharp increase of the spin-label mobility at pH 2.95. The change in spin mobility seemed to correspond to the spectrophotometric change known to occur with unmodified T2 DNA between pH 2.55 and 2.75.

Structural transitions of calf thymus DNA, rat spleen DNA, and spleen DNA of rats infected with erythromyclosis, all of which were spin labeled with the piperidine ethyleneimine **IX**, were analyzed by ESR (Mil' *et al.*, 1973). An Arrhenius plot showed the existence of several linear regions. It was suggested by the authors that structurally bound water plays an impor-

---

Fig. 14. The ESR spectra of spin-labeled tRNA's at ~ 22°C in 0.02 M Tris–HCl (pH 7.5) containing 10 mM MgCl$_2$. CSL-tRNA, tRNA spin labeled with piperidine carbodiimide **V**; BSL-tRNA, tRNA spin labeled with pyrrolidine bromoacetamide **III**; ASL-tRNA, tRNA spin labeled with pyrroline anhydride **XI**; MSL-tRNA, tRNA modified at the 3'-CCA end, where a morpholine spin-label derivative is believed to have formed; HSL-tRNA, tRNA spin labeled with pyrrolidine hydroxysuccinimide **XII**. [From Caron and Dugas (1976a); reprinted by permission of publisher.]

tant role in these DNA transitions. An intriguing reproducible transition at about 20°C takes place in calf thymus and rat spleen DNA that is characterized by an abrupt reduction in the rotational mobility of the spin label at this temperature. It is interesting that surface probing of DNA with TEMPOL (Sukhorukov and Kozlova, 1970) also revealed a low-temperature transition at about 20°C (for more details see Section II,C,2).

More recently, spin-labeled intercalating agents were used by two research groups primarily to probe DNA structures (Sinha and Chignell, 1975; Piette, 1974; Hong and Piette, 1976). Sinha and Chignell (1975) reported that in the presence of calf thymus DNA the ESR spectra of the aminoacridine spin probes **XVII** and **XVIII** were characteristic of highly immobilized nitroxide radicals. Such a broad asymmetric ESR spectrum is shown in Fig. 16 for a solution of spin probe **XVII** containing calf thymus DNA. A close examination of the spectrum reveals the presence of some probes that are still unbound (from Figs. 10 and 17 the peak positions of the fast-tumbling radicals are readily apparent). It is obvious from Fig. 16 that it is possible to monitor the concentration of the unbound spin probes by following the amplitude of the low-field line. This allowed the authors to determine the binding of probes **XVII** and **XVIII** over a wide range of nucleic acid building block to spin probe ratios and to analyze the binding of the intercalating agents to DNA with Scatchard plots. The results suggested the presence of two independent binding sites on DNA. Furthermore, an increase in ionic strength seemed to cause a decrease in the affinity of the

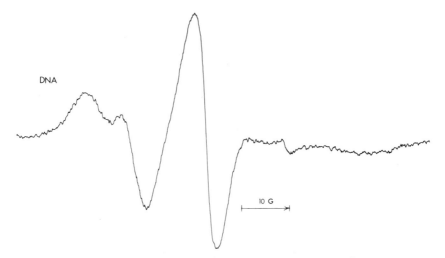

Fig. 16. The ESR spectrum of aminoacridine spin probe **XVII** (5×10^{-5} M) bound to calf thymus DNA (0.5 mg/ml) in the presence of sodium phosphate buffer (5 mM), pH 7.4. [From Sinha and Chignell (1975); reprinted by permission of publisher.]

aminoacridine spin probes for the high-affinity binding sites. The thermal melting curves of calf thymus DNA in the presence of the spin probes were determined by optical density at 260 nm or from the amplitude of the low-field line of the unbound probes. It was found that the T_m values determined by both approaches were in reasonably good agreement. Piette (1974) reported some preliminary melting studies with the ethidium bromide spin probe **XVI** and subsequently the data were published (Hong and Piette, 1976). In Fig. 17 the temperature dependence of the ESR spectrum of the ethidium bromide **XVI**–DNA complex is shown. No significant spectral changes are observed between 30° and 65°C. A dissociation of the probes becomes apparent above 65°C, resulting in an increase in peak b at the expense of peak a. An abrupt change occurs upon increasing the temperature from 73° to 74°C. The spectrum at 78°C is close to T_d, the dissociation transition temperature, and at temperatures above 90°C all the spin probes are believed to be dissociated in the system. A large difference between the T_m measured at 260 nm for the midpoint temperature of the unlabeled ethidium bromide–DNA complex and the T_d of the spin probe **XVI**–DNA complex was found. It is believed that the difference is due to a difference between labeled and unlabeled ethidium bromide in the extent of binding to DNA. Unfortunately, because of experimental limitations in the optical methods used by the authors, it was not possible to directly compare T_d and T_m for the spin probe **XVI**–DNA complex. However, the smooth cooperative dissociation of the spin probe from the complex as detected by ESR strongly suggests that most, if not all, of the labels were intercalated. Random nonspecific binding would not be expected to yield a cooperative dissociation of the labels upon thermal denaturation of the DNA. As earlier suggested by Pan and Bobst (1973) for a different spin-labeled nucleic acid system, the same temperature-dependent structural changes of DNA are monitored by UV and ESR spectroscopy, hypochromicity being observed on the one hand and spin probe mobility on the other. Finally, it is interesting to note from Fig. 17 that, upon renaturation of a melted ethidium bromide **XVI**–DNA complex, an ESR spectrum that suggests nonspecific binding and incomplete renaturation of DNA is obtained.

V. MOLECULAR ASSOCIATION STUDIES WITH SPIN-LABELED NUCLEIC ACIDS

A. Interaction with Small Molecules

So far, there have been a few studies on the interaction of spin-labeled nucleic acids with small, positively charged molecules. The same activation energies for ordered $(U)_n$ forms were observed in a buffer system containing either 5×10^{-4} M spermidine, 0.01 M phosphate, pH 7.4, or 0.42 M CsCl,

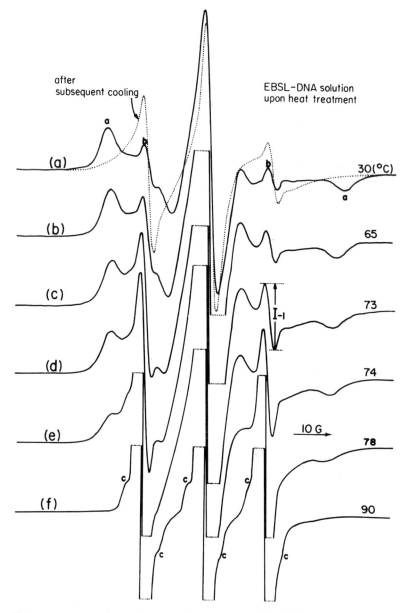

Fig. 17. Temperature dependence of the ESR spectrum of the DNA–EBSL (**XVI**) complex. The spectra (a)–(f) represent a composite consisting of bound probes designated by a, and unbound probes, b, at different temperatures. The peaks at c are due to ^{13}C peaks resulting from a natural abundance of this isotope on the carbon adjacent to the NO moiety. Dotted line indicates spectrum obtained after subsequent cooling from the state of complete dissociation at 90°C to the original temperature, 30°C. Final concentration of probe **XVI**, 1.33×10^{-4} M; that of DNA, 8.8×10^{-3} M. Dye/phosphate ratio, approximately 1.5×10^{-2} in 0.1 × SSC. [From Hong and Piette (1976); reprinted by permission of publisher.]

8×10^{-5} M EDTA, 8×10^{-3} M sodium cacodylate, pH 6.6 (Pan and Bobst, 1974). However, in the presence of spermidine the absolute mobility of the spin label in the ordered $(U)_n$ structure was slowed considerably, suggesting a decreased local mobility of the hairpin. This is not unexpected, since it is likely that the three positive charges on the oligoamine interact to a greater extent than a monovalent cation (Pan and Bobst, 1974). The ESR measurements on $(A)_n$ and $(dUfl)_n$, both spin labeled with compound VI, were carried out in the presence or absence of the amino acid L-Lys, the dipeptide L-Trp-L-Lys or the tripeptide L-Lys-L-Tyr-L-Lys, but detectable signal changes were observed only in the presence of the tripeptide (Pan, 1975). Sakai and Cohen (1976) explored the postulated binding site of spermidine to tRNA using tRNA labeled at s^4U residues with compound III. The ESR spectrum of this spin-labeled tRNA broadened and the signal heights decreased as the spermidine/tRNA ratio was increased. Spermidine, when added to spin-labeled tRNA at ratios greater than about 15, appeared to restrict the rotation of the spin label nearly twofold.

Zavriev et al. (1976) used DNA modified with the ethyleneimine spin-label derivative IX and the intercalating dyes ethidium bromide and acriflavine to study DNA–dye interactions. The oligopeptide hexalysine was used as a control, since it binds to DNA without the possibility of intercalation. An analysis of adsorption isotherms revealed that the nitroxide radicals are distributed randomly along the DNA and that each spin label covalently bound to DNA prevents three base pairs from binding with the dyes. An examination of the mean distance between the labels in the presence of ethidium bromide, acriflavine, or the oligopeptide hexalysine revealed an increased distance between the nitroxide radicals upon addition of the dyes.

B. Interaction with Polypeptides and Proteins

Attempts to clarify the question of whether conformational changes of DNA occur during the formation of a complex between RNA polymerase from E. coli and T2 DNA also were made with the aid of spin labeling (Kamzolova et al., 1973; Sukhorukov et al., 1973). For this purpose T2 DNA with three to four single breaks was spin labeled with the pyrroline imidazole VIII, and it was assumed that the nitroxide radical was attached to the OH groups of glucose (in T2 DNA 80% of the 5-hydroxymethylcytidine is glucosylated). Such labeling is believed to have little effect, if any, on the physical properties of the secondary structure. It also was noted that the labeling does not affect the activity of the DNA as template with RNA polymerase. Namely, the incorporation of $[^{14}C]UTP$ into RNA was 1124 counts per 100 sec with an unmodified T2 DNA and 1164 counts per 100 sec with spin-labeled T2 DNA as template under the same conditions of RNA synthesis. The ESR measurements revealed that, upon addition of RNA polymerase to spin-labeled T2 DNA at 37°C, the spin label becomes more mobile, whereas formation of the RNA polymerase–T2 DNA complex at 2°C results in no change of the ESR spectrum of the nitroxide label. By taking into account results obtained from nitrocellulose filtration of ^{14}C-labeled T2 DNA–RNA polymerase complexes formed at various temperatures, the authors interpreted the ESR data as suggesting that the DNA contained in a complex with RNA polymerase undergoes a conformational change above 20°C. The authors believe that the spin label reflects a transition of DNA that allows "passive" DNA to be converted to "active" DNA (Sukhorukov, 1973). It is interesting that the ratio of spin-labeled bases in contact with RNA polymerase was about 1% (Sukhorukov, 1973), suggesting that the local conformational changes taking place at the DNA–enzyme binding sites initiate a cooperative structural change of the whole duplex. This increased spin mobility (by a factor of about 2, which gives the ESR spectrum of an essentially free spin label) upon complex formation is a provocative finding. However, since rapid hydrolysis of the T2 DNA containing three to four breaks by a nuclease at 37°C could also explain the increase in spin mobility as well as the small retention of ^{14}C-labeled T2 DNA on nitrocellulose, one must be cautious in interpreting the results obtained with T2 DNA and RNA polymerase.

Zavriev et al. (1974b) studied DNA–RNA polymerase interaction with a different template and a different spin label. They used DNA and RNA polymerase from E. coli, the former having been spin labeled with the ethylene IX. In contrast to the results with spin-labeled T2 DNA, the RNA polymerase activity as measured by its ability to incorporate radiolabeled nucleotides into RNA was strongly sensitive to the presence of the spin label. This should not be surprising since spin labeling of E. coli DNA does not offer the opportunity to label glucose residues far removed from the helix axis of the duplex. A spin label/base pair ratio of 50–60 reduced the biosynthesis of RNA by more than a factor of 2. It is, however, interesting that 1–2% bound label IX did not strongly affect the physical properties of a double-helical structure. Incidentally, it was with this label attached to calf thymus DNA that a reproducible transition was shown to occur in DNA at about 20°C (Section IV,B) (Mil' et al., 1973). The loss of RNA polymerase activity when E. coli DNA spin labeled with IX was used as template can be explained by the following two mechanisms (Zavriev et al., 1974b). (a) The presence of the label hinders the motion of the polymerase along the template in a purely mechanical way. (b) The label prevents a B → A transition of the DNA and, since it is believed that RNA polymerase interacts with the DNA in its A form, the suppression of a B → A transition prevents transcription. It is apparent that these are extremely complex systems, and considerably more work is needed to obtain a better understanding of RNA polymerase–template interactions through ESR, but the potential of using spin labels to gain some insight into these processes should be evident.

Figure 18 is an example illustrating the advantage of using ESR to monitor nucleic acid–protein interactions. For this purpose, a simple model system consisting of $(U)_n$ or spin-labeled $(U)_n$ and the polypeptide $(Lys)_n$ was selected, and the ESR results (Fig. 18A) were compared with those obtained by circular dichroism or optical density spectroscopy. The studies were carried out not only in a buffer system that is convenient for OD or CD measurements, i.e., contains only an aqueous solution, but also in a buffer containing glycerol that can significantly protect a protein from denaturation. The titration of spin-labeled $(U)_n$ by $(Lys)_n$ is monitored by ESR in two buffer systems, one of which contains glycerol (Fig. 18A). The ESR spectral changes of spin-labeled $(U)_n$ upon addition of small aliquots of $(Lys)_n$ are followed by plotting h_{+1}/h_0 versus $[Lys]/[U]$. The h_{+1}/h_0 ratio decreases in a nonlinear manner, and at a point in the vicinity of electrostatic equivalence of the polymer complex no further change in the ratio h_{+1}/h_0 is observed. In the presence of glycerol, there is a slight delay in reaching the point at which the ratio remains constant. However, if the same study is done by CD or OD measurements the results show a strong dependence on the buffer system used (Fig. 18B). Earlier studies showed that addition of $(Lys)_n$ to a $(U)_n$ solution containing no glycerol results in a maximum of the CD band at 265 nm ($\Delta\varepsilon_{265\,nm}$, left ordinate) at a $[Lys]/[U]$ ratio of about 0.7. Such a titration also reveals the formation of a highly turbid solution when electrostatic equivalence is achieved by OD measurements at 340 nm (right ordinate). On the other hand, the presence of glycerol in the buffer has a completely different effect. From the beginning the OD at 265 nm increases, indicating the formation of turbidity even at an early stage of the titration, and the CD shows only a small decrease in $\Delta\varepsilon_{265\,nm}$ upon addition of $(Lys)_n$. Thus, this example should make it clear that the magnetic resonance approach described here is of great value when one is studying interactions in opalescent or turbid solutions.

A convenient procedure for determining the relative affinity of proteins for nucleic acids by ESR was reported (Bobst and Pan, 1975). The assay was tested by determining the affinity of T4 gene 32 protein (P32), an unwinding protein, for various single-stranded nucleic acids. This was achieved by carrying out competition experiments between P32 and spin-labeled and unlabeled polynucleotides. For this purpose, a limiting amount of protein was complexed with a slight excess of spin-labeled nucleic acid. Upon addition of an unlabeled polynucleotide with, for instance, a strong affinity for P32 in an equimolar amount to spin-labeled nucleic acid, the latter would be completely displaced from the complex. The advantage of using ESR is that such

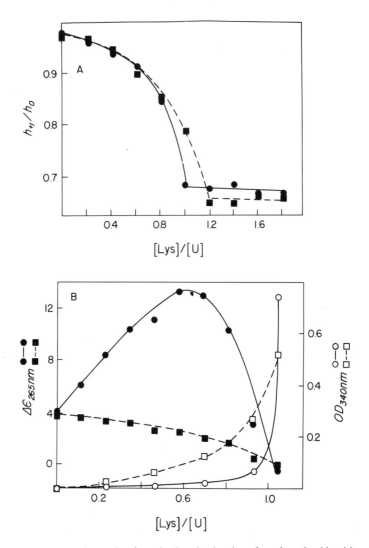

Fig. 18. A comparative study of monitoring the titration of a polynucleotide with poly-L-lysine by ESR, CD, and UV spectroscopy in two different buffer systems at 24°C. Key: ○ or ●, in 0.001 M Na-cacodylate, 0.005 M NaCl, pH 7; □, or ■ in 0.02 M Tris, pH 8.1, 0.05 M NaCl, 0.001 M EDTA, 10% (w/v) glycerol. (A) ESR titration of 23.4 nmoles of $(U)_n$ spin labeled with piperidine iodoacetamide **VI** to an extent of about one nitroxide per 100 bases, with aliquots of a 2.36×10^{-3} M poly-L-lysine solution. (B) CD (left ordinate) and UV (right ordinate) titration of 396 nmoles $(U)_n$ with aliquots of a 2.36×10^{-2} M poly-L-lysine solution.

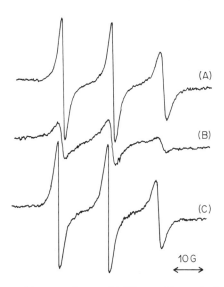

Fig. 19. An ESR competition experiment with T4 gene 32 protein (P32) and $(RUGT,U)_n$ (Section II,B) and $(dUfl)_n$. (A) 10.5 nmoles $(RUGT,U)_n$; (B) $(RUGT,U)_n$–P32 complex [nucleotide/P32 = 9.5; 10.5 nmoles $(RUGT,U)_n$]; (C) addition of 10.6 nmoles $(dUfl)_n$ to $(RUGT,U)_n$–P32 complex characterized under (B). [From Bobst and Torrence (1978) reprinted by permission of publisher.]

a displacement can be monitored directly, as is shown in Fig. 19. By using this strategy, it was possible to show unequivocally that a significant difference must exist in the binding constants of various P32–nucleic acid complexes. The following P32 affinity relationship was determined (Bobst and Pan, 1975):

$$(dT)_n > (dUfl)_n > (A)_n > (U)_n \approx (dA)_n$$

A typical displacement reaction is shown in Fig. 19, where $(RUGT, U)_n$ obtained via enzymatic synthesis (Section II,B) was used to corroborate the results obtained earlier (Bobst and Pan, 1975) with $(U)_n$ spin labeled site nonspecifically with the piperidine iodoacetamide **VI**. Spectrum A of Fig. 19 is the ESR spectrum of $(RUGT, U)_n$ in its single-stranded form, and spectrum B shows the effect of P32 on the ESR spectrum of $(RUGT, U)_n$. It is apparent from spectrum C that $(RUGT, U)_n$ in the $(RUGT, U)_n$–P32 complex is completely displaced by $(dUfl)_n$. Incidentally, the detection of a definite degree of nucleic acid specificity in the case of P32 was unexpected, since it generally was assumed that unwinding proteins belong to the class of "nonspecific site binding" proteins. In the meantime, however, it has been recognized that the standard error of the binding parameter measurements could be large enough to give parameter differences for different polynucleotides, a situation which would explain the differences in the relative affinity of various polynucleotides for P32 as determined by ESR competition experiments (Kelly *et al.*, 1976).

C. Interaction with Mammalian Cell Monolayers

The interaction of Vero cell monolayers with the nucleic acids $(A)_n$, $(dUfl)_n$, $(U)_n$, and $(A)_n \cdot (U)_n$, all of which were spin labeled with the piperidine iodoacetamide **VI** to an extent of about one nitroxide per 100 bases, was examined by ESR at various temperatures (Kouidou *et al.*, 1978). Vero cells CCCL-81, American Type Culture Collection, were grown in monolayers at 37°C, and the procedure outlined in Fig. 20 was used. Cell monolayers for the ESR cavity were obtained by placing one or two quartz plates (carefully treated with chromic acid cleaning solution for at least 16 hr, rinsed with distilled water, and then boiled for a half-hour in distilled water) into a plastic Falcon petri dish with 15 ml of MEM (minimal essential medium containing Hank's balanced salt solution with penicillin and streptomycin), and then approximately 5×10^5 cells were suspended in the solution. After 1 day of incubation, about 2.5×10^5 cells were attached to the surface of each quartz plate (Fig. 20A). For some experiments, twice as many cells were added, resulting in 5×10^5 cells per quartz plate. The cell viability was always 95% or better. On inspection under a microscope, the cells grown on quartz plates showed the same shape as those grown on

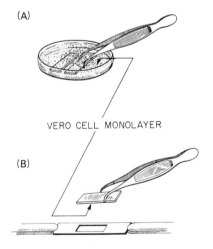

Fig. 20. Procedure illustrating the approach taken to prepare Vero cell monolayers in a suitable form for ESR measurements in a multipurpose ESR cavity.

traditional (plastic) surfaces. A quartz tissue cell with cavity well dimensions of 23 × 7 × 0.5 mm was used for the ESR measurements (Fig. 20B). A fresh aliquot of culture medium was preincubated under constant CO_2 pressure for about 2 hr and then utilized to fill the cavity. Small aliquots of spin-labeled nucleic acids were then added to the well before it was closed with the quartz plate populated with the Vero cell monolayer.

The time at which the cells were brought into contact with the cavity well medium was taken as $t = 0$. The temperature of the experiments was not controlled as previously described in Section III,A since a tissue cell does not fit into the Dewar of a variable-temperature control unit. In addition, it was inadvisable to expose the rather delicate quartz plate positioned on top of the well to an intense nitrogen stream. Thus, a completely different approach for proper temperature control was adopted. This approach was successful because of the location of the Varian E-4. The E-4 unit was placed in a small, windowless, well-insulated room that essentially served the purpose of a walk-in incubator. With appropriate cooling and heating, the temperature of the room was regulated carefully. A thermometer placed in the immediate vicinity of the ESR cavity was used to continuously monitor the temperature, which remained constant within $\pm 1°C$. It also was established with two thermocouples that the temperatures inside and in the immediate vicinity of the cavity in tune position differed by less than 0.2°C.

Figure 21 shows the kinetics of the normalized center-line height of spin-labeled $(A)_n$ exposed to about 50% of the coverplate containing approximately 5×10^5 Vero cells per plate at 20° and 30°C, respectively. For the normalization procedure, an external Cr^{3+} standard attached to the back of the tissue cell was used. The error limits of the normalized center heights, a and b in Fig. 21, were calculated with the t test for confidence limits of 98%. It is apparent that at the lower temperature there is no decrease in the spin concentration over a 1-hr period. At 26°C, however, the spin concentration remains unchanged only for about 15–20 min before decaying for the next 45 min. A similar observation was made when the reaction was followed at 20° and 30°C with 28 nmoles of spin-labeled $(A)_n$ in the presence of 1.2×10^5 exponentially growing cells. The line shapes of all the ESR signals were examined by calculating the ratios h_0/h_{-1} and h_0/h_{+1}, and it was found that these parameters were not affected by the presence of Vero cells. No change in the h_0/h_{-1} and h_0/h_{+1} ratios during the course of these experiments implies that the mobility of the nitroxide radicals covalently attached to the nucleic acids does not undergo a change in the motional narrowing region time scale. The signal decrease can be attributed either to a decrease in the concentration of nitroxide radicals resulting from reduction by *in vivo* cell respiratory activity at the membrane surface or to considerable broadening of the signal caused by the formation of strongly immobilized nitroxide

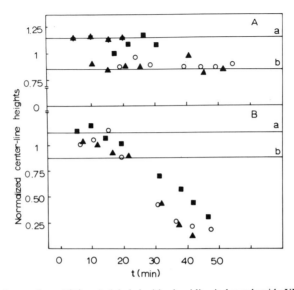

Fig. 21. Interaction of $(A)_n$ spin labeled with piperidine iodoacedamide **VI** to an extent of about one nitroxide per 100 bases at 20°C (A) and 26°C (B) with Vero cells. Approximately 2.5×10^5 exponentially growing cells were in contact with either 19 nmoles (A) or 37 nmoles (B) of spin-labeled $(A)_n$. Error limits of the normalized center-line heights are indicated by **a** and **b**.

radicals. Although the exact cause of the decrease in ESR signal intensity is not known at present, it should be stressed that the actual cause does not have to be known in order to state that a definite temperature barrier in the interaction pattern exists between spin-labeled nucleic acids and Vero cells. By examining all the results obtained with the spin-labeled nucleic acids listed above, it was concluded that the phenomenon is independent of the nature of the single strands tested. In addition, the duplex showed an ESR signal decay at 26°C, which occurred after 35–40 min of contact between cavity well medium and Vero cell surface instead after 15–20 min as observed with the single strands. Finally, a set of control experiments with TEMPOL (Section II,C,2) and nitroxide radicals covalently bound to Sepharose 4B made it possible to conclusively establish that the temperature barrier phenomenon is nucleic acid related.

At present, a definite explanation for the observed temperature effect existing in the interaction between spin-labeled nucleic acids and cell monolayers cannot be given in molecular terms. The following two models give a reasonable explanation of the data. Either the spin-labeled polynucleotides interact with the cell surface at the higher temperatures due to a temperature-dependent phase transition of the interacting membrane components or the cell surface activates products at the higher temperatures which are released from the cell surface to interact with spin-labeled nucleic acids.

D. Bioassay for Detecting $(A)_n$ Tracts in RNA's

Polymer $(A)_n$ tracts are found in eukaryotic messenger RNA's and in 70 S RNA of tumor viruses. These tracts are located at the 3'-OH end of the polynucleotide and can be detected by means of hybridization experiments. Usually, hybridization is detected by isolating the hybrids by filtration onto membranes or fiberglass or by hydroxyapatite chromatography.

A novel bioassay was reported for detecting $(A)_n$ tracts in RNA's with ESR (Bobst, 1974; Bobst et al., 1975). The assay is based on monitoring the complex formation between the spin-labeled $(U)_n$ or spin-labeled $(dUfl)_n$ (nitroxide/nucleotide $\simeq 1/75$) and the $(A)_n$ tail by ESR. As shown in Section III,B,1, hybridization of a spin-labeled single-stranded polynucleotide with a complementary single strand has a pronounced effect on the ESR line shape. The mobility of the spin label decreases, and a characteristic anisotropic motion of the nitroxide radical becomes apparent. The observed significant line shape change allows one to directly follow the titration of spin-labeled $(U)_n$ or $(dUfl)_n$ with $(A)_n$. The titration is followed by plotting the ratio h_{-1}/h_0 as a function of the number of nanomoles of $(A)_n$ added. Figure 22 shows two typical examples of such titration experiments. In the first case (Fig. 22a), 2.8 nmoles of spin-labeled $(dUfl)_n$ is titrated with a mixture of R17 RNA and $(A)_n$ at a molar ratio of 9 : 1. The R17 RNA was added as a control since it contains no known long $(A)_n$ tracts and therefore should not interact with the spin-labeled polynucleotide. Figure 22a shows that h_{-1}/h_0 first decreases linearly and then remains constant. The intercept of the two resulting line segments corresponds to 2.8 nmoles $(A)_n$, indicating that R17 RNA does not bind to the spin-labeled polynucleotide. In Fig. 22b, the titration experiment is repeated in the absence of R17 RNA, and the same titration result is observed as in Fig. 22a. Thus, it was firmly established that for spin-labeled $(dUfl)_n$ as well as for spin-labeled $(U)_n$ the ratio of the two hyperfine components remains constant when equimolar amounts of the components of the hybridization product are present, and RNA's containing no long $(A)_n$ stretches do not interfere with the assay.

In order to trap an $(A)_n$ tail in a viral or eukaryotic mRNA, one must ascertain that the tail is freely accessible for hybridization, i.e., does not interact strongly, if at all, with other regions of the RNA molecule. That this is the case was demonstrated by CD for rabbit globin mRNA, of which 20% or less did not bind to oligo cellulose (dT). Changes in ESR line shape of spin-labeled $(dUfl)_n$ upon addition of small amounts of rabbit globin mRNA revealed the presence of a sharp break point (Bobst et al., 1975). Although

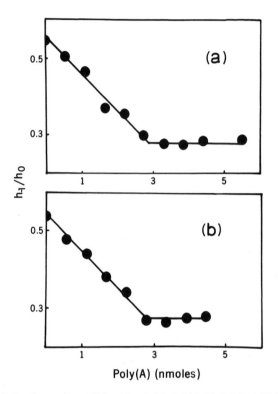

Fig. 22. ESR titration studies of $(A)_n$ with spin labeled $(dUfl)_n$ in 0.01 M Tris–HCl, 0.01 M NaCl, 0.001 M EDTA, pH 7.5, at 26°C. (a) Variation of the ratio of high-field (h_{-1}) to center-field (h_0) hyperfine components of 2.8 nmoles of spin-labeled $(dUfl)_n$ upon titration with a mixture consisting of R17 RNA and $(A)_n$ at a molar ratio of 9 : 1. (b) Variation of the ratio of h_{-1} to h_0 of 2.8 nmoles of spin-labeled $(dUfl)_n$ upon titration with $(A)_n$. [From Bobst *et al.* (1975); copyright © 1977 by the American Association for the Advancement of Science.]

the spin assay for detecting $(A)_n$ tracts in RNA's gives satisfactory results with spin-labeled $(U)_n$ or $(dUfl)_n$, it was advantageous to use the polynucleotide analogue $(dUfl)_n$ for titration experiments because the analogue is more stable toward hydrolysis than $(U)_n$. From the titration plot, it was determined that rabbit globin mRNA has an average $(A)_n$ content that can be hybridized with spin labeled $(dUfl)_n$ of about 6.6%. Circular dichroism and gel electrophoresis studies gave a slightly higher $(A)_n$ content of about 8%. It seems likely that a steric constraint effect present in a hybridization assay gives a systematically lower $(A)_n$ content in viral or eukaryotic mRNA. [This comment also should be applicable to hybridization results obtained with radioactively labeled $(U)_n$.] Finally, it should be noted that the bioassay for detecting $(A)_n$ tracts in RNA's requires no more than 35 μg of

(A)$_n$-containing RNA and can be used for direct quantitation of any (A)$_n$-containing RNA, provided that correction for any systematic errors in the bioassay is made.

VI. CONCLUSION

In this chapter, we have discussed the practical and theoretical aspects of nucleic acids that are spin labeled or interact with spin probes. Because of the complexity of nucleic acid chemistry progress in the use of nitroxide labels or probes to study the structure and function of nucleic acids has not been as rapid as in other areas of biochemistry. However, the advancement over the last several years has firmly established the potential of applying the spin-label technique to nucleic acids. It has been shown conclusively that nitroxide radicals can serve as sensitive microscopic seismic stations to detect subtle conformational changes in nucleic acids otherwise often overlooked by more conventional techniques. Despite the success of the method, the reader should keep in mind that the technique is limited by the fact that it makes use of a reporter group of a finite size, which can cause undesirable perturbations. Therefore, it is imperative to take some precautionary steps including measurement of the structural properties and, if possible, the biological activity of the spin-labeled systems in order to minimize the chance of the information obtained being artifactual due to the introduction of the nitroxide radical. Recently, it was discovered that a spin-labeled nucleic acid building block analogue could be synthesized and enzymatically incorporated into a nucleic acid matrix. The novel spin-labeled polynucleotide produced shows a great deal of promise for acquiring unique information on the structure and function of nucleic acids because of the specificity of the nitroxide labeling with respect to the nucleic acid residue and because the incorporated label is not readily hydrolyzed. Further systematic studies on the application of spin labeling to nucleic acids should also result in the design of many other spin bioassays. The creative potential of this application has not yet been fully exploited. Chemical synthesis of new spin labels and spin probes, as well as of spin-labeled nucleoside phosphates that can be incorporated into polynucleotides by nucleic acid polymerases, offers an almost unlimited resource to solve nucleic acid-related problems in biological systems by ESR.

ACKNOWLEDGMENT

I wish to thank Mrs. E. V. Bobst for her technical assistance and gratefully acknowledge helpful discussions with Dr. Stephen A. Goldman and Dr. Patricia G. Warwick. This work was supported in part by grant CA 15717, awarded by the National Cancer Institute, DHEW.

REFERENCES

Artyukh, R. I., G. B. Postnikova, B. I. Sukhorukov, and S. G. Kamzolova (1972). Study of spin-labeled DNA. Chemical modification of DNA by means of stable free radical 2,2,6,6-tetramethyl-4-bromoacetoxypiperidine-1-oxyl. *Biokhimiya* **37**, 902–906.

Bobst, A. M. (1971). Studies on spin-labeled polyriboadenylic acid. *Proc. Int. Congr. Pure Appl. Chem.* **23**, 1228–1235.

Bobst, A. M. (1972). Studies on spin-labeled polyriboadenylic acid. *Biopolymers* **11**, 1421–1433.

Bobst, A. M. (1974). A spin label assay for detecting poly (A) tracts in RNA's. *Fed. Proc., Fed. Am. Soc. Exp. Biol.* **33**, Abstr. No. 1481.

Bobst, A. M. (1977). A spin probe approach for characterizing molecular associations in biological systems. *Chimia* **31**, 141–142.

Bobst, A. M., and Y. C. E. Pan (1975). A spin probe approach for measuring the nucleic acid affinity of gene 32 protein. *Biochem. Biophys. Res. Commun.* **67**, 562–570.

Bobst, A. M., and P. F. Torrence (1978). Incorporation of spin probes into polynucleotides by enzymatic polymerization. *Polymer* **19**, 115–117.

Bobst, A. M., T. K. Sinha, and Y. C. E. Pan (1975). Electron spin resonance for detecting polyadenylate tracts in RNA's. *Science* **188**, 153–155.

Caron, M., and H. Dugas (1976a). Specific spin-labeling of transfer ribonucleic acid molecules. *Nucleic Acids Res.* **3**, 19–34.

Caron, M., and H. Dugas (1976b). A Spin label study of the thermal unfolding of secondary and tertiary structure in *E. coli* transfer RNAs. *Nucleic Acids Res.* **3**, 35–47.

Caron, M., N. Brisson, and H. Dugas (1976). Evidence for a conformational change in tRNA[Phe] upon aminoacylation. *J. Biol. Chem.* **251**, 1529–1530.

Caspary, W. J., J. J. Greene, L. M. Stempel, and P. O. P. Ts'o (1976). Spin labeled nucleic acids. *Nucleic Acids Res.* **3**, 847–861.

Danchin, A., and M. Guéron (1970). Cooperative binding of manganese (II) to transfer RNA. *Eur. J. Biochem.* **16**, 532–536.

Daniel, W. E., Jr., and M. Cohn (1975). Proton nuclear magnetic resonance of spin-labeled *E. coli* tRNA$_{f1}^{Met}$. *Proc. Natl. Acad. Sci. U.S.A.* **72**, 2582–2586.

Daniel, W. E., Jr., and M. Cohn (1976). Changes in tertiary structure accompanying a single base change in transfer RNA. Proton magnetic resonance and aminoacylation studies of *E. coli* tRNA$_{f1}^{Met}$ and tRNA$_{f3}^{Met}$ and their spin-labeled (s^{4}U8) derivatives. *Biochemistry* **15**, 3917–3924.

Dugas, H. (1977). Spin-labeled nucleic acids. *Acc. Chem. Res.* **10**, 47–54.

Freed, J. H. (1976). Theory of slow tumbling ESR spectra for nitroxides. *In* "Spin Labeling: Theory and Applications" (L. J. Berliner, ed.), Vol. 1, pp. 53–132. Academic Press, New York.

Freed, J. H., G. V. Bruno, and C. F. Polnaszek (1971). Electron spin resonance lineshapes and saturation in the slow motional region. *J. Phys. Chem.* **75**, 3385–3399.

Gaffney, B. J. (1975). The chemistry of spin labels. *In* "Spin Labeling: Theory and Application" (L. J. Berliner, ed.), Vol. 1, pp. 183–238. Academic Press, New York.

Girshovich, A. S., M. A. Grachev, D. G. Knorre, V. P. Kumarev, and V. I. Levintahl (1971). Reaction of a spin-labeled carbodiimide with nucleosides, poly U and tRNA. *FEBS Lett.* **14**, 199–202.

Goldman, S. A., G. V. Bruno, C. F. Polnaszek, and J. H. Freed (1972). An ESR study of anisotropic rotational reorientation and slow tumbling in liquid and frozen media. *J. Chem. Phys.* **56**, 716–735.

Hara, H., T. Horiuchi, M. Saneyoshi, and S. Nishimura (1970). 4-Thiouridine-specific spin-labeling of *E. coli* transfer RNA. *Biochem. Biophys. Res. Commun.* **38**, 305–311.

Hoffman, B. M., P. Schofield, and A. Rich (1969). Spin-labeled transfer RNA. *Proc. Natl. Acad. Sci. U.S.A.* **62**, 1195–1202.

Hong, S. J., and L. H. Piette (1976). Electron spin resonance spin-label studies of intercalation of ethidium bromide and aromatic amine carcinogens in DNA. *Cancer Res.* **36**, 1159–1171.

Ishizu, K., H. H. Dearman, M. T. Huang, and J. R. White (1969). Interaction of the 5-methylphenazinium cation radical with deoxyribonucleic acid. *Biochemistry* **8**, 1238–1246.

Jost, P., and O. H. Griffith (1976). Instrumental aspect of spin labeling. *In* "Spin Labeling: Theory and Applications" (L. J. Berliner, ed.), Vol. 1, pp. 251–272. Academic Press, New York.

Jouve, H., E. Melgar, and B. Lizarraga (1975). A study of the binding of Mn^{2+} to bovine pancreatic deoxyribonuclease I and to deoxyribonucleic acid by electron paramagnetic resonance. *J. Biol. Chem.* **250**, 6631–6635.

Kabat, D., B. Hoffman, and A. Rich (1970). Synthesis and characterization of a spin-labeled aminoacyl transfer ribonucleic acid. *Biopolymers* **9**, 95–101.

Kamzolova, S. G., A. I. Kolontarov, L. I. Elfimova, and B. I. Sukhorukov (1973). Study of structural transitions of an RNA polymerase T2 DNA complex from *E. coli* B. *Dokl. Akad. Nauk SSSR* **208**, 245–247.

Kelly, R. C., D. E. Jensen, and P. H. von Hippel (1976). DNA "melting" proteins. IV. Fluorescence measurements of binding parameters for bacteriophage T4 gene 32-protein to mono-, oligo-, and polynucleotides. *J. Biol. Chem.* **251**, 7240–7250.

Kouidou, S., T. K. Sinha, B. Janik, and A. M. Bobst (1978). Nucleic acid interaction with Vero cells: A temperature barrier in the interaction pattern. *Biochim. Biophys. Acta* **520**, 210–218.

Kumarev, V. P., and D. G. Knorre (1970). Water-soluble carbodimides with a spin label. *Dokl. Akad. Nauk SSSR* **193**, 103–105.

Leigh, J. S., Jr. (1970). ESR rigid-lattice line shape in a system of two interacting spins. *J. Chem. Phys.* **52**, 2608–2612.

Leventahl, V. I., J. M. Backer, Y. N. Molin, V. P. Kumarev, M. A. Grachev, and D. G. Knorre (1972). Spin-exchange interaction in polyuridylic acid modified with a spin-labeled carbodiimide. *FEBS Lett.* **24**, 149–152.

McIntosh, A. R., M. Caron, and H. Dugas (1973). A specific spin labeling of the anticodon of *E. coli* tRNA^{Glu}. *Biochem. Biophys. Res. Commun.* **55**, 1356–1363.

Mil', E. M., S. K. Zavriev, G. L. Grigoryan, and K. E. Kruglyakova (1973). An investigation of structural transitions of spin-labeled DNA. *Dokl. Akad. Nauk SSSR* **209**, 217–220.

Ohnishi, S., and H. M. McConnell (1965). Interaction of the radical ion of chloropromazine with deoxyribonucleic acid. *J. Am. Chem. Soc.* **87**, 2293.

Pan, Y. C. E. (1975). A spin label approach for studying nucleic acid-nucleic acid and nucleic acid-protein interactions. Ph.D. Thesis, University of Cincinnati, Cincinnati, Ohio.

Pan, Y. C. E., and A. M. Bobst (1973). Melting studies on spin-labeled polyadenylic-polyuridylic complexes. *Biopolymers* **12**, 367–371.

Pan, Y. C. E., and A. M. Bobst (1974). Spin-labeled polyuridylic acid. Formation of an intermediate structure. *Biopolymers* **13**, 1079–1083.

Petrov, A. I., and B. I. Sukhorukov (1975). Spin-labeled nucleosides, nucleoside-5'-mono-, di- and triphosphates. *Biofizika* **20**, 965–966.

Petrov, A. I., G. B. Postnikova, and B. I. Sukhorukov (1972). Application of *N*-(2,2,5,5-tetramethyl-3-carbonylpyrollin-1-oxyl)-imidazole to obtain spin labeled nucleic acids. *Izv. Akad. Nauk SSSR, Ser. Khim.* **6**, 1453–1454.

Piette, L. H. (1974). ESR-spin label studies of ethidium bromide and carcinogen intercalation in DNA. *Fed. Proc., Fed. Am. Soc. Exp. Biol.* **33**, Abstr. No. 832.

Porumb, T., and E. F. Slade (1976). Electron-spin-resonance studies of a chloropromazine derivative bound to DNA fibres. *Eur. J. Biochem.* **65**, 21–24.

Reuben, J., and E. J. Gabbay (1975). Binding of manganese (II) to DNA and the competitive effects of metal ions and organic cations. An electron paramagnetic resonance study. *Biochemistry* **14**, 1230–1235.

Sakai, T. T., and S. S. Cohen (1976). Effects of polyamines on the structure and reactivity of tRNA. *Prog. Nucleic Acid Res. Mol. Biol.* **17**, 15–42.

Schofield, P., B. M. Hoffman, and A. Rich (1970). Spin-labeling studies of aminoacyl transfer ribonucleic acid. *Biochemistry* **9**, 2525–2533.

Singer, B. (1975). Methylation and ethylation of uridylic acid and thymidylic acid. Reactivity of the ring and phosphate as a function of pH and alkyl group. *Biochemistry* **14**, 4353–4357.

Sinha, B. K., and C. F. Chignell (1975). Acridine spin labels as probes for nucleic acids. *Life Sci.* **17**, 1829–1836.

Sinha, B. K., R. L. Cysyk, D. B. Millar, and C. F. Chignell (1976). Synthesis and biological properties of some spin-labeled 9-aminoacridines. *J. Med. Chem.* **19**, 994–998.

Sinha, T. K. (1977). ESR as a tool for detecting nitroxide radicals covalently bound to nucleic acids. M.S. Thesis, University of Cincinnati, Cincinnati, Ohio.

Smith, I. C. P., and T. Yamane (1967). Spin-labeled nucleic acids. *Proc. Natl. Acad. Sci. U.S.A.* **58**, 884–887.

Smith, I. C. P., and T. Yamane (1968). Spin label studies of polynucleotide conformations. *In* "Recent Developments of Magnetic Resonance in Biological Systems" (S. Fujiwara and L. H. Piette, eds.), pp. 95–100. Hirokawa Publ. Co., Tokyo.

Snipes, W., J. Cupp, G. Cohn, and A. Keith (1974). Electron spin resonance analysis of the nitroxide spin label 2,2,6,6-tetramethylpiperidone-N-oxyl (Tempone) in single crystals of the reduced tempone matrix. *Biophys. J.* **14**, 20–32.

Sprinzl, M., E. Krämer, and D. Stehlik (1974). On the structure of phenylalanine tRNA from yeast. *Eur. J. Biochem.* **49**, 595–605.

Stone, T. J., T. Buckman, P. L. Nordio, and H. M. McConnell (1965). Spin-labeled biomolecules. *Proc. Natl. Acad. Sci. U.S.A.* **54**, 1010–1017.

Sukhorukov, B. I. (1973). Spin label studies of DNA. *Stud. Biophys.* **40**, 33–40.

Sukhorukov, B. I., and L. A. Kozlova (1970). Detection by the method of paramagnetic probe of new spectral transformations in the system DNA-water with change in temperature. *Biofizika* **15**, 539–541.

Sukhorukov, B. I., A. M. Vasserman, L. A. Kozlova, and A. L. Buchachenko (1967). The use of the paramagnetic probe method to investigate a nucleic acid-water system. *Dokl. Akad. Nauk SSSR* **177**, 454–457.

Sukhorukov, B. I., S. G. Kamzolova, A. I. Kolontarov, and A. I. Petrov (1973). Study of the structural conversions of DNA in complex with RNA-polymerase by the method of spin labels. *Biofizika* **18**, 377–378.

Vasserman, A. M., A. N. Kuznetsov, A. L. Kovarskii, and A. L. Buchachenko (1971). Anisotropic rotational diffusion of stable azoxy radicals. *Zh. Strukt. Khim.* **12**, 609–616.

Vocel, S. V., I. A. Slepneva, and J. M. Backer (1975). Influence of Mn^{2+} ion coordination on tRNA^{Val}1 macrostructure and determination of some coordination sites of Mn^{2+} ions in tRNA^{Val}1. *Biopolymers* **14**, 2445–2456.

Wallach, D. (1967). Effect of internal rotation on angular correlation functions. *J. Chem. Phys.* **47**, 5258–5268.

Zavriev, S. K., G. L. Grigoryan, I. Krilova, and P. A. Karmanov (1973). A spin-label for RNA study. *Biochem. Biophys. Res. Commun.* **54**, 123–126.

Zavriev, S. K., G. L. Grigoryan, P. A. Karmanov, G. A. Rtveliashvili, and E. G. Rozantsev (1974a). An investigation of rRNA and synthetic polyribonucleotides by the spin-label method. *Mol. Biol. (Moscow)* **8**, 302–309.

Zavriev, S. K., G. L. Grigoryan, L. E. Minchenkova, and V. I. Ivanov (1974b). Study of conformational transitions within the double-stranded state of spin-labeled DNA. *Mol. Biol. (Moscow)* **8**, 775–783.

Zavriev, S. K., G. L. Grigoryan, and L. E. Minchenkova (1976). The study of DNA-dye interaction by the method of spin-labels. *Mol. Biol. (Moscow)* **10**, 1387–1393.

APPENDIX I

Since September, 1977, the following pertinent papers have been published concerning the topic of nitroxide chemistry as discussed in Chapter 3.

Alcock, N. W., B. T. Golding, P. V. Ioannou, and J. F. Sawyer (1977). Preparation, crystal structure and reactions of a new spin-labeling reagent, cis-3,5-dibromo-4-oxo-2,2,6,6.-tetramethylpiperidin-1-yloxy. *Tetrahedron* **33**, 2969–2980.

Anderson, D. R., and T. H. Koch (1977). Upper excited state photochemistry of di-*tert*-butyl nitroxide. *Tetrahedron Lett.*, pp. 3015–3018.

Aplin, J. D., and L. D. Hall (1977). A spin labeling study of a polysaccharide support matrix for affinity chromatography. *J. Am. Chem. Soc.* **99**, 4162–4163.

Bartolin, G., and L. Piette (1978). Synthesis of intercalating and carcinogenic nitroxide spin labels. *J. Chem. Soc. Chem. Commun.*, pp. 215–216.

Benson, W. R., M. Maienthal, G. C. Yang, E. B. Sheinin, and C. W. Chung (1977). Synthesis of spin-labeled nitroxyl esters of steroids. *J. Med. Chem.* **20**, 1308–1312.

Cable, M. B., J. Jacobus, and G. L. Powell (1978). Cardiolipin: A stereospecifically spin-labeled analogue and its specific enzymic hydrolysis. *Proc. Natl. Acad. Sci. U.S.A.* **75**, 1227–1231.

Cella, J. A., and J. A. Kelley (1977). Preparation of spin-labeled opiates: morphine and codeine. *J. Pharm. Sci.* **66**, 1054–1056.

Chan, T. W., and Bruice, T. C. (1977). Reaction of nitroxides with 1,5-dihydroflavins and $N^{3,5}$-dimethyl-1,5-dihydrolumiflavin. *J. Am. Chem. Soc.* **99**, 7287–7291.

Dvolaitzki, M., and C. Taupin (1977). Synthesis and use of some spin-labeled long chain quarternary ammonium salts. *Nouv. J. Chim.* **1**, 355–356.

Espie, J.-C., R. Ramasseul, and A. Rassat (1978). Nitroxydes LXXXV: Nitroxydes azetidiniques. *Tetrahedron Lett.*, 795–796.

Goldberg, J. S., E. J. Rauchman, and G. M. Rosen (1977). Bioreduction of nitroxides by *Staphylococcus aureus. Biochem. Biophys. Res. Commun.* **79**, 198–202.

Gupta, C. M., R. Radhakrishnan, and H. G. Korana (1977). Glycerophospholipid synthesis: Improved general method and new analogues containing photoactivable groups. *Proc. Natl. Acad. Sci. U.S.A.* **74**, 4315–4319.

Guyot, N., and Dvolaitzky (1977). Synthesis of new substituted N-oxy-pyrrolidines as spin labels. *J. Chem. Res.*, 1734–1747.

Hideg, K., O. H. Hankovszky, and J. Tigyi (1977). Nitroxides, I. Sulfonic esters of 4-hydroxy-2,2,6,6-tetramethyl-piperidine 1-oxyl and their reaction with nucleophilic reagents. *Acta Chim. Acad. Scientiarum Hungaricae* **92**, 85–87.

Parmon, V. N., A. I. Kokorin, and G. M. Zhidomir (1977). Conformational structure of nitroxide biradicals. Use of biradicals as spin probes. *J. Struct. Chem.* **18**, 104–147.

Rauchman, E. J., G. M. Rosen, and W. W. Hord (1977). Use of sodium cyanoborohydride in the synthesis of biradical nitroxides. *Org. Prep. Proced. Inst.* **9**, 53–56.

Rosen, G. M., and E. J. Rauchman (1978). A new route to the synthesis of nitroxide carboxylic acids. *Org. Prep. Proced. Inst.* **10**, 17–20.

Rosen, G. M., E. J. Rauchman, and K. W. Hanck (1977). Selective bioreduction of nitroxides by rat liver microsomes. *Toxicology Lett.* **1**, 71–74.

Sheats, J. R., and H. M. McConnell (1977). Photochemical reaction of alkylpentacyanocobaltates with nitroxides. A new biophysical tool. *J. Am. Chem. Soc.* **99**, 7091–7092.

Sosnovsky, G., and M. Konieczny (1977). Transphosphorylative spin-labeling utilizing diimidazolides of pentavalent phosphorus. *Z. Naturforsch., Teil B* **32**, 1048–1059.

Sukhanov, V. A., V. I. Shvets, R. I. Zhdanov, and R. P. Evstigne (1977). Chemical synthesis of spin-labeled phospholipids. *Dokl. Akad. Nauk SSSR* **236**, 1386–1389.

Warner, T. G., and A. A. Benson (1977). An improved method for the preparation of unsaturated phosphatidylcholines: acylation of *sn*-glycero-3-phosphorylcholine in the presence of sodium methyl-sulfinylmethide. *J. Lipid Res.* **18**, 548–552.

Wenzel, H. R., G. Becker, and E. von Goldammer (1978). Synthese eines bifunktionellen Spin-Labels zur Quervernetzung (cross-linking) von Proteinen. *Chem. Ber.* **111**, 2453–2454.

Wu, W. Y., L. G. Abood, M. Gates, and R. W. Kreilick (1977). Spin-labeled narcotics. *Mol. Pharmacol.* **13**, 766–773.

Yamaguchi, T., A. Yamauchi, and H. Kimizuka (1978). Synthesis and properties of a spin labeled sodium dodecyl sulfate. *Chemistry Lett. (Japan)*, pp. 941–942.

APPENDIX II—Updated List of Commercial Sources of Spin Labels and Nitroxide Precursors

Since the publication of the first volume some additional suppliers of spin labels and nitroxide precursors have entered the market. This listing comprises all of the most common sources to date and is by no means intended to be complete.

Aldrich Chemical Co., Inc., 940 W. St. Paul Ave., Milwaukee, Wisconsin 53233
Cilag-Chemie, Schaffhausen, Switzerland
Eastman Organic Chemicals, Eastman Kodak Co., 343 State St., Rochester, New York, 14650
Fluka A. G., (U.S. Representative) Tridom Chemical, Inc., 255 Oser Ave., Hauppauge, New York 11787
Frinton Labs., P.O. Box 301, Grant Ave., South Vineland, New Jersey 08360
Molecular Probes, Inc., 849 J Place, Suite B, Plano, Texas 75074
New England Nuclear, 549 Albany St., Boston, Massachusetts 02118
Reanal, P.O. Box 54, 1441 Budapest 70, Hungary (exported by Medimpex, P.O. Box 126, Budapest 5, Hungary)
Stohler Isotope Chem, 49 Jones Rd., Waltham, Massachucetts 02154
SYVA, 3220 Porter Dr., Palo Alto, California 94304

Index

A

Acetazolamide spin probe, 232
Acetylcholinesterase, spin labeled inhibitors, 232
Actin, 43, 45
Adiabatic rapid passage, 7, 15
Adenylyl cyclase, spin labeled antagonists, 233
2-Aminoacetyl fluorene spin probe
  in nucleic acids, 269
  in rat liver microsomal membranes, 266
Aminoacridine spin probes, 307
  intercalating DNA, 322
Aminoanthracene spin probe, and nucleic acids, 269, 274
Aminochrysene spin probe, and nucleic acids, 269, 274, 276
Anesthetics, parethoxycaine spin probes, 241
Angular diffusion, 6
Anhydride spin label and t-RNA^{Glu}, 301–302
Anisotropic motion, 58, 62
  in an anisotropic environment, 65
  in DNA, 13
  in an isotropic environment, 63
  in labeled polynucleotides, 314
  in lipid bilayers, 55
  microwave frequency, 60
Anticancer spin labeled drugs, 226
Anti-DNP serum, in SMIA, 262
Arrhenius plots, 319–320, 323
Avidin, see Biotin spin probes
10-Azethoxyleicosanoic acid, synthesis, 161
8-Azethoxyl-1-iodooctadecane, synthesis, 159

Azethoxy nitroxides
  chemical stability, 135
  synthesis, 134, 158–162
8-Azethoxyloctadecanol, synthesis, 159
8-Azethoxyloctadecanol tetrahydropyranyl ether, synthesis, 158

B

Binding tightness
  label-protein, 12
  protein conformational fluctuations, 13
Biotin spin probes, 234, 239
Biradicals, 152
  synthesis, 153
  deca- or hexa-methonium, 233
Bloch equation, 22
Bredt's rule, 119
Bromoacetamide spin labels
  pyrrolidine nitroxide and RNA, t-RNA, 294
  with 4-thiouridine in t-RNA, 300
Bromoacetoxypiperidine spin label
  and calf thymus DNA, 299
  and T2 phage, 299
Brownian diffusion, 6, 313

C

Carbodiimide spin label
  mononucleotides, 296
  piperidine nitroxide, 296
  t-RNA, 301
Carbonic anhydrase and spin labeled drugs, 231, 235–238

Carcinogenic spin probes, *see* aminochyr-
    sene, aminoacetylfluorene, ethidium
    bromide, or estradiol spin probes
Chlorpromazine cation radical and DNA,
    223, 306
Complementary probes
    definition, 94
    Cu^{2+}, 95
    Gd^{3+}, 101
    Mn^{2+}, 98
    VO^{2+}, 100
Computer simulation of ESR spectra
    ST-EPR, 19, 22, 27
    motional narrowing region, 313, 315
Contractile proteins
    actin, 45
    myosin, 45
    tropomyosin, 45
Correlation time
    in labeled polymers, 188–189
    in labeled polynucleotides, 312
    translational, 82
Cu^{2+}, 95
    monothiosemicarbazone, 97
Cytochrome P-450
    and metapyrone spin probes, 227, 240
    reductase and piperidinol nitroxide or
        5-doxylstearic acid, 229
    and spin trapping, 279

D

Decamethonium spin probe as acetyl-
    cholinesterase inhibitor, 233, 240
Density matrix, 22
Dichloroisoproterenol spin probe, as
    adenylyl cyclase antagonists, 233
Diffusion, Brownian, 6, 313
    angular, 6
    anisotropic, 65
    spectral, 76
2,5-Dimethyl-5-nonyl-Δ^1-pyrroline   and
    N-oxide synthesis, 157
2,5-Dimethyl-Δ^1-pyrroline and N-oxide syn-
    thesis, 157
5,5'-(Dimethyl-1-pyrroline-1-oxide), 281
Dinitroxide spin labels, 152
    decamethonium spin probes, 233
    hexamethonium spin probes, 233
    synthesis, 153

Diphenylhydantoin spin probe, 225
Dipolar Hamiltonian
    A component, 74, 77
    B-F components, 78, 81
    cross relaxation (B component), 82
    in liquid phase, 87
Dispersion signal detection, 17
Disproportionation
    of nitroxides, mechanisms, 118, 124
    kinetics, 119
DNA
    Mn(II) complexes, 305
    melting, 276
    spin labeled, 295, 325–327
    spin probe intercalation 13, 223, 267–271,
        273, 328
    ST-EPR, 39
DNA–RNA polymerase, 331–336
Doped polymers, 184
Doxyl nitroxides
    oxaziridine intermediate, 131
    from oxazolines, 131
    stereochemical considerations, 130
    synthetic procedures, 131
Doxyl spin probes
    actractyloside, 54
    choline ester, 53
    5-doxylpalmitic acid, 51
    5-doxylstearic acid and cyt. P-450 reduc-
        tase, 299
    N-O-maleimidoester, 51
DPPC-DMPE, 54
Drug binding in plasma
    diphenylhydantoin spin probe, 225
    distribution, 226
    morphine spin probes, 225
    triethylenethiophosphoramide spin probe,
        226
Dy^{3+}, 82, 106

E

EDTA, ESR signals from, 126
Electron–electron double resonance (EL-
    DOR), 14, 18, 20, 62
End-labeled polymers, 182
Enzyme catalyzed spin labeled nucleic acid
    synthesis, 302–304
    ppRUGT, 302

ESR sensitivity, experimental considerations, 308–310
Estradiol spin probes, 265
Ethidium bromide spin probe, and DNA, 267, 273, 308, 318
Ethyleneimine spin label, and DNA, 299

F

Free radicals, interactions with metal ions, 106

G

Gadolinium, 91, 101
Glass transition, spin probe detection, 202–203

H

Heisenberg exchange, 83
  Coulomb repulsion, 86
  interspecies, 84
  strong, 85, 106
  weak, 85
Hemoglobin
  maleimide spin labeled, 20, 25
  S, 45
Hexamethonium spin probe, as acetylcholinesterase inhibitor, 233
N-Hydroxyamine of nitroxide, 118
N-Hydroxysuccinimide spin label, pyrroline nitroxide and val-t-RNA, 296, 300–302

I

Imidazolide spin label (pyrroline carboxylic acid imidazolide nitroxide), and T2 phage DNA, 299
Imidazoline nitroxide, 137
  2,5-dimethyl-Δ^1-pyrroline N-oxide synthesis, 157
  N-hydroxy-Δ^3, 139
  Δ^1:derived (nitronyl nitroxides), 140
  2-Undecyl-5,5-dimethyl-Δ^1-pyrroline N-oxide, synthesis, 154
In vivo ESR
  in rat liver, 286
  sensitivity, 285
  traveling wave helices, 285

Inhomogeneous broadening, 27
Iodoacetamide spin label
  4(2-iodoacetamido)-2,2,6,6,-tetramethylpiperdien-1-oxyl, 41
  homopolynucleotides, 296, 331
  t-RNA, 301
  pyrrolidine nitroxide
    and nucleic acids, 299
    and cys-t-RNA^{Cys}, 300

L

Labeling synthetic polymers
  copolymerization
    with nitroxide monomer, 175
    with nitroxide precursor, 180
  end-labeled polymers, 182
  labeling a preformed polymer, 181
  randomly labeled, 175
  with spin probes, 184
Lanthanides, 87
  as biochemical probes, 91
  chemical properties, 90
  magnetic properties, 90
  as NMR shift reagents, 92
  spin lattice relaxation times, 91
Leigh theory, 72, 74, 104, 319
  Q band, 77
Lipid bilayers, 54
Lipid peroxidation, 282

M

Magnetic resonance theory
  Bloch equation formalism, 21
  density matrix formalism, 21
Maleimide spin label
  labeled albumin, 226
  4-maleimido 2,2,6,6-tetramethylpiperidine-1-oxyl, 25, 32, 43
  labeled RNA and t-RNA, 294
Matrix inversion, Gauss–Jordan (Gauss–Pivot) method, 25
Melting curves
  critical temperature, T_{crit}, 323
  of labeled polynucleotides, 310
  spin denaturation temperature, T_m^{sp}, 322, 325
  of spin probe-DNA complexes, 329
Membranes, permeability, 105

Metal ions, measuring binding, 107
Metapyrone spin probes, 227
Methyl phenazinium cation, 307
Minimum perturbation spin labels (azethoxyl), 134
Mn²⁺, 98
  creatine kinase complexes, 99
  Mn(II)-DNA complexes, 305
  spin probes of nucleic acids, 304, 317, 319
Morphine spin probe, 225, 255
Motion
  in bulk polymers, 195
  comparison of spin label to other techniques, 195, 208
Motional narrowing
  in labeled polymers, 191
  in labeled polynucleotides, 311
Myofibrils, 43
Myosin, 45

N

NADPH mixed function oxidases, spin trapping, 281
Nitrone, 117, 119
  5-5′-(dimethyl-1-pyrroline 1-oxide), 281
  conversion to proxylnitroxide, 132
  2-methyl, condensation, 137
  phenyltertiary butyl (PBN), 280
  as spin traps, 280
Nitronyl nitroxides, 140
  NAD analog, 141
Nitroxide derivatives
  N-acetoxy, 128
  bicyclic, 119
  characterization, 126
  conjugated, stable, 121–123
  decomposition, 118
  dimers of thionitroxides, 121
  dinitroxide, 152
  disproportionation, 117
  imidazoline, 137
  IR spectra, 129
  mass spectra, 129
  ¹⁵N substituted, 129
  NMR spectra, 126–127, 129
  overoxidation, 147
  oxazolidine, 119
  oxidation of secondary amine precursor, 144

  proline, 119
  reduction, 148
  stability, 116–117
    proxyl vs. doxyl, 133
  tetrahydrooxazine, 141–142
  thio, 121
  unstable, 124
Nitroxides
  acetazolamide spin probe, 232
  acylimidazole, 147
  alcohol, 147
  aldehydo, 146
  amino, 146
  2-aminoacetyl fluorene spin probe, 266, 269
  aminoacridine spin probes, 307, 328
  aminoanthracene spin probe, 269, 274
  aminochrysene spin probe, 269, 274, 276
  anhydride spin label, 301–302
  aryl-tert-butyl, 124
  azethoxyl, 134, 158–162
  10-azethoxyleicosanoic acid, synthesis, 161
  8-azethoxyl-1-iodooctadecane, synthesis, 159
  8-azethoxyloctadecanol, synthesis, 159
  8-azethoxyloctadecanol tetrahydropyranyl ether, synthesis, 158
  biotin spin probes, 234, 239
  bromoacetamide spin labels, 294, 300
  bromoacetoxypiperidine spin label, 299
  carbodiimide spin label, 296, 301
  decamethonium spin probe, 233, 240
  dichloroisoproterenol spin probe, 233
  2,5-dimethyl-5-nonyl-Δ¹-pyrroline N-oxide, synthesis, 157
  2,5-Dimethyl-Δ¹-pyrroline N-oxide, synthesis, 157
  diphenylhydantoin spin probe, 225
  di-tert-butyl, 128
  doxyl spin probes, 130
  5-doxylpalmitic acid, 51–54
  5-doxylstearic acid, 229
  estradiol spin probes, 265
  ethidium bromide spin probe, 267, 273, 308, 318
  ethyleneimine spin label, 299
  hexamethonium spin probe, 233
  N-hydroxysuccinimide spin label, 296, 300–302

imidazolide spin label (pyrroline carboxylic acid imidazolide nitroxide), 299

4(2-iodoacetamido)-2,2,6,6-tetramethylpiperidine-1-oxyl, 41

iodoacetamide spin label, 41, 294

isoquinuclidine, 143

maleimide spin label, 25, 38, 43, 226, 294

4-maleimido 2,2,6,6-tetramethylpiperidine-1-oxyl, 25, 32, 143

metapyrone spin probes, 227

N-methyl-N-tri-*tert*-butylphenyl, 126

minimum perturbation spin labels (azethoxyl), 134

mixed anhydride spin label, 147

morphine spin probe, 225, 255

N-(1-oxyl-2,2,6,6-tetramethyl-4-piperidinyl)-O-[1-β-D-ribofuranosyluracil-5-yl]-glycolamide (RUGT), 302, 335

1- and 2-naphthyl-*tert*-butyl, 124

1-palmitoyl-2-(*trans*-10-azethoxyleicosoyl)-glycerophosphatidylcholine, synthesis, 161

parethoxycaine spin probes, 241

phosphatidylcholine, 134

phosphatidyl serine, 134

phosphoryl, 147

propanolol spin probe, 233

proxyl, 132

7-proxyloctadecanal synthesis, 156

12-proxylpalmityl alcohol, 127

7-proxylstearic acid synthesis, 156

7-proxylstearyl alcohol, synthesis, 154

  methanesulfonate, synthesis, 155

  trimethylammonium methanesulfonate, 155, synthesis

sulfanilamide spin probes, 231

TEMPOcholine spin probe, 260

TEMPOL (*see* 2,2,6,6-tetramethylpiperidine-1-oxyl-4-ol)

tert-butylisopropyl, 112

2,2,6,6-tetramethylpiperidine-1-oxyl-4-ol (TEMPOL, TANOL), 8, 10, 120, 305

triethylenethiophosphoramide spin probe, 225

Nucleic acids, *see* Spin labeling of nucleic acids

O

Oxaziridine, intermediate in doxyl (oxazolidine) nitroxide synthesis, 131

Oxazolidine nitroxide, *see* Doxyl nitroxide

Oxazolines, general synthesis, 160

N-(1-Oxyl-2,2,6,6-tetramethyl-4-piperidinyl)-O-[1-β-D-ribofuranosyluracil-5-yl]glycolamide (RUGT), 302, 335

P

1-Palmitoyl-2-(*trans*-10-azethoxyleicosoyl)-glycerophosphatidylcholine, synthesis, 161

Parethoxycaine spinprobes, 241

2,2-Pentamethylene-4,4,5,5-tetramethylimidazolidine, synthesis, 163

  -1-oxyl derivative, 163

  -N-acetoxy, 163–164

  -1-hydroxy, 164

  -1,2 dioxyl, 165

Phenyltertiarybutyl nitrone as spin trap, 280

Photochemical

  nitroxide destruction by, 151

  nitroxide synthesis, 143–144

Polyamide, 180, 183, 206–207

Polybenzylglutamate, 177, 182, 184, 188, 189, 190, 191, 194, 205, 212

Poly(N-butyl methacrylate), 177, 181, 196, 197

Polyester, 180, 183, 206

Polyether, 180, 182, 201, 202

Polyethylene, 176, 180, 199, 200, 205, 208

Poly(N-2-hydroxypropylmethacrylamide), 177, 181, 193, 205, 207, 208

Polylysine, 177, 182, 205

Polymer solutions

  concentrated, 209–212

  dilute, 204–209

Poly(methylacrylate), 176, 181, 196, 197, 209

Poly(methyl methacrylate), 175, 176, 181, 183, 186, 187, 188, 195, 196, 197, 205–214

Polynucleotides, *see* Spin labeling of nucleic acids

Polystyrene, 176, 181, 182, 198, 205, 206, 208–210, 212–214

Polyurethane, 202

Poly(vinylacetate), 176, 182, 202, 203, 205, 206, 207, 215
Poly(vinylpyrrolidone), 177, 215
Progressive saturation, 14, 79, 81
Propanolol spin probe, adenylyl cyclase antagonists, 233
Protein–lipid interactions, 47
Proxyl nitroxides
    from N-hydroxyamine intermediate, 133
    from nitrones, 132
7-Proxyloctadecanal, synthesis, 156
7-Proxylstearic acid, synthesis, 156
7-Proxylstearyl alcohol, synthesis, 154
    methane sulfonate, synthesis, 155
    trimethyl ammonium methanesulfonate, synthesis, 155
Pyrrolidine nitroxides, side chain substituted, see Proxyl nitroxides

                            R

Receptors Ca-ATPase, 39, 48
    Torpedo marmorata, 39
Reduction of nitroxides
    carboxylic acid nitroxides, 152
    catalytic, 128
    by cyt. P-450 reductase, 229
    by Grignard reagents, 148
    by phenylhydrazine, 122
Restoring potential, 65
RNA, spin labeled, 320
Rotation correlation times
    methods of determination, 14
    in labeled polymers, 195–214
Runge–Kutta method, 27
Rod outer segments, 39, 51

                            S

Saturation
    progressive, 14
    recovery experiment, 79, 81
Saturation-transfer spectroscopy, experimental methods, 61
    computer simulated, 19, 22, 27
    EPR spectra acquisition, 22
    EPR spectral components, 22
    theory, 3
Slow motion, in polymer systems, 187

Solid–liquid interface, absorbed polymers, 214
Solid polymers, 203
Spectral diffusion of saturation, 4, 6
Spin assay
    FRAT, 249
    human serum albumin, 249
    nucleic acids, 298
    for poly-A tracts, 339
    reserve bilirubin binding capacity (RBBC), 249
    spin membrane immunoassays, 255
Spin immunoassay, 224, 255
Spin labeling of nucleic acids (see also DNA)
    by chemical modification, 293
    enzymatically, 302
    homo-polynucleotides, 296
    and interaction with other ligands, 331
    monitoring polynucleotide-polypeptide complexes, 298, 331
    site-nonspecific labels, 295
    site-specific labeling, 300
Spin lattice relaxation, 8
    anisotropic electron-Zeeman interaction, 9
    anisotropic hyperfine interaction, 9
    electronic dipolar in solids, 77
    spin–rotation interactions, 9
    two phonon Raman process, 10
Spin membrane immunoassay (SMIA)
    complement, 261
    liposomes, 261
    sensitizers, 261
    technique, 259
    thyroxine assay, 263
Spin probing nucleic acids
    Mn(II), 304
    TEMPOL, 205
Spin–spin relaxation mechanisms
    electron–electron, 11
    electron–nuclear, 10
    librational motions, 12
Spin trapping
    with cyt. P-450
    of drug metabolites, 229
    of free radicals in carcinogenesis, 279
    in lipid peroxidation, 282
    with monooxygenase enzymes, 279
    technique, 280

Subfragment-1, 40
Sulfanilamide spin probes, 231
Surface catalysts, 56
Synthetic polymers, 39, 173–221
    adsorbed at interfaces, 214–216
    effect of diluents, 204–214
    glass transition, 202–203
    labeling, 174–185
    motion in bulk, 195–204
    spectral analysis, 185–194
Synthetic polynucleotide, *see* Spin labeling
    of nucleic acids

T

TEMPOcholine spin probe in SMIA, 260
TEMPOL (*see* 2,2,6,6-Tetramethylpiperi-
    dine-1-oxyl-4-ol)
Tetrahydrooxazine derived nitroxides, 141
2,2,6,6-Tetramethylpiperidine-1-oxyl-4-ol
    (TEMPOL, TANOL) 8, 10, 120
    as nucleic acid spin probe, 305
Thyroxine assay by SMIA, 263
Transfer RNA
    Mn(II) complexes, 305

pyrroline   hydroxysuccinimide   spin
    labeled, 300
    spin labeled 294–396, 300–302, 326
    spin melting studies, 325–327
Transition metal ions, properties, 93
Trapezoidal field, modulation, 62
Triethylenethiophosphoramide spin probe,
    225
Tris, ESR signals from, 126
Tropomypsin, 45

U

Undecyl-5,5-dimethyl-Δ^1-pyrroline-N-oxide,
    154
Unstable nitroxides, 124

V

Vero cell monolayers
    interaction with polynucleotides, 336–339
    techniques, 336
Very slow tumbling, 3
VO^{2+}, 100

Molecular Biology

An International Series of Monographs and Textbooks

Editors

BERNARD HORECKER

Roche Institute of Molecular Biology
Nutley, New Jersey

NATHAN O. KAPLAN

Department of Chemistry
University of California
At San Diego
La Jolla, California

JULIUS MARMUR

Department of Biochemistry
Albert Einstein College of Medicine
Yeshiva University
Bronx, New York

HAROLD A. SCHERAGA

Department of Chemistry
Cornell University
Ithaca, New York

HAROLD A. SCHERAGA. Protein Structure. 1961

STUART A. RICE AND MITSURU NAGASAWA. Polyelectrolyte Solutions: A Theoretical Introduction, *with a contribution by Herbert Morawetz.* 1961

SIDNEY UDENFRIEND. Fluorescence Assay in Biology and Medicine. Volume I—1962. Volume II—1969

J. HERBERT TAYLOR (Editor). Molecular Genetics. Part I—1963. Part II—1967

ARTHUR VEIS. The Macromolecular Chemistry of Gelatin. 1964

M. JOLY. A Physico-chemical Approach to the Denaturation of Proteins. 1965

SYDNEY J. LEACH (Editor). Physical Principles and Techniques of Protein Chemistry. Part A—1969. Part B—1970. Part C—1973

KENDRIC C. SMITH AND PHILIP C. HANAWALT. Molecular Photobiology: Inactivation and Recovery. 1969

RONALD BENTLEY. Molecular Asymmetry in Biology. Volume I—1969. Volume II—1970

JACINTO STEINHARDT AND JACQUELINE A. REYNOLDS. Multiple Equilibria in Protein. 1969

DOUGLAS POLAND AND HAROLD A. SCHERAGA. Theory of Helix-Coil Transitions in Biopolymers. 1970

JOHN R. CANN. Interacting Macromolecules: The Theory and Practice of Their Electrophoresis, Ultracentrifugation, and Chromatography. 1970

WALTER W. WAINIO. The Mammalian Mitochondrial Respiratory Chain. 1970

LAWRENCE I. ROTHFIELD (Editor). Structure and Function of Biological Membranes. 1971

ALAN G. WALTON AND JOHN BLACKWELL. Biopolymers. 1973

WALTER LOVENBERG (Editor). Iron-Sulfur Proteins. Volume I, Biological Properties—1973. Volume II, Molecular Properties—1973. Volume III, Structure and Metabolic Mechanisms—1977

A. J. HOPFINGER. Conformational Properties of Macromolecules. 1973

R. D. B. FRASER AND T. P. MACRAE. Conformation in Fibrous Proteins. 1973

OSAMU HAYAISHI (Editor). Molecular Mechanisms of Oxygen Activation. 1974

FUMIO OOSAWA AND SHO ASAKURA. Thermodynamics of the Polymerization of Protein. 1975

LAWRENCE J. BERLINER (Editor). Spin Labeling: Theory and Applications. Volume I, 1976. Volume II, 1978

T. BLUNDELL AND L. JOHNSON. Protein Crystallography. 1976

HERBERT WEISSBACH AND SIDNEY PESTKA (Editors). Molecular Mechanisms of Protein Biosynthesis. 1977

J. HERBERT TAYLOR (Editor). Molecular Genetics, Part III, Chromosome Structure